Electric Vehicles in Energy Systems

Ali Ahmadian • Behnam Mohammadi-ivatloo
Ali Elkamel
Editors

Electric Vehicles in Energy Systems

Modelling, Integration, Analysis, and Optimization

Editors
Ali Ahmadian
Department of Electrical Engineering
University of Bonab
Bonab, Iran

Behnam Mohammadi-ivatloo
Faculty of Electrical and Computer Engineering
University of Tabriz
Tabriz, Iran

Ali Elkamel
Department of Chemical Engineering
University of Waterloo
Waterloo, ON, Canada

College of Engineering
Khalifa University of Science and
Technology, The Petroleum Institute
Abu Dhabi, UAE

ISBN 978-3-030-34450-4   ISBN 978-3-030-34448-1   (eBook)
https://doi.org/10.1007/978-3-030-34448-1

© Springer Nature Switzerland AG 2020
This work is subject to copyright. All rights are reserved by the Publisher, whether the whole or part of the material is concerned, specifically the rights of translation, reprinting, reuse of illustrations, recitation, broadcasting, reproduction on microfilms or in any other physical way, and transmission or information storage and retrieval, electronic adaptation, computer software, or by similar or dissimilar methodology now known or hereafter developed.
The use of general descriptive names, registered names, trademarks, service marks, etc. in this publication does not imply, even in the absence of a specific statement, that such names are exempt from the relevant protective laws and regulations and therefore free for general use.
The publisher, the authors, and the editors are safe to assume that the advice and information in this book are believed to be true and accurate at the date of publication. Neither the publisher nor the authors or the editors give a warranty, express or implied, with respect to the material contained herein or for any errors or omissions that may have been made. The publisher remains neutral with regard to jurisdictional claims in published maps and institutional affiliations.

This Springer imprint is published by the registered company Springer Nature Switzerland AG
The registered company address is: Gewerbestrasse 11, 6330 Cham, Switzerland

Preface

The transportation sector utilizes a considerable amount of energy worldwide; therefore, it has a significant impact on today's energy systems. In recent years, electric vehicles (EVs), as a technical solution, have been given great attention in order to address the environmental concerns in modern cities. Meanwhile, the EVs impact on electric grids and energy systems should be investigated in both planning and operation studies. Different charging strategies including uncoordinated and coordinated charging methods have individual impact on energy systems. In addition, the vehicle to grid (V2G) and vehicle to home (V2H) abilities of EVs can improve the efficiency of energy utilization in smart energy systems. Therefore, it is necessary to study the impact of EVs on energy systems from various points of view.

In this book, a comprehensive study about EVs is done so that technical, economic, and environmental aspects are taken into account. The EVs load modeling techniques, EVs integration with renewable energies, and energy optimization in the future energy systems are discussed completely.

This book contains 15 chapters in which numerous researchers and experts from academia and industries collaborated. The breakdown of the chapters is as follows:

- Chapter 1 reviews the challenges and opportunities of the electric vehicles in energy systems. The transportation electrification as one of the pillars of intelligent transportation systems is investigated in details.
- Chapter 2 models the electric vehicles' travel behavior using artificial intelligence-based approach. The data mining of electric vehicles is investigated using deep learning method.
- Chapter 3 describes the role of off-board electric vehicles battery chargers in smart home and smart grid applications. The energy storage system and renewable energies are also taken into account.
- Chapter 4 presents a nonlinear bi-level model for optimal operation of smart distribution network. The electric vehicles can operate in both grid to vehicle and vehicle to grid charge modes.

v

- Chapter 5 manages the energy of a typical microgrid considering solar energy and electric vehicles. The chapter develops a charging management program that increases the renewable energy's penetration.
- Chapter 6 presents an optimal operation model for electric vehicles. The wind energy is integrated in the studied case, and the impact of coordinated and uncoordinated operations of vehicles on stochastic generation of wind energy is investigated.
- Chapter 7 presents the distributed charging management of electric vehicles in smart microgrids. The proposed cooperative control system is introduced to accomplish a wide range of auxiliary services.
- Chapter 8 proposes an energy and reserve management model for a distribution network considering electric vehicles, renewable energy recourses, and distributed generations. In this chapter, an operation scheme for the electric vehicles aggregator is accomplished with main objective function of decreasing operation costs of distribution network.
- Chapter 9 presents a bidding strategy model for electric vehicles' parking lots to participate in the electricity market.
- Chapter 10 introduces battery sweep stations in microgrids and proposes a stochastic model to participate in electricity market.
- Chapter 11 presents the integration of electric vehicles charge stations with renewable energy resource in order to participate in electricity market.
- Chapter 12 models the impact of electric vehicles charging demand on residential energy hubs.
- Chapter 13 presents a stochastic model for electric vehicles battery replacement stations. The renewable energy recourses are taken into account.
- Chapter 14 models the participation of electric vehicles in demand response program.
- Chapter 15 optimizes the charge demand of electric vehicles in smart homes. The load demand profiles of smart homes are modified by the proposed model.

The editors of the book warmly thank all the contributors for their valuable works. Also, they would like to thank the respected reviewers who improved the quality of the book by their valuable and important comments.

Bonab, Iran	Ali Ahmadian
Tabriz, Iran	Behnam Mohammadi-ivatloo
Abu Dhabi, UAE	Ali Elkamel

Contents

1   Why Electric Vehicles? 1
Hamidreza Jahangir, Masoud Aliakbar Golkar, Ali Ahmadian,
and Ali Elkamel

**2   Artificial Intelligence-based Approach For Electric Vehicle
Travel Behavior Modeling** 21
Hamidreza Jahangir, Masoud Aliakbar Golkar, Ali Ahmadian,
and Ali Elkamel

**3   The Role of Off-Board EV Battery Chargers in Smart Homes
and Smart Grids: Operation with Renewables
and Energy Storage Systems** 47
Vitor Monteiro, Jose Afonso, Tiago Sousa, and Joao L. Afonso

**4   Optimal Charge Scheduling of Electric Vehicles
in Solar Energy Integrated Power Systems Considering
the Uncertainties** 73
S. Muhammad Bagher Sadati, Jamal Moshtagh,
Miadreza Shafie-Khah, Abdollah Rastgou,
and João P. S. Catalão

**5   Optimal Utilization of Solar Energy for Electric Vehicles
Charging in a Typical Microgrid** 129
Mohammad Saadatmandi and Seyed Mehdi Hakimi

**6   Integration of Electric Vehicles and Wind Energy
in Power Systems** 165
Morteza Shafiekhani and Ali Zangeneh

**7   Distributed Charging Management of Electric Vehicles
in Smart Microgrids** 183
Reza Jalilzadeh Hamidi

vii

8 Optimal Energy and Reserve Management of the Electric Vehicles Aggregator in Electrical Energy Networks Considering Distributed Energy Sources and Demand Side Management 211
Mehrdad Ghahramani, Morteza Nazari-Heris, Kazem Zare, and Behnam Mohammadi-ivatloo

9 An Interactive Model for the Participation of Electric Vehicles in the Competitive Electricity Market 233
Mohammad Reza Fallahzadeh and Ali Zangeneh

10 Optimal Scheduling of Smart Microgrid in Presence of Battery Swapping Station of Electrical Vehicles 249
Mohammad Hemmati, Mehdi Abapour, and Behnam Mohammadi-ivatloo

11 Risk-Based Long Term Integration of PEV Charge Stations and CHP Units Concerning Demand Response Participation of Customers in an Equilibrium Constrained Modeling Framework 269
Pouya Salyani, Mehdi Abapour, and Kazem Zare

12 Modelling the Impact of Uncontrolled Electric Vehicles Charging Demand on the Optimal Operation of Residential Energy Hubs 289
Azadeh Maroufmashat, Q. Kong, Ali Elkamel, and Michael Fowler

13 Optimal Operation of Electric Vehicle's Battery Replacement Stations with Taking into Account Uncertainties 313
Babak Mardan, Sahar Seyyedeh Barhagh, Behnam Mohammadi-ivatloo, Ali Ahmadian, and Ali Elkamel

14 Participation of Aggregated Electric Vehicles in Demand Response Programs 327
Maedeh Yazdandoust and Masoud Aliakbar Golkar

15 Optimal Charge Scheduling of Electric Vehicles in Smart Homes 359
Arezoo Hasankhani and Seyed Mehdi Hakimi

Index 385

Chapter 1
Why Electric Vehicles?

Hamidreza Jahangir, Masoud Aliakbar Golkar, Ali Ahmadian, and Ali Elkamel

1.1 Introduction

Recently, automobile manufacture companies began to produce electric vehicles (EVs) and new developments were made every day. At first, the progress speed of the vehicle electrification was very slow, but nowadays some manufactures have spent the time to completely change their vehicle from petrol to electric, and some countries are preparing laws to ban petrol vehicles [1–3]. The problem of air pollution is also widespread these days; in this regard, policy makers in modern societies focus on electrification of the transportation fleet more than before and, in near future, the next vehicle of every person all around the world will be electric [4, 5]. In this chapter, an introduction to the motivations of implementing the EVs in transportation fleet from different points including environmental, economic, political and other aspects is presented. The incentives for the purchase of EVs in different developed countries including America, Canada and Japan have been considered and the most important concerns about EVs from the customers' point of view have been studied. Furthermore, the challenges imposed on the power system, which are aggravated by increasing the penetration of the EVs in transportation fleet, are also explained.

H. Jahangir (✉) · M. A. Golkar
Faculty of Electrical Engineering, K. N. Toosi University of Technology, Tehran, Iran
e-mail: h.r.jahangir@email.kntu.ac.ir; golkar@kntu.ac.ir

A. Ahmadian
Department of Electrical Engineering, University of Bonab, Bonab, Iran
e-mail: ahmadian@bonabu.ac.ir

A. Elkamel
Department of Chemical Engineering, University of Waterloo, Waterloo, ON, Canada

College of Engineering, Khalifa University of Science and Technology, The Petroleum Institute, Abu Dhabi, UAE
e-mail: aelkamel@uwaterloo.ca

© Springer Nature Switzerland AG 2020
A. Ahmadian et al. (eds.), *Electric Vehicles in Energy Systems*,
https://doi.org/10.1007/978-3-030-34448-1_1

1.2 The Motivations for Increasing the Penetration Rate of EVs

Today, the transportation sector consumes a significant amount of energy around the world [6]. For example, the transportation sector uses about one-third of the US energy consumption annually (27.5 quadrillion in 2010), and fossil fuels are the main sources of the transportation fleet [7–9]. In addition, as shown in Fig. 1.1, energy consumption by the transport sector has increased dramatically since 1950. This fact is confirmed by a review of NHTS database information released in 2011. According to NHTS, the number of cars per family has increased dramatically in recent years.

Also, despite the increase in car efficiency, as shown in Fig. 1.2, fuel consumption increased from 1980 to 2010; although the increased truck sales somewhat reduced the efficiency of the transportation system, the overall efficiency increased each year. Furthermore, the increase in vehicle miles travelled (VMT) per person shows a significant increase in energy consumption pattern. As shown in Fig. 1.3, the per capita distance traveled per vehicle is lower than the per capita distance traveled by each household due to the increasing number of the vehicles purchased by households [10–12]. Therefore, this information well suggests that the number of vehicles in modern societies is steadily increasing. This increase in the number of vehicles, despite their efficiency, will increase the need for fossil fuels.

Increasing vehicle fuel consumption has a significant effect on the greenhouse gas emissions and air pollution. The vehicle share of the fossil fuels consumption in the United States in 2011 includes about 53% of CO, 31% of NOX, 24% of VOCs, and 1.7% of PM2.5 [13, 14]. The transport sector also accounts for a large portion of

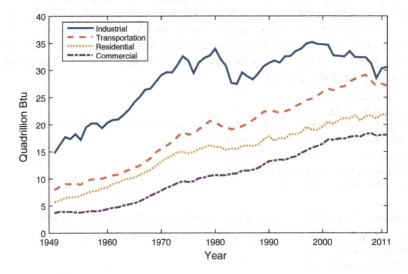

Fig. 1.1 Energy consumption trends in various parts of the United States from 1950 to 2011

1 Why Electric Vehicles? 3

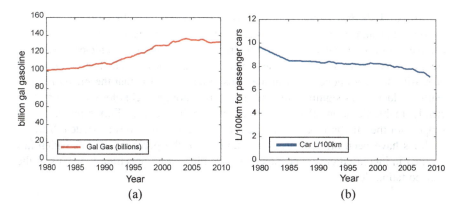

Fig. 1.2 The efficiency of passenger transportation and fossil fuels in the United States. (**a**): Fossil fuel consumption pattern; (**b**): Travel distance pattern

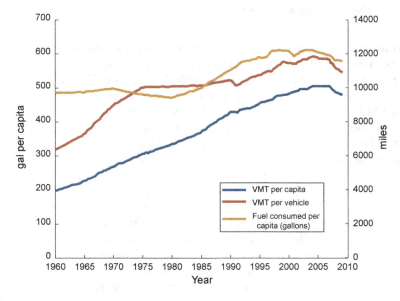

Fig. 1.3 The trend of increasing per capita fuel consumption, the distance traveled per car and the per capita distance traveled per household

CO2 production, with a share of 31% in 2009 [15]. Therefore, statesmen as well as environmental activists are constantly striving to overcome this great challenge in modern societies. In addition, concerns over the limitation of the fossil fuel energies have doubled the incentive for governments to address this problem [16]. There are various technical solutions to reduce fossil fuel consumption, for example, the policy for increasing the fossil fuel prices, efforts to increase consumer efficiency and obtain fines for air pollution from the consumers, implementation of traffic control policies such as blocking single-vehicle traffic, traffic plans, and so on. Meanwhile, the use of EVs has been suggested as an effective way to reduce fossil fuel consumption and environmental concerns. However, given the high cost of EVs compared to traditional vehicles and the risk of charging these vehicles, customers are not very keen on buying these vehicles right now [17]. The solution for increasing the EVs penetration in the transportation fleet is that the price of petroleum products rises significantly and the energy storage technology improves efficiently. In this situation, EVs can compete with traditional cars. Furthermore, special support for the purchase of EVs in the form of various facilities is needed. Various facilities have been offered in many countries for the purchase of EVs which has resulted in increased customer satisfaction and increased penetration of EVs in the transportation fleet.

1.2.1 Facilities in Different Countries to Encourage the Purchase of EVs

In many countries, various facilities such as tax exemptions, discounts for environmental protection organizations, and so on are offered to buy Evs [18–22]. In Iran, for example, in 2014, the government abolished the taxes tariff on importing EVs into the country, thereby providing a good incentive to interest consumers. In the United States, in 2008, the law about EVs tax provided special discounts for EV buyers to increase market penetration. In Japan, the first incentive program began in 1996 and improved in 1998. Relevant facilities in Japan included subsidies and a tax rebate on the purchase of EVs. The plan employed in Japan would subsidize up to 50% of the extra costs of an EV compared to a conventional vehicle, which is about $ 8500. This may have caused Japan to have the largest number of EVs after the United States, so that about 95,000 EVs were sold in Japan from 2009 to 2014. In Canada, there is a $ 5000 discount for EVs up to 4 kW and $ 8500 for EVs up to 17 kW. Furthermore, EVs in Canada also have a special license plate called green license plate, with the benefits of being allowed to travel on routes that only vehicles with more than one passenger are permitted, or charging and parking in special parking lots. These incentives lead to an increase in the penetration of EVs in transportation fleet. The most important infrastructure needed in this regard is the availability of the installed charging stations in homes, personal business centers, and public stations across the city and along the roads.

1.2.2 The Cost of the Charging Infrastructure of EVs

To increase the penetration of EVs in transportation fleet, we need to consider the infrastructure cost for electrifying vehicles. The cost of EVs charging infrastructure will vary depending on the type, size and location of charging stations

Table 1.1 Cost of different charging stations

Charging station model	Lower bound price ($)	Base price ($)	Upper bound price ($)
Home EV charger 1.4 kW	25	75	550
Home EV charger 7.7 kW	500	1125	4000
Public EV charger 1.4 kW	1050	3000	9000
Public EV charger 7.7 kW	2500	5000	15,000
Public EV charger 38.4 kW	11,000	20,000	50,000

[23]. However, based on some assumptions, the cost of EVs' infrastructure can be estimated. One hypothesis is that the EV charging system should be uniform throughout the world; in other words, it should have the same standard [24]. This assumption makes a low charge cost, as anyone with any type of EV can charge at any station at any time. If this uniformity is not respected in the manufacture of charging equipment, in addition to causing problems in the charging process, the charging cost for EV owners will also be higher [25]. EV charging infrastructure costs include equipment costs and installation costs. To estimate the cost of the equipment, in general, the charge is assumed to be the same in all areas, and all equipment are installed by one company. In this case, the base price is extracted and the real price can be more or less dependent on the circumstances. It will be cheaper to install charging equipment at homes with robust wiring infrastructure. Installing a payment tool for credit cards or other ways will raise the price of this equipment. Generally, installing charging equipment in public places is more expensive than installing it at home [26]. As for home installation, the homeowner would have to pay more if the home has not strong wiring infrastructure. In the case of public charging, including charging in commercial areas, the cost of operating such as the cost of lighting and so on will also affect the final cost. The basic cost of EV charging infrastructure, along with its lower limit and upper limit, is shown in Table 1.1 [27].

From the points outlined in the preceding sections, it can be concluded that with the support of governments and the investment in appropriate infrastructures, EVs share will be increased significantly in transportation fleet by near future. It should be noticed that by increasing the penetration of EVs in transportation fleet, the electric load demand of EVs will be increased dramatically and we need new power plants to meet this need. If these new power plants are constructed based on fossil fuels, the environmental and pollutants concerns about conventional vehicles will lead to electricity generation part. Also, this increase in electrical load demand by EVs will put considerable stress on power systems, especially on electricity distribution networks, and network operators will have to build or reinforce transmission lines and other parts of the power system. To overcome this problem, researchers in EVs energy management field have proposed coordinated or smart charging strategies to reduce the adverse effects of EVs on power systems, and recommended various approaches for optimal smart charging producers [28]. Other researchers have also suggested the use of vehicle to grid (V2G) operation mode of EVs through smart charging in which EVs provide ancillary services to reduce the power system stress.

1.3 Different Types of Electric Vehicles

Generally, based on the technology used in EVs and type of their connection to the power grid, EVs are categorized in three main types including all-electric vehicles, hybrid electric vehicles (HEV), and plug-in electric vehicles (PEV) [29]. It is predicted that EVs with the ability to connect to the power grid will have a better future than others. In this part, different kinds of EVs are explained in details.

1.3.1 All-electric Vehicles

All-electric vehicles are the first generation of EVs which use the energy stored in batteries to power electric motors and provide propulsion power. The propulsion power of the vehicle is provided solely by electrical energy that is free of pollution and, therefore, they are known as non-polluting vehicles (zero pollution). The first generation of the this EV had limited battery capacity and was not capable of long distances. This, coupled with the high cost of batteries and lack of possibility for these vehicles to be charged by other sources, made the all-electric vehicles uneconomical and led them to be rejected by the customers. Recently, according to the significant improvements in battery technology, the prices of these vehicles have dropped significantly and it caused these EVs to receive high welcome from buyers. Nowadays, EVs are defined as plug-in electric vehicles (PEVs) and can be charged by power grid either at homes or in public places. This feature makes the all-electric vehicle as a good choice for customers and improves the reliability of its charging procedure. The main advantages of these vehicles are defined as follows:
- Completely free of greenhouse gas emissions.
- Much higher efficiency than internal combustion engines.
- Can be recharged with renewable energy sources.

1.3.2 Hybrid Electric Vehicles (HEVs)

These vehicles have both a fuel engine (internal combustion engine) and an electric motor with sufficient battery capacity to save energy from the fuel engine and brakes. Batteries come in handy when needed to produce auxiliary power, or at low speeds, by turning off the fuel motor to provide the driving force. About 1.5 million HEVs have been sold in the last decade. In the developed countries such as the United States, about 3% of existing vehicles are hybrid. Disadvantages of these vehicles are listed as follows:

- Unable to charge batteries over the network.
- Dependence on fossil fuel consumption engine.

1 Why Electric Vehicles? 7

Table 1.2 Specifications of common EVs

Vehicle model	EV kind	Charging rate (kW)	Charging time (h)	Trip length (km)	Battery capacity (kWh)
Mitsubishi iMiEV	EV	3.1	7	100	16
Nissan leaf	EV	3.3	8	118	24
Tesla model S	EV	11	8.5	425	85
Chevrolet volt	PHEV	3.3	3	61	16.5
Toyota Prius	PHEV	3.3	1.5	18	5.2
Ford fusion	PHEV	3.3	2.5	34	7.5

1.3.3 Plug-in Hybrid Electric Vehicles (PHEVs)

The plug-in hybrid electric vehicles (PHEVs) are combinations of the two previous types, and have been designed to eliminate the disadvantages of them. As they have rechargeability by the power grid, they require batteries with more capacity than HEVs. The major difference between the batteries of these two types of EVs (PHEVs and HEVs) is that the PHEV battery must be capable of fast discharge and fast recharge, while HEV batteries operate in near-full charge and discharge rarely occurs. Since the PHEVs can be recharged with a power grid at homes and in public places, the volume of batteries used in PHEVs has increased. It is therefore possible to travel longer distances than HEVs using the non-polluting mode (electric mode). The most important feature of PHEVs or PEVs is their ability to connect to the power grid and the possibility of bi-directional power exchange— grid to vehicle and vehicle to grid— which are known as G2V and V2G, respectively. It should be noted that with the increasing number of PHEVs and PEVs in the future, there will be a significant volume of energy storage which can help the operation of the power system by providing the ancillary services. The specifications of some of the most popular EVs in the world are shown in Table 1.2 [28].

1.4 Different Charging Rates of EVs

To assess and evaluate the impact of EVs and power system on each other, the charging level of the EVs must be known. The researchers in this field, by considering the charging infrastructure capability, have introduced three levels of charging as follows [30–32]:

- Level 1 (about 2 kW charging rate)
- Level 2 (about 7 kW charging rate)
- Level 3 (about 50 kW charging rate)

In this part, we are going to describe these charging levels in details.

1.4.1 Charging Level 1

Level 1 (Slow Charging): This level of charge is usually AC and single phase and is known as the standard charging level. Charging rates vary across countries; in many European countries, it is around (230 V, 16 A and 3.7 KW). In some other European countries, such as England and Switzerland, the standard charging levels are lower, (230 V and 13 A) and (230 V and 10 A), respectively.

1.4.2 Charging Level 2

Level 2 (fast charge): This charge level is in one or three phases and beyond the standard charge level; these conditions may be available in residential and commercial environments. There are different charging rates in different countries, ranging from 10 to 20 kW.

1.4.3 Charging Level 3

Level3 (very fast charge): This charge level comes in three phases AC and DC. This charging surface requires an external charging, and depending on the level of charging, it requires power equipment for cooling electronic devices. Charging rate is above 50 kW. This level of charging is usually used by drivers in emergency situations that require low charging time. In this case, the time factor is very important.

However, when the penetration of EVs is high, usually charging level 3 can refer to special charging stations like gas stations, where the EV can get 50% of its charge in 10–15 minutes. A brief description about various charging levels are illustrated in Table 1.3.

1.5 Charging of EVs

Charging of EVs is done for different purposes with different structures. Depending on the power system conditions, the purpose of the optimal charging process and its implementation structure will be determined. In this regard, in this part, different goals of EVs charging and then different structures of smart charging implementation are introduced.

1 Why Electric Vehicles?

Table 1.3 various charging levels [33]

Charging specifications	Charging level 1	Charging level 2	Charging level 3
Voltage (V)	120 V 1-phase AC	208 V DC or 240 V 1-phase AC	208 V DC or 480 V 3-phase AC
AMPS (A)	12–16	12–80	<125
Charging rates (kW)	1.4–1.9	2.5–19.2	<90
Charging time for vehicle	10–20 miles of range per hour	10–20 miles of range per hour	80% charge in 20–30 minutes

1.5.1 Different Purposes of Optimal Charging of EVs

The EVs' smart charging strategy can be set for different purposes. Each of these goals has its own advantages and disadvantages which are given in Table 1.4 [34–36]. As shown in Table 1.4, as we move from uncoordinated charging to smart charging, the EVs' charging implementation algorithm becomes more sophisticated and more infrastructures are required; however, smart charging algorithms bring more benefits for the power grid operator and EV owners. It should be mentioned that if all the EVs charge in an uncoordinated manner without specific control and scheduling, and each EV does its charging as soon as it arrives home, subscribers will have more freedom to operate; however, power grid will experience sharp peak loads which have harmful effects on the power system stability and operation. So, it is better to use smart charging procedure.

1.5.2 Different Structures for Implementation of Smart Charging

To implement smart charging of EVs, the infrastructure needs to communicate with different components of this problem including subscribers, aggregators and network operators. Following, we discuss different structures for implementing smart charging procedure.

1.5.2.1 Centralized Charging Control Structure

This structure is known as direct charging. As shown in Fig. 1.4, in this arrangement, the aggregator is directly responsible for charging all EVs in its area. The aggregator

Table 1.4 Advantages and disadvantages of different charging approaches

Charging mode	Load profile	Advantages	Disadvantages
Uncoordinated charging	0 2 4 6 8 10 12 14 16 18 20 22	Easy implementation Comfortable for EV owners	Overloading in power transformers and distribution feeders. Increasing the peak load of the distribution network Increasing the electricity cost Increasing the need to reinforce the grid
Charging in off-peak time	0 2 4 6 8 10 12 14 16 18 20 22	Easy implementation Flattening the load demand profile Improving the integration of renewable energies at off-peak hours Delaying the grid investments	Easy implementation Unbalancing the load demand profile due to the sharp increase of EVs demand Possibility for voltage deviations Decreasing the customers' welfare
Optimal charging for valley filling	0 2 4 6 8 10 12 14 16 18 20 22	Providing the ancillary services for power system Flattening the load demand profile Improving the integration of renewable energies at off-peak hours Delaying the grid investments	Complex implementation Requiring the ICT infrastructures Decreasing the customers' welfare
Optimal charging for peak shaving	0 2 4 6 8 10 12 14 16 18 20 22	Providing the ancillary services for power system Reducing the peak load Improving the integration of renewable energies at off-peak hours Delaying the investments for the power system reinforcements	Too complex implementation Requiring the ICT infrastructures Degradation of the customers' batteries during the V2G Energy loss in different operation modes

can control other organizations such as charging stations, where each charging station is responsible for charging the EVs inside its territory. In this structure, the aggregator is connected to the distribution network operator and the transmission system operator to control the charging procedure of the EVs according to technical and economic conditions of the power system. In this way, the aggregator first

1 Why Electric Vehicles?

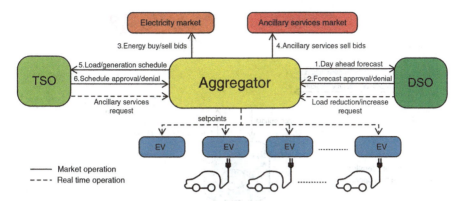

Fig. 1.4 The overall structure of the centralized charging control method [37, 38]

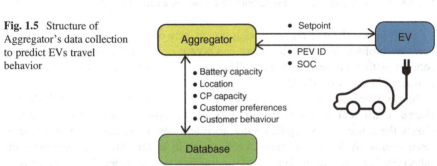

Fig. 1.5 Structure of Aggregator's data collection to predict EVs travel behavior

estimates the load demand of EVs according to the predicted subscribers' travel behavior, then, informs the distribution and transmission network operator, and after confirming them, finally, the aggregator submits the predicted load demand in the electricity market.

Under these circumstances, the technical constraints of the upstream network have also been considered. The aggregator, in coordination with the distribution and transmission network, can also participate in the electricity market for ancillary services such as frequency control and reservation [39, 40]. The aggregator has to collect various information about the subscribers' travel behavior to predict the load demand of EVs as shown in Fig. 1.5. In this structure, the aggregator sends the charging signal to the EV owners and controls the charging demand of the EVs by considering the technical limitations of the power system for different purposes including frequency and voltage regulations, power loss reduction, minimizing charging costs, and so on.

1.5.2.2 Decentralized Charging Control Structure

This structure is known as an indirect charging; in this case, rather than an intermediary entity (aggregator) determines the charging procedure, EVs receive different signals from the upstream network or aggregators and participate in various schemes

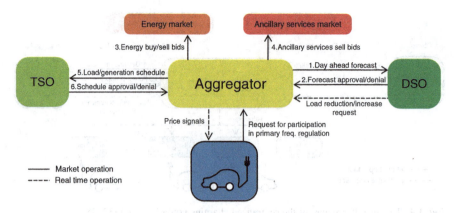

Fig. 1.6 The overall structure of the decentralized charging control method

by their choice. In this regard, the upstream network only suggests the optimal charging profiles and the subscribers can accept or reject the suggested charging patterns with their own judgment. The main configuration of this structure is illustrated in Fig. 1.6 [41, 42].

In fact, in this structure, it is up to the customer to decide how much and when to charge; in different situations, the aggregator or the upstream network operator only directs the subscriber by updating the price and control signals. This structure is more commonly known as multi-agent approach. In this structure, because each subscriber is a decision-maker, smart communication equipment has to be considered for each EV or private charging places which creates additional cost in the charging infrastructure part.

1.5.2.3 Comparison Between Centralized and Decentralized Smart Charging Structures

To choose the right smart charging structure, these two general methods need to be compared.

In the centralized method, all the charging procedures are performed by the aggregator and it is the aggregator that adjusts the charging profile of all the EVs by considering different charging demand of the subscribers and the upstream network condition. By increasing the penetration of EVs in transportation fleet, centralized method requires a large database and a powerful processor in the aggregator's compiler center to determine the optimal charging procedure of EVs [43]. This structure has higher reliability than decentralized structure because, here, the charging method is determined by the aggregator and the subscribers (EV owners) must obey the charging signals most of the time. In this regard, the upstream network operator can count on the ancillary services such as frequency control by EVs.

1 Why Electric Vehicles?

Table 1.5 Specifications of common electric vehicles

Smart charging mode	Advantages	Disadvantages
Centralized optimal charging	Clear and well-known configuration Optimal usage of the power grid capacity Providing a reliable way for ancillary services	Robust data processing server is needed Complex communication services for charging signals are required Customers privacy is not considered
Decentralized optimal charging	Improving the customers' welfare Charging control in user side Lower fault tolerance	High uncertainty in the final results Inability to provide reliable uncertainty services Forecasting customers reaction is necessary Simultaneous behavior of the customers may be happened which causes avalanche effects

In the decentralized structure, the upstream network operator or aggregator only transmits different charging commands or price signals to the subscribers depending on the power systems conditions, and it is the EV owners who execute these signals according to their situations, and they may not participate in some programs. In this way, each subscriber must have its own smart charging program that requires more intelligent equipment. Furthermore, in this case, EVs will be less likely to participate in ancillary services such as frequency control and reservation backups.

Considering different studies in this field, each of the above structures has its own advantages and disadvantages. A general comparison of these two structures is presented in Table 1.5 which implies that the centralized structure is more applicable with high penetration rate of EVs in transportation fleet.

1.6 EVs as a Big Storage Unit in Power System by V2G

EVs, when connected to the power grid (plug-in EVs), offer a wide range of applications. If they are equipped with the right hardware and software, EVs will be able to transfer power in bi-directional ways including charging (G2V[1]) and discharging (V2G[2]) [44, 45]. Indeed, in V2G mode, EV acts as a storage unit and by increasing the penetration of EVs in transportation fleet, we have a large capacity of batteries. In this regard, V2G mode is an important ability of EVs which can help power system operation in different ancillary services. In this part, we are going to discuss various purposes of operating EVs in V2G mode.

---

[1]Grid to vehicle

[2]Vehicle to grid

1.6.1 Application of the V2G Mode as Virtual Power Plants

The storage capacity of a bunch of EVs can be used as virtual power plants as shown in Fig. 1.7. This approach is commonly applied by V2G operation mode in the power generation and transmission sector. Minimizing energy costs and emissions by focusing on the integration of large-scale renewable resources is the most important issue in this regard [46, 47]. Virtual power plants can be used to balance generation and load demand, reduce fossil power generation capacity and replace expensive power plants, especially during peak hours. In this application, the energy of each EV is not considerable, but the total available capacity of EVs' batteries which provide possible ancillary services is calculated. In fact, the concept of virtual power plants has been taken into account because of the small size of the EVs' batteries in power system scale. From this point of view, this application can be applied to all levels of the power system with the presence of other available resources. Similarly, the concept of virtual power plants can be used in micro-grids, in standalone operation mode, to control the fluctuations of renewable energies [48].

1.6.2 Application of V2G to Improve Power System Security and Resiliency

The optimal location of EV charging stations is important for the power system security, particularly, when this approach is implemented in the distribution

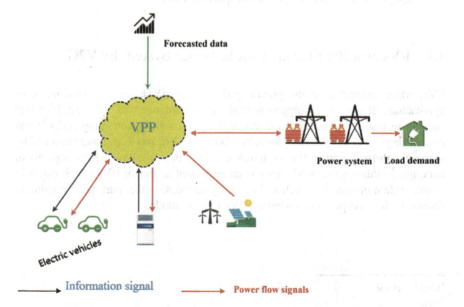

Fig. 1.7 VPP configuration with different energy resources

networks and microgrids with different types of configurations, such as radial or ring grids with different voltage levels. The V2G is useful in different contingency situations such as outage of a power system equipment. This highlights the importance of finding the optimal location of EVs parking lots to provide the needed energy in contingency situations by V2G. Optimal location of EVs parking lots is so useful at medium voltage distribution levels.

1.6.3 Application of V2G for Implementation of the Microgrids

A microgrid is defined as the set of the various distributed generation energy sources and storage units. Most of the valuable energy resources in the microgrid are renewable resources, so using the capacity of EVs as a storage unit can be so effective in damping their natural power fluctuations [49]. Besides, V2G can be used for services such as voltage control and frequency regulation in the microgrid. It should be mentioned that EVs are high-performance resources with flexible behavior and quick response, and smart V2G can also improve demand-side management approaches in microgrids.

1.6.4 Application of V2G for Ancillary Services in the Distribution Network

The importance of EV charging units in the electricity market has been further enhanced by the increasing tendency to use renewable energy sources in combination with a reliable and flexible energy source for better regulation and reservation. The number of available batteries is important for quick response applications in short-term markets. The best market for EVs is the ancillary services market [50]. The main auxiliary services in the wholesale market, where EVs can play a key role in them, are the frequency regulation and the spinning reserve markets. The frequency regulation market is an efficient market for V2G, and advances in fast-charging infrastructures can make it even more operational. EVs can provide both one-way and two-way auxiliary services [51]. In one-sided performance, EVs only participate in pricing the frequency regulation and spinning reserve in the ancillary service market. They cannot inject the energy stored in the batteries into the grid, so a V2G concept is provided for this goal. V2G, in a demand-side management system, can technically lead to balance the power in the smart grid structure. In addition, V2G capabilities can be used to manage load demand profile and reduce the peak load as an alternative to expensive generation units. V2G services have a significant impact on the peak load of the small networks such as microgrids. It can be assumed that EV owners have fully charged the batteries at work and injected them into the power grid during the peak hours. In this approach, the available battery capacity for

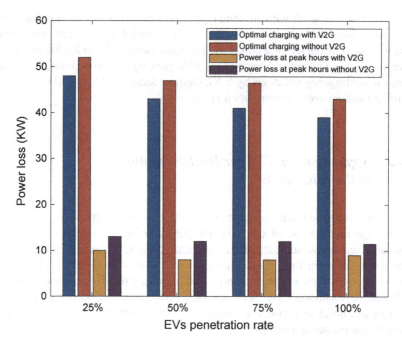

Fig. 1.8 Residential distribution network power losses with different charging modes

V2G will be obtained through the driving pattern between home and work. It is a safe assumption that this will be possible with the expansion of smart parking lots in the commercial industrial areas, universities and so on. As shown in Fig. 1.8, V2G has successfully reduced the power loss at peak hours by half. The relationship between different charging strategies and power loss of the distribution network is illustrated in Fig. 1.9 [52].

1.7 Conclusion

In this chapter, a brief discussion about various features of EVs is presented. The modern societies are moving toward electrification of the transportation fleet. This happens due to the limitation of fossil fuel resources and environment pollution in the high crowded cities. EVs are expensive and they have some limitations which are not in conventional internal combustion vehicles such as several charging for long distances and so on; in this regard, policymakers in different developed countries consider various incentives such as tax reduction and low-priced public services to encourage the people to buy EVs. In other points, as we know, the electrification of the transportation fleet is the best solution to solve these problems, but it should be noticed that by increasing the penetration of EVs, we are facing large electrical power demand in power system. This can cause some problems for power system operation such as voltage deviations, increasing the power loss and so on. To solve this problem,

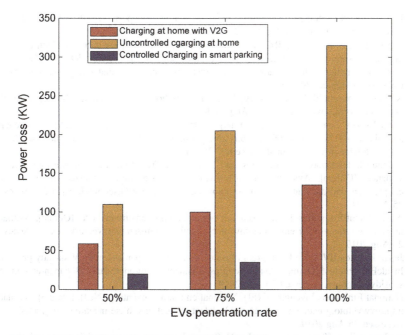

Fig. 1.9 Residential distribution network power losses for different charging scenarios with V2G at home and controlled charging in smart parking

intermediary units, which are known as aggregators, are considered to do the charging procedure of EVs by considering the power system constraints and minimizing the EV owners charging cost as much as possible. In this way, various smart charging approaches are introduced to handle the aggregator's job such as increasing the penetration rate of EVs and improving the efficiency of the charging procedure of EV; but now, we are at the first steps of the electrification of the power system and to achieve a complete electric transportation, which is the target of the developed countries, we have a long way to go and more studies are needed in this regard.

References

1. X. Li, P. Chen, X. Wang, Impacts of renewables and socioeconomic factors on electric vehicle demands–panel data studies across 14 countries. Energy Policy **109**, 473–478 (2017)
2. H. Tayarani, H. Jahangir, R. Nadafianshahamabadi, M. Aliakbar Golkar, A. Ahmadian, A. Elkamel, Optimal charging of plug-in electric vehicle: Considering travel behavior uncertainties and battery degradation. Appl. Sci. **9**(16), 3420 (2019)
3. H. Kheradmand-Khanekehdani, M. Gitizadeh, Well-being analysis of distribution network in the presence of electric vehicles. Energy **155**, 610–619 (2018)
4. W. Su, H. Eichi, W. Zeng, M.-Y. Chow, A survey on the electrification of transportation in a smart grid environment. IEEE Trans. Industr. Inform. **8**(1), 1–10 (2011)

5. K.J. Dyke, N. Schofield, M. Barnes, The impact of transport electrification on electrical networks. IEEE Trans. Ind. Electron. **57**(12), 3917–3926 (2010)
6. J. Aghaei, A.E. Nezhad, A. Rabiee, E. Rahimi, Contribution of plug-in hybrid electric vehicles in power system uncertainty management. Renew. Sustain. Energy Rev. **59**, 450–458 (2016)
7. E. U. Information Administration – Department of Energy, Annual Energy Review 2011 – Released September 2012, (2011)
8. National Household Travel Survey, National household travel survey. [Online]. Available https://nhts.ornl.gov/. Accessed 25 Aug 2019
9. Total Energy Annual Data – U.S. Energy Information Administration (EIA), Total energy annual data – U.S. Energy Information Administration (EIA). [Online]. Available https://www.eia.gov/totalenergy/data/annual/. Accessed 25 Aug 2019
10. Bureau of, Transportation, Statistics, and U. S. D. of Transportation, National transportation statistics. [Online]. Available https://www.bts.gov/sites/bts.dot.gov/files/docs/browse-statistical-products-and-data/national-transportation-statistics/220806/ntsentire2018q1.pdf. Accessed 25 Aug 2019
11. Air Pollutant Emissions Trends Data, Air pollutant emissions trends data. [Online]. Available https://www.epa.gov/air-emissions-inventories/national-emissions-inventory-nei. Accessed 25 Aug 2019
12. Japan Extends EV Subsidy Program | InsideEVs Photos, Japan extends EV subsidy program | InsideEVs Photos. [Online]. Available https://insideevs.com/photos/657686/japan-extends-ev-subsidy-program/. Accessed 25 Aug 2019
13. National Emissions Inventory (NEI), National emissions inventory (NEI). [Online]. Available http://www.mto.gov.on.ca/english/vehicles/electric/electric-vehicle-incentive-program.shtml. Accessed 25 Aug 2019
14. P. Campbell, Y. Zhang, F. Yan, Z. Lu, D. Streets, Impacts of transportation sector emissions on future US air quality in a changing climate. Part I: Projected emissions, simulation design, and model evaluation. Environ. Pollut. **238**, 903–917 (2018)
15. S. Wang, J. Wang, J. Li, J. Wang, L. Liang, Policy implications for promoting the adoption of electric vehicles: Do consumer's knowledge, perceived risk and financial incentive policy matter? Transp. Res. Part A Policy Pract. **117**, 58–69 (2018)
16. P. Campbell, Y. Zhang, F. Yan, Z. Lu, D. Streets, Impacts of transportation sector emissions on future US air quality in a changing climate. Part II: Air quality projections and the interplay between emissions and climate change. Environ. Pollut. **238**, 918–930 (2018)
17. J. Zhao, C. Wan, Z. Xu, J. Wang, Risk-based day-ahead scheduling of electric vehicle aggregator using information gap decision theory. IEEE Trans. Smart Grid **8**(4), 1609–1618 (2015)
18. S. Ma, P. Gao, H. Tan, The impact of subsidies and charging facilities on demand for electric vehicles in China. Environ. Urban. ASIA **8**(2), 230–242 (2017)
19. M.A. Aasness, J. Odeck, The increase of electric vehicle usage in Norway—Incentives and adverse effects. Eur. Transp. Res. Rev. **7**(4), 34 (2015)
20. E. Morganti, M. Browne, Technical and operational obstacles to the adoption of electric vans in France and the UK: An operator perspective. Transp. Policy **63**, 90–97 (2018)
21. M.A. Brown, A. Soni, Expert perceptions of enhancing grid resilience with electric vehicles in the United States. Energy Res. Soc. Sci. **57**, 101241 (2019)
22. D. Lopez-Behar, M. Tran, J.R. Mayaud, T. Froese, O.E. Herrera, W. Merida, Putting electric vehicles on the map: A policy agenda for residential charging infrastructure in Canada. Energy Res. Soc. Sci. **50**, 29–37 (2019)
23. S. Davidov, M. Pantoš, Planning of electric vehicle infrastructure based on charging reliability and quality of service. Energy **118**, 1156–1167 (2017)
24. I. Rahman, P.M. Vasant, B.S.M. Singh, M. Abdullah-Al-Wadud, N. Adnan, Review of recent trends in optimization techniques for plug-in hybrid, and electric vehicle charging infrastructures. Renew. Sustain. Energy Rev. **58**, 1039–1047 (2016)

25. T.D. Chen, K.M. Kockelman, J.P. Hanna, Operations of a shared, autonomous, electric vehicle fleet: Implications of vehicle & charging infrastructure decisions. Transp. Res. Part A Policy Pract **94**, 243–254 (2016)
26. F. Ahmad, M.S. Alam, M. Asaad, Developments in xEVs charging infrastructure and energy management system for smart microgrids including xEVs. Sustain. Cities Soc. **35**, 552–564 (2017)
27. A. Ahmadian, M. Sedghi, M. Aliakbar-Golkar, Fuzzy load modeling of plug-in electric vehicles for optimal storage and dg planning in active distribution network. IEEE Trans. Veh. Technol. **66**, 3622–3631 (2017)
28. H. Jahangir et al., Charging demand of plug-in electric vehicles: Forecasting travel behavior based on a novel rough artificial neural network approach. J. Clean. Prod. **229**, 1029–1044 (2019)
29. K.Y. Bjerkan, T.E. Nørbech, M.E. Nordtømme, Incentives for promoting battery electric vehicle (BEV) adoption in Norway. Transp. Res. Part D Transp. Environ **43**, 169–180 (2016)
30. G. Binetti, A. Davoudi, D. Naso, B. Turchiano, F.L. Lewis, Scalable real-time electric vehicles charging with discrete charging rates. IEEE Trans. Smart Grid **6**, 2211–2220 (2015)
31. F.V. Cerna, M. Pourakbari-Kasmaei, R.A. Romero, M.J. Rider, Optimal delivery scheduling and charging of EVs in the navigation of a city map. IEEE Trans. Smart Grid **9**, 4815–4827 (2018)
32. M. Aziz, T. Oda, M. Ito, Battery-assisted charging system for simultaneous charging of electric vehicles. Energy **100**, 82–90 (2016)
33. Sunlight Solar Energy, Sunlight solar energy – Electric Vehicle (EV) chargers with solar energy. [Online]. Available http://sunlightsolar.com/learning-center/ev-charging/. Accessed 25 Aug 2019
34. J. García-Villalobos, I. Zamora, J.I. San Martín, F.J. Asensio, V. Aperribay, Plug-in electric vehicles in electric distribution networks: A review of smart charging approaches. Renew. Sustain. Energy Rev. **38**, 717–731 (2014)
35. Y. Zheng, S. Niu, Y. Shang, Z. Shao, L. Jian, Integrating plug-in electric vehicles into power grids: A comprehensive review on power interaction mode, scheduling methodology and mathematical foundation. Renew. Sustain. Energy Rev. **112**, 424–439 (2019)
36. R. Wang, P. Wang, G. Xiao, Two-stage mechanism for massive electric vehicle charging involving renewable energy. IEEE Trans. Veh. Technol. **65**, 4159–4171 (2016)
37. M.D. Galus, M.G. Vayá, T. Krause, G. Andersson, The role of electric vehicles in smart grids. Wiley Interdiscip. Rev.: Energy Environ. **2**(4), 384–400 (2013)
38. J.A.P. Lopes, F.J. Soares, P.M.R. Almeida, Integration of electric vehicles in the electric power system. Proc. IEEE **99**, 168–183 (2011)
39. S.I. Vagropoulos, A.G. Bakirtzis, Optimal bidding strategy for electric vehicle aggregators in electricity markets. IEEE Trans. Power Syst. **28**, 4031–4041 (2013)
40. Y. He, B. Venkatesh, L. Guan, Optimal scheduling for charging and discharging of electric vehicles. IEEE Trans. Smart Grid **3**, 1095–1105 (2012)
41. X. Xi, R. Sioshansi, Using Price-based signals to control plug-in electric vehicle Fleet charging. IEEE Trans. Smart Grid **5**, 1451–1464 (2014)
42. L. Gan, U. Topcu, S.H. Low, Optimal decentralized protocol for electric vehicle charging. IEEE Trans. Power Syst. **28**, 940–951 (2013)
43. H. Jahangir et al., A novel electricity price forecasting approach based on dimension reduction strategy and rough artificial neural networks. IEEE Trans. Industr. Inform. **99**, 1–1 (2019)
44. S. Habib, M. Kamran, U. Rashid, Impact analysis of vehicle-to-grid technology and charging strategies of electric vehicles on distribution networks – A review. J. Power Sources **277**, 205–214 (2015)
45. M.H. Abbasi, M. Taki, A. Rajabi, L. Li, J. Zhang, Coordinated operation of electric vehicle charging and wind power generation as a virtual power plant: A multi-stage risk constrained approach. Appl. Energy **239**, 1294–1307 (2019)

46. H. Jahangir, A. Ahmadian, M. Aliakbar-Golkar, M. Fowler, A. Elkamel, Optimal design of standalone micro-grid considering reliability and investment costs, in *IET Conference Publications* (2016)
47. H. Tayarani, S. Baghali, H. Jahangir, M.A. Golkar, A. Fereidunian, Travel behavior and system objectives uncertainties in electric vehicle optimal charging, in *2018 Smart Grid Conference (SGC)* (2018), pp. 1–6
48. J. Aghaei, M. Barani, M. Shafie-Khah, A.A. Sanchez De La Nieta, J.P.S. Catalao, Risk-constrained offering strategy for aggregated hybrid power plant including wind power producer and demand response provider. IEEE Trans. Sustain. Energy **7**, 513–525 (2016)
49. S. Dinkhah, C.A. Negri, M. He, S.B. Bayne, V2G for reliable microgrid operations: Voltage/frequency regulation with virtual inertia emulation, in *2019 IEEE Transportation Electrification Conference and Expo (ITEC)* (2019), pp. 1–6
50. B.K. Sovacool, R.F. Hirsh, Beyond batteries: An examination of the benefits and barriers to plug-in hybrid electric vehicles (PHEVs) and a vehicle-to-grid (V2G) transition. Energy Policy **37**(3), 1095–1103 (2009)
51. L. Noel, G. Zarazua de Rubens, J. Kester, B.K. Sovacool, Navigating expert skepticism and consumer distrust: Rethinking the barriers to vehicle-to-grid (V2G) in the Nordic region. Transp. Policy **76**, 67–77 (2019)
52. H. Turton, F. Moura, Vehicle-to-grid systems for sustainable development: An integrated energy analysis. Technol. Forecast. Soc. Change **75**, 1091–1108 (2008)

Chapter 2
Artificial Intelligence-based Approach For Electric Vehicle Travel Behavior Modeling

Hamidreza Jahangir, Masoud Aliakbar Golkar, Ali Ahmadian, and Ali Elkamel

2.1 Introduction

With the increase of the penetration of the electric vehicles (EVs) in the transportation fleet, finding an accurate and precise method for modeling EV travel behavior is a vital challenge in the optimal charging procedure of these vehicles. In fact, if we do not have a suitable model of the EVs travel behavior, we cannot estimate their electric load demand profile accurately which has a solid effect on the optimal charging results. Forecasting the travel behavior profile of EVs is really a complicated task because these vehicles have a high intermittent behavior which cannot be forecasted easily. In this way, we need a robust forecasting technique to estimate the intermittent behavior of various drivers in an accurate manner. The aggregator, which is an internal unit between EV owners and power system operator, should forecast the EVs behavior accurately to increase its profit and decrease the charging cost of EV owners as much as possible [1]. In this way, the aggregator can attract more customers, which will be needed soon when the number of aggregators increases dramatically. It should be noticed that in modeling the EV travel behavior problem, we are facing various travels with different travel purposes which impose high uncertainty in this problem. The best solution for this problem—with a large

---

H. Jahangir (✉) · M. A. Golkar
Faculty of Electrical Engineering, K. N. Toosi University of Technology, Tehran, Iran
e-mail: h.r.jahangir@email.kntu.ac.ir; golkar@kntu.ac.ir

A. Ahmadian
Department of Electrical Engineering, University of Bonab, Bonab, Iran
e-mail: ahmadian@bonabu.ac.ir

A. Elkamel
Department of Chemical Engineering, University of Waterloo, Waterloo, ON, Canada

College of Engineering, Khalifa University of Science and Technology, The Petroleum Institute, Abu Dhabi, UAE
e-mail: aelkamel@uwaterloo.ca

© Springer Nature Switzerland AG 2020
A. Ahmadian et al. (eds.), *Electric Vehicles in Energy Systems*,
https://doi.org/10.1007/978-3-030-34448-1_2

volume of data and high uncertainty—is the artificial intelligence-based approaches [2]. These approaches use historical data to find the pattern of the travels and have a good performance in large dimension problems. Recently, artificial intelligence-based approaches are constructed based on a new concept which is known as Deep Learning. Deep learning is the artificial intelligence-based method with various number of representation layers to handle data in a very precise manner. Deep learning has been employed in various studies such as image processing, pattern recognition, and classification tasks [3]. In transportation systems such as EV travel behavior modeling, we need an artificial-intelligence-based network with multi representation layers. In this chapter, we are going to introduce the implementation of data engineering-based approach with deep artificial neural networks in forecasting the travel behavior of EVs and study the effects of accurate EV travel behavior forecasting on the aggregator's incomes.

The overall structure of this chapter is presented as follows:

First, in Sect. 2.2, various EV travel behavior modeling approaches are presented with their pros and cons. After that, in Sect. 2.3, the artificial intelligence approach based on artificial neural networks is explained with the formulations. In Sect. 2.4, the optimal charging process of EVs is presented. Some numerical results about the forecasting of the EV's travel behavior are given in Sect. 2.5. Section 2.6 concludes the findings of this chapter.

2.2 Different Approaches to EV Travel Behavior Modeling

Various studies have been done on optimal charging of EVs and different methods were applied to handle the EV's modeling problem. We can categorize them in five main approaches as follows:

- Monte Carlo simulation
- Markov Chain
- Queuing theory
- Trip chain and origin-destination
- Artificial intelligence

In this section, these methods as well as different studies which have been done in this field will be discussed by explaining the deficiencies of each model.

2.2.1 Monte Carlo Simulation method

The Monte Carlo simulation (MCS), which is the benchmarking method in this field [4–10], is implemented to model EV travel behavior based on generating different scenarios. The precision of the MCS method highly depends on the number of generated scenarios [11]. In this way, the computational cost will increase expressively, and the proposed optimal charging algorithm based on the MCS method is pointless for the complex problem with a large number of EVs. The low number of generated scenarios will also significantly reduce the correctness of the numerical results [12]. Most existing works by MCS method, [13–15], employed normal probability distribution function for forecasting all the travel behavior parameters such as departure time, arrival time, and trip length which reduces the precision of forecasting results.

Moreover, the main deficiency of implementing the MCS method for modeling the travel behavior of EVs is that the correlation between various travel behavior parameters such as arrival time, departure time and trip length is not employed in generating scenarios because they are generated by different probability distribution functions separately. Thus, the forecasting result may have some impossible trips. For instance, it is possible to generate scenarios in which the trip length may not tie with the departure and arrival times.

2.2.2 Markov Chain Theory

Some studies have employed the Markov chain theory to forecast EVs travel behavior and their electric load demand. For instance, Sun et al., [14, 16] have implemented a conventional Markov chain method to model EV travel behavior which considers various drivers' actions as the states of the transition matrix of the Markov chain. Considering the transition matrix as the probability of transition between different steps is another study which has employed the Markov chain to model EVs travel behavior [17]. In another work, Zhou et al. [18] implemented grey-Markov chain theory to forecast EVs travel behavior based on transition matrix among different statues of EVs optimal charging process. Sun et al. [14] have combined MCS with Markov chain theory to generate various scenarios. In all these studies, the Markov transition matrix is calculated separately for various travel parameters, and the correlation between various travel parameters such as departure time, arrival time, and trip length is neglected. Furthermore, to increase the accuracy of the Markov chain method, we need a large number of states which increase the transition matrix dimension and the computational cost dramatically. These short-comings have limited the application of Markov chain method in real case studies which have large number of EVs.

2.2.3 Queuing Theory

The queuing theory is another approach in EV travel behavior modeling. This method is more complicated than MCS and Markov chain methods and considers an intermittent behavior of various drivers by applying a homogeneous Poisson technique to find the departure and arrival times of EVs [19–23]. For example, [24] has integrated M/M/s queuing topology by employing the traffic flaw instructions to forecast EVs load demand in the highway and commercial centers charging stations with various charging rates. This method needs many assumptions, and for implementing this approach for modeling the EVs travel behavior, many parameters must be selected based on the operator experience, which is the main drawback for large scale projects.

Furthermore, in queuing theory, travel behavior and optimal charging procedure are determined altogether and, in this way, it cannot be operative in an optimal charging process with considering various charging constrains. In another work, Hafez et al. [25], have presented a new queuing-based approach for optimal charging of EVs by employing a non-homogeneous Poisson function for modeling travel behavior of different drivers. This method presents more accurate results in comparison with previous studies in which queuing theory has been employed; however, still it does not consider the correlation between various travel behavior parameters which are so important in the final numerical results of the optimal charging procedure.

2.2.4 Trip Chain and Origin-Destination Methods

In most of the studies by transportation researchers, the trip chain and origin-destination (O-D) are employed. For instance, in [26], a trip chain modeling approach has been presented for forecasting the trip length, arrival, and departure times of EVs travel behavior. The main focus of this method is to find the transition between different driving states. This method can be combined with other approaches such as the MCS method. In [27], the authors have introduced a hybrid method based on MCS and trip chain technique by employing the probability distribution of drivers' travel data. Mu et al. [28] have proposed a novel model for forecasting the load demand of the EVs by employing the O-D technique. This approach, however, is more appropriate for modeling the conventional private vehicles which have simpler travel behavior than EVs because they do not need various charging during a week.

2.2.5 Artificial Intelligence

Artificial intelligence-based approaches such as artificial neural networks (ANNs) are the best solution for the problems with large dimension input data and high stochastic behavior. Indeed, in modeling the EVs travel behavior, we need a robust approach in handling the high uncertainty of the input data. Furthermore, the input data regarding travel behaviors of various drivers, which are taken from various sources such as GPS, may contain a bad data, and, in this way, omitting this bad data by denoising techniques should be considered. In this regard, recently, some studies have been presented for modeling the EVs travel behavior by ANNs. In one of the few studies in this field, in [29], ANNs have been employed in the conventional form to forecast EVs travel behavior, including departure and arrival times and trip length. The findings of this work are not accurate because a sample form of ANN is employed with limited data.

Furthermore, this study used MCS to increase the training data which has a negative effect on the training procedure. The conventional forms of ANNs have shallow structures and do not present an accurate result because they have weak ability in feature extraction task. To handle this problem, ANNs with deep structure, which are robust data engineering methods in large dimension tasks such as EVs travel behavior forecasting, are employed. EVs travel behavior has high uncertainty which needs to be controlled by a strong data engineering method such as rough ANNs. Rough ANNs have acceptable performance in handling the input data uncertainty and are implemented in various tasks such as short-term wind speed forecasting [30], signature recognition with large and noisy data [31], short-term electrical load demand forecasting [32], and presented acceptable results. Rough networks improved the forecasting results with high uncertainties by implementing interval upper and lower bounds weights in hidden layers. Rough neuron-based ANNs with various hidden layers are promising tools to handle great stochastic problems [33].

Based on the mentioned survey of various approaches in modeling travel behavior of EVs, the ANN-based approaches are the best solution for this goal and can implement the correlation between various travel behavior parameters carefully with denoising input data. Considering the excellent performance of ANNs with rough neurons, in this chapter, we are going to introduce these techniques for EVs travel behavior forecasting task. More details about these networks are given in Sect. 2.3.

2.3 Modeling of EVs by ANNs

ANNs are promising tools in handling large dimension tasks with high uncertainty such as EVs travel behavior forecasting. ANNs learn the EVs travel behavior by their historical data, and, in this way, we need to find the appropriate input data parameters to forecast the target data. In ANN's training procedure, we divide historical data into three parts, including training, validation, and test. Indeed, ANN learns the data behavior from training data sets, and after that, the training accuracy is evaluated by validation data sets during training. Finally, when training and validation processes are terminated, the final test of the ANN performance is done by the test data set. In this way, we can guarantee the ANN performance with other data sets.

It should be noticed that the structure of the ANN is made based on the level of difficulty of the case study. EVs travel behaviors have intermittent character, and, accordingly, we need to employ the robust form of the ANNs to forecast travel behavior of the EVs. In this regard, we have introduced robust training form of

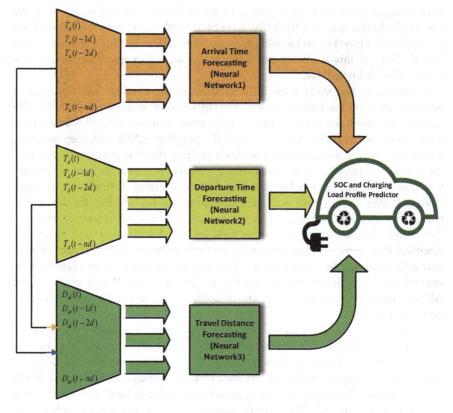

Fig. 2.1 The overall configuration of the ANN-based method

ANNs such as Levenberg-Marquardt (LM), which is developed based on second-order derivative method, and robust form of neurons such as rough structure-based neurons which have high ability in handling the uncertainty of input data.

We are going to present the ANN-based approach in this section, and in this manner, it is supposed that communication services between EV owners and optimal charging operators (aggregators) are considered for sending charging signals. By this communication service, EV's travel data has been updated to train the ANN every hour. The main configuration of the proposed method is shown in Fig. 2.1. To employ the ANN for forecasting different travel data, we use departure and arrival times as time-series. In this way, we employ the previous data of these profiles to forecast the future data. However, in forecasting traveled distance of every EV, the departure and arrival times are considered as input data. In this regard, we consider the correlation of various travel parameters which results a rational pattern for the forecasted travels. To improve the accuracy of the forecasting results, we employ the K-means method, which is an unsupervised classification method, to cluster the input data of various drivers in different groups and special forecasting network is implemented for each group in forecasting task.

In this section, first, we explain various forms of ANNs in complete details, and present optimal charging procedure of the EVs.

2.3.1 Conventional ANN

In this section, we categorize the conventional ANN into two groups as follows:

- Multilayer perceptron ANNs with the conventional form of error back propagation learning method
- Multilayer perceptron ANNs with the Levenberg-Marquardt learning method

More details about these approaches are presented as follow.

2.3.1.1 Multilayer Perceptron ANN with Error Back Propagation Learning Method

The initial form of ANNs is known as multilayer perceptron (MLP) network. The most common type of training procedure of ANN is the conventional error back propagation (CEBP) algorithm which was implemented for ANN training by Hinton [34], based on gradient descent theory. A simple ANN with two hidden layers is shown in Fig. 2.2. The activation functions of the hidden layer and output layer are considered as sigmoid and linear, respectively. To illustrate the performance of this ANN, the Feed-forward equations are presented as follows:

$$net_1^1(k) = \left(w_1^1(k)\right)^T . X \tag{2.1}$$

$$net_2^1(k) = \left(w_2^1(k)\right)^T . X \tag{2.2}$$

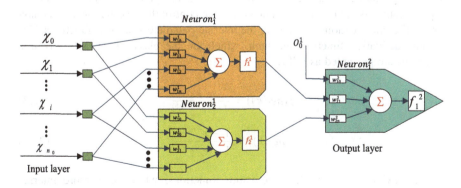

Fig. 2.2 Overall configuration of MLP ANN

$$O^1(k) = \left[O_0^1(k), f_1^1\left(net_1^1(k)\right), f_2^1\left(net_2^1(k)\right) \right]^T \tag{2.3}$$

$$net_1^2(k) = \left(w_1^2(k)\right)^T . O^1(k) \tag{2.4}$$

$$O_1^2(k) = f_1^2\left(net_1^2(k)\right) \tag{2.5}$$

Based on the gradient descent algorithm, the back-propagation equations, which present the learning approach, are defined as follows:

$$E(k) = \frac{1}{2}\left(e_1^2(k)\right)^2 = \frac{1}{2}\left(d(k) - o_1^2(k)\right)^2 = \frac{1}{2}\left(d(k) - f\left(net_1^2(k)\right)\right)^2 \tag{2.6}$$

$$\nabla w_1^2(E(k)) = \frac{\partial E(k)}{\partial w_1^2(k)} = \frac{\partial E}{\partial net_1^2} \times \frac{\partial net_1^2}{\partial w_1^2}(k) \tag{2.7}$$

$$\nabla w_1^1(E(k)) = \frac{\partial E(k)}{\partial w_1^1(k)} = \frac{\partial E}{\partial net_1^2} \times \frac{\partial net_1^2}{\partial o_1^1} \times \frac{\partial o_1^1}{\partial net_1^1} \times \frac{\partial net_1^1}{\partial w_1^1}(k) \tag{2.8}$$

$$\nabla w_2^1(E(k)) = \frac{\partial E(k)}{\partial w_2^1(k)} = \frac{\partial E}{\partial net_1^2} \times \frac{\partial net_1^2}{\partial o_1^1} \times \frac{\partial o_1^1}{\partial net_2^1} \times \frac{\partial net_2^1}{\partial w_2^1}(k) \tag{2.9}$$

Finally, the network's weights are trained as follows:

$$\Delta w_j^s(k) = -\eta_w \nabla w_j^s(E(k)) \tag{2.10}$$

$$\Delta w_j^s(k) = w_j^s(k+1) - w_j^s(k) = -\eta_w \nabla w_j^s(E(k)) \tag{2.11}$$

$$w_j^s(k+1) = w_j^s(k) - \eta_w \nabla w_j^s(E(k)) \tag{2.12}$$

The most common activation function in ANN literature is the sigmoid activation function. However, it should be noticed that the sigmoid activation function will be saturated by large input data sets, and this problem disrupts the training process of the ANN [35]. The saturation state of neurons has a significant effect on the forecasting results. To handle this problem, flexible activation function with internal hyper parameters are employed. Indeed, we can control the flexible function behavior in different situations by training the internal parameters of these functions. The sigmoid activation function and tanh, which are the classic forms of activation functions, are presented as follows:

$$f(net(k)) = \frac{1}{1 + e^{-net(k)}} \tag{2.13}$$

$$f(net(k)) = \frac{1 - e^{-net(k)}}{1 + e^{-net(k)}} \tag{2.14}$$

The flexible forms of these activation functions, which have more internal parameters, are expressed as follows:

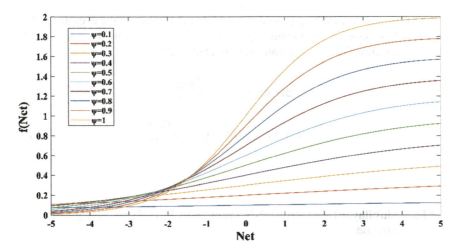

Fig. 2.3 The overall behavior of flexible Sigmoid function by different values of ψ

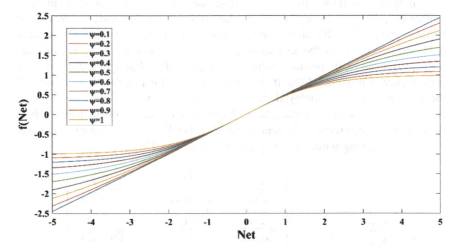

Fig. 2.4 The overall behavior of Flexible tanh function by different values of ψ

$$f(net(k), \psi(k)) = \frac{2|\psi(k)|}{1 + e^{-net(k) \times \psi(k)}} \quad (2.15)$$

$$f(net(k), \psi(k)) = \frac{1}{\psi(k)} \times \frac{1 - e^{-net(k) \times \psi(k)}}{1 + e^{-net(k) \times \psi(k)}} \quad (2.16)$$

The behaviors of sigmoid and tanh activation function with various internal parameters are shown in Figs. 2.3 and 2.4, respectively.

In this way, in the training procedure of ANNs, in addition to the neuron weights, the activation functions parameter (ψ) is trained. The training procedure of the flexible activation function variables is presented as follows:

$$\Delta\psi(k) = -\eta_\psi \times \frac{\partial E(k)}{\partial \psi(k)} \tag{2.17}$$

$$\psi(k+1) = \psi(k) + \eta_\psi \times \frac{\partial E(k)}{\partial \psi(k)} \tag{2.18}$$

2.3.1.2 Multilayer Perceptron ANN Training by Levenberg–Marquardt Method

The first-order derivative methods are the most common training procedures of the back propagation algorithm. The first-order derivative method is slow in large dimension problems. By implementing the second-order derivative approach, the training speed of ANNs significantly improved [36]. The Levenberg-Marquardt (LM) method employs the second-order derivative procedure by combining the backpropagation with newton techniques. Indeed, LM has quick and accurate performance in training ANNs. It should be mentioned that in the Newton method, the Hessian matrix is employed, but the Hessian matrix calculation causes high computational cost. In this regard, the LM approach uses the Jacobian matrix, which is the approximated form of Hessian matrix [37]. The Hessian and Jacobian matrixes in the ANN training procedure are presented as follows [38]:

$$H(k) = \frac{\partial^2 E}{\partial w^2} = \nabla^2 E(k) = \begin{bmatrix} \dfrac{\partial^2 E}{\partial w_1^2} & \dfrac{\partial^2 E}{\partial w_2 \partial w_1} & \cdots & \dfrac{\partial^2 E}{\partial w_{N_w} \partial w_1} \\[2ex] \dfrac{\partial^2 E}{\partial w_1 \partial w_2} & \dfrac{\partial^2 E}{\partial w_2^2} & \cdots & \dfrac{\partial^2 E}{\partial w_{N_w} \partial w_2} \\[2ex] \vdots & \vdots & & \vdots \\[2ex] \dfrac{\partial^2 E}{\partial w_1 \partial w_{N_w}} & \dfrac{\partial^2 E}{\partial w_2 \partial w_{N_w}} & \cdots & \dfrac{\partial^2 E}{\partial w_{N_w}^2} \end{bmatrix} \tag{2.19}$$

$$H(k) \approx J(k)^T J(k) \tag{2.20}$$

$$J(k) = \begin{bmatrix} \dfrac{\partial e_{11}}{\partial w_1} & \dfrac{\partial e_{11}}{\partial w_2} & \cdots & \dfrac{\partial e_{11}}{\partial w_{N_w}} \\[2mm] \dfrac{\partial e_{12}}{\partial w_1} & \dfrac{\partial e_{12}}{\partial w_2} & \cdots & \dfrac{\partial e_{12}}{\partial w_{N_w}} \\[1mm] \cdots & \cdots & \cdots & \cdots \\[1mm] \dfrac{\partial e_{1M}}{\partial w_1} & \dfrac{\partial e_{1M}}{\partial w_2} & \cdots & \dfrac{\partial e_{1M}}{\partial w_{N_w}} \\[1mm] \cdots & \cdots & \cdots & \cdots \\[1mm] \dfrac{\partial e_{j1}}{\partial w_1} & \dfrac{\partial e_{j1}}{\partial w_2} & \cdots & \dfrac{\partial e_{j1}}{\partial w_{N_w}} \\[2mm] \dfrac{\partial e_{j2}}{\partial w_1} & \dfrac{\partial e_{j2}}{\partial w_2} & \cdots & \dfrac{\partial e_{j2}}{\partial w_{N_w}} \\[1mm] \cdots & \cdots & \cdots & \cdots \\[1mm] \dfrac{\partial e_{N_T M}}{\partial w_1} & \dfrac{\partial N_T M}{\partial w_2} & \cdots & \dfrac{\partial e_{N_T M}}{\partial w_{N_w}} \end{bmatrix} \quad e_{L-M}(k) = \begin{bmatrix} e_{11} \\ e_{12} \\ \cdots \\ e_{1M} \\ e_{j1} \\ e_{j2} \\ \cdots \\ e_{jM} \\ \cdots \\ e_{N_T 1} \\ e_{N_T 2} \\ \cdots \\ e_{N_T M} \end{bmatrix} \tag{2.21}$$

Based on the steepest descent and Newton methods, the updating procedure of weights in the LM method is defined as follows [37]:

$$\Delta w(k) = \left(J(k)^T J(k) + \mu(k)I(k)\right)^{-1} J(k)^T e_{L-M}(k) \tag{2.22}$$

$$w(k+1) = w(k) - \eta_w \left(\left(J^T(k)J(k) + \mu(k)I\right)^{-1} J^T(k) e_{L-M}(k)\right) \tag{2.23}$$

The learning procedure in the LM method is a hybrid approach which switches between Newton and gradient descent methods. In this procedure, $\mu(k)$ is the key parameter. When $\mu(k)$ goes near zero, the LM shifts to the Gauss-Newton approach, and by increasing the $\mu(k)$ value, the LM shifts to the gradient descent method [39]. In this method, the training procedure starts with small $\mu(k)$ to employ the Newton technique and profits from its good convergence speed. After that, if we don't find a smaller error value, we do the training steps with a higher value of $\mu(k)$ to employ the gradient descent algorithm and find a better solution.

2.3.2 ANN with Rough Neurons

In the real world, there are uncertainty and noise in input data and they have a significant effect on data engineering-based forecasting approaches such as ANNs. In fact, to guarantee the performance of the ANNs in a real case study, we must employ the robust neurons which can handle the uncertainty of the input data. The rough neuron-based ANN was introduced by Lingras [40]. In this structure, neurons of the hidden layers are structured based on Rough theory. The rough neuron is defined with a pair of neurons which are named as upper and lower bounds. In rough neurons structure, when x implies a variable, x and $\bar{x}$ presents the lower and upper bounds of the variable x, respectively. In this section, we introduce the CEBP and LM algorithms for rough neuron-based ANNs.

2.3.2.1 Multilayer Perceptron Neural Network with Rough Neurons and Back Propagation Learning Approach

A simple form of a rough neuron is shown in Fig. 2.5. This structure is designed to omit the input data noise by interval weights, including upper bound and lower bound weights.

The Feed forward equations for a rough neuron are defined as follows [41]:

$$net^s_{L_j}(k) = \left(w^{s-1}_{Lj}(k)\right)^T . X \qquad (2.24)$$

$$net^s_{U_j}(k) = \left(w^{s-1}_{Uj}(k)\right)^T . X \qquad (2.25)$$

$$O^s_{L_j}(k) = \min\left(f^s_j\left(net^s_{L_j}(k)\right), f^s_j\left(net^s_{U_j}(k)\right)\right) \qquad (2.26)$$

$$O^s_{U_j}(k) = \max\left(f^s_j\left(net^s_{L_j}(k)\right), f^s_j\left(net^s_{U_j}(k)\right)\right) \qquad (2.27)$$

$$O^s_j(k) = \gamma^s_j O^s_{L_j}(k) + \lambda^s_j O^s_{U_j}(k) \qquad (2.28)$$

As shown in the above equations, in rough neurons, we have two ways of the information flow, one from upper bound and another from lower bound weights. This configuration improves the robustness of the neurons in high uncertainty data sets. After rough neurons, two specific coefficients are allocated for upper bound and lower bound outputs which are named as γ and λ, respectively. The learning procedure in rough neurons is the same as in simple neurons which is defined based on the gradient descent method, but with two different ways for back propagation process.

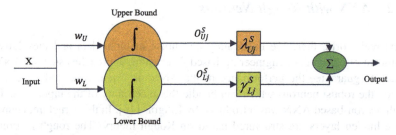

Fig. 2.5 A simple form of the rough neuron

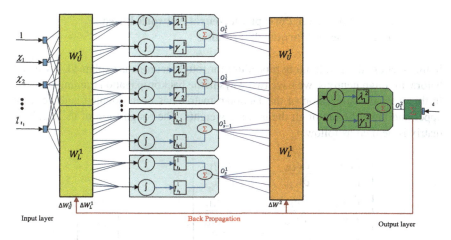

Fig. 2.6 R-ANN with error back propagation learning method

To show the rough neurons-based networks in a clear way, a rough ANN (R-ANN) with rough hidden layer is shown in Fig. 2.6. The training process of neurons' weights are defined as follows:

The output layer has conventional neurons, and the training procedure of this layer is the same as previous [42]:

$$\Delta w_{ji}^2(k) = -\eta_w \frac{\partial E}{\partial w_{ji}^2}(k) \quad (2.29)$$

The hidden layer is considered with rough neurons, and the training procedures of lower bound and upper bound neurons are defined as follows:

$$\Delta w_{L_j}^1 = -\eta_w \frac{\partial E}{\partial w_{L_j}^1}(k)$$

$$= -\eta_w \times \frac{\partial E}{\partial e^2} \times \frac{\partial e^2}{\partial o^2} \times \frac{\partial o^2}{\partial net^2} \times \frac{\partial net^2}{\partial o_j^1} \times \frac{\partial o_j^1}{\partial o_{L_j}^1} \times \frac{\partial o_{L_j}^1}{\partial net_{L_j}^1} \times \frac{\partial net_{L_j}^1}{\partial w_{L_j}^1}(k) \quad (2.30)$$

$$\Delta w_{U_j}^1 = -\eta_w \frac{\partial E}{\partial w_{U_j}^1}(k)$$

$$= -\eta_w \times \frac{\partial E}{\partial e^2} \times \frac{\partial e^2}{\partial o^2} \times \frac{\partial o^2}{\partial net^2} \times \frac{\partial net^2}{\partial o_j^1} \times \frac{\partial o_j^1}{\partial o_{U_j}^1} \times \frac{\partial o_{U_j}^1}{\partial net_{U_j}^1} \times \frac{\partial net_{U_j}^1}{\partial w_{U_j}^1}$$

$$\times (k) \quad (2.31)$$

2.3.2.2 Rough Multilayer Perceptron Neural Network Based on Levenberg–Marquardt Training

In this section, the LM training procedure for rough neurons is explained. As stated before, in rough neurons, we have two ways in the backpropagation training process. In this way, the dimension of the Jacobian matrix has been doubled and we have upper and lower bound of all the training weights. The rough form of the Jacobian matrix is defined as follows:

$$
J(k) = \begin{bmatrix}
\dfrac{\partial e_{11_u}}{\partial w_{1u}} & \dfrac{\partial e_{11_u}}{\partial w_{2u}} & \cdots & \dfrac{\partial e_{11_u}}{\partial w_{N_w u}} \\[2ex]
\dfrac{\partial e_{11_l}}{\partial w_{1l}} & \dfrac{\partial e_{11_l}}{\partial w_{2l}} & \cdots & \dfrac{\partial e_{11_l}}{\partial w_{N_w l}} \\[1ex]
\cdots & \cdots & \cdots & \cdots \\[1ex]
\dfrac{\partial e_{1M_u}}{\partial w_{1u}} & \dfrac{\partial e_{1M_u}}{\partial w_{2u}} & \cdots & \dfrac{\partial e_{1M_u}}{\partial w_{N_w u}} \\[2ex]
\dfrac{\partial e_{1M_l}}{\partial w_{1l}} & \dfrac{\partial e_{1M_l}}{\partial w_{2l}} & \cdots & \dfrac{\partial e_{1M_l}}{\partial w_{N_w l}} \\[1ex]
\cdots & \cdots & \cdots & \cdots \\[1ex]
\dfrac{\partial e_{N_T M_u}}{\partial w_{1u}} & \dfrac{\partial e_{N_T M_u}}{\partial w_{2u}} & \cdots & \dfrac{\partial e_{N_T M_u}}{\partial w_{N_w u}} \\[2ex]
\dfrac{\partial e_{N_T M_l}}{\partial w_{1L}} & \dfrac{\partial e_{N_T M_u}}{\partial w_{2u}} & \cdots & \dfrac{\partial e_{N_T M_l}}{w_{N_w L}}
\end{bmatrix}
\quad
e_{L-M}(k) = \begin{bmatrix}
e_{11_u} \\[1ex]
e_{11_l} \\[1ex]
\cdots \\[1ex]
e_{1M_u} \\[1ex]
e_{1M_l} \\[1ex]
\cdots \\[1ex]
e_{N_T M_u} \\[1ex]
e_{N_T M_l}
\end{bmatrix}
\tag{2.32}
$$

2.4 Optimal Charging of EVs

To handle the challenges of EVs on the distribution network, we cannot allow the EV owners to charge their vehicle in an uncoordinated manner. In this regard, the aggregators, which are intermediary management units between the power system operator and EVs' owners, do the optimal charging procedure by considering power system operation constraints such as power loss, operation cost (OC), and voltage magnitudes [43]. The optimal charging equations are defined as follows:

$$
OC = \sum_{t=1}^{24} \{ca(t) \times Ra(t) + Cr(t) \times Rr(t)\} + C_{inf}
\tag{2.33}
$$

Power balance equations are defined as follows [25, 44]:

2 Artificial Intelligence-based Approach For Electric Vehicle Travel... 35

$$\sum_{t}^{24} Ra(t) = \sum_{t}^{24} La(t) + \sum_{t}^{24} PEVa(t) + \sum_{t}^{24} Plossa(t) \tag{2.34}$$

$$\sum_{t}^{24} Rr(t) = \sum_{t}^{24} Lr(t) + \sum_{t}^{24} PEVr(t) + \sum_{t}^{24} Plossr(t) \tag{2.35}$$

$$Plossa(t) = \sum_{p=1}^{n} V(p,q,t) \times \sum_{q=1}^{n} V(p,q,t)$$
$$\times \left\{ G(p,q) \times \cos\left(\theta_{p,t} - \theta_{q,t}\right) + B(p,q) \times \sin\left(\theta_{p,t} - \theta_{q,t}\right) \right\} \tag{2.36}$$

$$Plossr(t) = \sum_{p=1}^{n} V(p,q,t) \times \sum_{q=1}^{n} V(p,q,t)$$
$$\times \left\{ G(p,q) \times \cos\left(\theta_{p,t} - \theta_{q,t}\right) - B(p,q) \times \sin\left(\theta_{p,t} - \theta_{q,t}\right) \right\} \tag{2.37}$$

Power system operation constraints are defined as follows:

$$P^u(t) \le P_{max}^u, \quad Q^u(t) \le Q_{max}^u \quad u = 1, 2, \dots, n_{eq} \tag{2.38}$$

$$V_{min} \le V_p(t) \le V_{max} \quad, p = 1, 2, \dots, n, \quad t = 1, 2, \dots, 24 \tag{2.39}$$

The State of Charge (SOC) of EV's batteries, which is determined based on forecasted travel behavior parameters, is presented as follows [13, 15, 45]:

$$SOC_{init,l} = 100 - \frac{Tl_l}{C_{eff} \times Cap_{bat,l}} \times 100 \tag{2.40}$$

The C_{eff} and $Cap_{bat, p}$, which are EV's parameters, are obtained from [33]. The SOC at departure time is given as follows [13, 46]:

$$SOC_l(t) = SOC_l(t - 1) + P_l^{chr}(t) \times \rho_{chr} \tag{2.41}$$

2.5 Numerical Study

In this section, EVs data and forecasting results are presented.

2.5.1 EVs and Power System Data

In this study, we use the EV's travel behavior of the 2017 NHTS database [47] which has complete information of departure time, arrival time, and travel distance.

Table 2.1 Data of various EVs [48]

PEV model	Total number	Battery capacity (kWh)	Max. charging rate (kWh)	PEV model	Total number	Battery capacity (kWh)	Max. charging rate (kWh)
LEAF Nissan	103,578	30	6.6	e-Golf VW	4589	26.5	6.6
Model S Tesla	93,277	100	17.2	Class E Mercedes B-	3312	36	10
i3 BMW	24,721	42	7.4	Soul EV Kia	2993	30.5	6.6
500E Fiat	10,229	24	6.6	i-MiEV Mitsubishi	2098	16	3.6
Spark Chevrolet	7369	19	7.6	Fit EV Honda	1071	20	6.6
Focus EV Ford	6839	33.5	6.6	Active E BMW	965	33	6.4

Various EVs are considered in this study, and the penetration rate of each EV in the US market is obtained from [48] and is shown in Table 2.1. In this work, the battery capacity and the charging rate of every EVs are selected based on the model of EV [49].

The power system topology (feeders information) and operation data (electricity price, load demand) are obtained from [50, 51], respectively.

2.5.2 Evaluation Criteria

To verify the robustness of the ANN-based approach, various error criteria including the Root Mean Square Error (RMSE), the Mean absolute error (MAE), the Mean Absolute Percentage Error (MAPE), and [52] are implemented as follows:

$$MAE = \frac{1}{n_0} \sum_{g=1}^{n_0} \left(\left| \widehat{Y_g} - Y_g \right| \right) \tag{2.42}$$

$$MAPE = \frac{1}{n_0} \sum_{g=1}^{n_0} \left(\left| \widehat{Y_g} - Y_g \right| \right) \tag{2.43}$$

$$RMSE = \sqrt{\frac{1}{n_0} \sum_{g=1}^{n_0} \left(\left| \widehat{Y_g} - Y_g \right| \right)^2} \tag{2.44}$$

For further confirmation, the numerical results are also evaluated with the R-squared criterion which is a popular performance metric [38, 53]. R-square states the correlation of real and forecasted data, and the higher R-squared value indicates more precise results. The R-squared formulation is presented as follows [54]:

$$R^2 = 1 - \frac{\sum_{g=1}^{n_0}\left(Y_g - \widehat{Y}_g\right)^2}{\sum_{g=1}^{n_0}\left(Y_g - \overline{Y}_g\right)^2} \tag{2.45}$$

2.5.3 Forecasting Results

The forecasting results of different ANNs and various training procedures are presented in this section. As we know, the initial values of neuron weights are selected in a random manner. Accordingly, we have done training procedure more than 100 times to find the best results based on the error criteria. In this study, we consider 50 hidden layers, and this large number of hidden layers, which presents the deep learning approach, is too sensitive during the training procedure. To avoid overlearning, during the training procedure, the dropout and L2 regulation techniques have been used in this study [55].

First, we employ the conventional ANNs with first-order and second-order derivative error back propagations which are considered as CEBP and LM techniques, respectively. In the training procedure of ANNs, we use 80% of input data for training, 10% of input data for validation, and 10% of input data for the test. The numerical results of test data are shown in Table 2.2.

As illustrated in Table 2.2, the LM method presents more accurate results. However, the accuracy of the forecasting results needs to be improved because aggregators' financial profit or loss is related to the precision of the forecasting results.

To increase the accuracy of the forecasting results, we employ the rough neuron-based networks, and the forecasting results are presented in Table 2.3. Division of input data into training, validation, and testing parts is considered as conventional ANNs.

As shown in Table 2.3, the accuracy of the forecasting results has been improved significantly, and it implies the high ability of rough neurons in handling the uncertainty of input data. Indeed, R-ANNs employ the interval weights (upper bound and lower bound weights) which are so effective in forecasting profiles with high stochastic behavior.

To verify the correlation between various travel behavior parameters in the forecasting results by ANNs, we compare the simulation results of conventional and rough ANNs with MCS method, which is a benchmark approach in modeling travel behavior parameters. The comparison results are shown in Figs. 2.7 and 2.8.

Table 2.2 Error criteria for conventional ANNs

	CEBP			LM		
Error criteria	Arrival time (h)	Departure time (h)	Trip length (mile)	Arrival time (h)	Departure time (h)	Trip length (mile)
MAE	5.33	3.68	11.33	4.43	2.69	9.09
RMSE	6.35	4.49	15.90	5.16	3.18	12.56
MAPE (%)	31.86	34.77	37.28	26.45	25.44	29.90

Table 2.3 Error criteria for Rough ANNs

	R-CEBP			R-LM		
Error criteria	Arrival time (h)	Departure time (h)	Trip length (mile)	Arrival time (h)	Departure time (h)	Trip length (mile)
MAE	4.12	2.53	7.84	1.81	2.90	5.87
RMSE	4.80	3.20	11.27	2.17	2.60	8.11
MAPE (%)	24.63	23.92	25.82	17.12	17.34	19.32

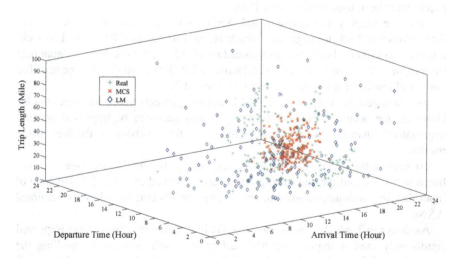

Fig. 2.7 EV' travel behavior forecasting results for conventional LM, MCS and real data

The MCS method does not consider the correlation between various parameters, and all the forecasting result are around the mean values (as shown by Figs. 2.7 and 2.8). In this way, the forecasted travels are so similar to each other, and the variety of travels is very low. In this regard, MCS result has a high error rate in modeling EVs

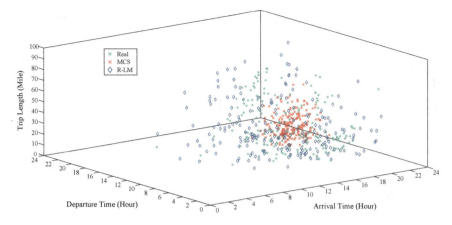

Fig. 2.8 EV's travel behavior forecasting results for rough neuron-based LM, MCS and real data

travel behavior, and aggregators cannot forecast the actual electrical demand of their customers accurately. In ANN-based approaches (conventional neurons and rough neurons), the correlation between departure time, arrival time, and traveled distance have been considered.

To evaluate the forecasting results of various ANN-based approaches mentioned in this chapter, the R-squared values of different methods are presented in Fig. 2.9.

As shown in Fig. 2.9, R-LM method outperforms other ANN-based approaches and presents the 0.944 R-squared value which implies the effect of the rough neurons and LM method. As mentioned before, accurate forecasting of EVs travel behavior has a significant effect on aggregator's financial loss and profit which is presented in Table 2.4.

As shown in Table 2.4, the aggregator's financial loss has been significantly reduced by implementing the ANN-based approaches in comparison with MCS. Indeed, ANN-based approaches improved the forecasting accuracy of various travel parameters about 23%. This is so important in optimal charging procedure because the aggregator would have to pay extra money to upstream network or EV owners based on the difference between actual load demand and forecasted load demand. In ANN-based approaches, R-ANN has the best performance because of the interval weights which have high ability in handling the input data uncertainty. It should be noticed that the presented results are just for 210 EVs in a day, and by rising the penetration of EVs in transportation fleet, the effect of accurate forecasting has been significantly increased in the aggregator's financial loss.

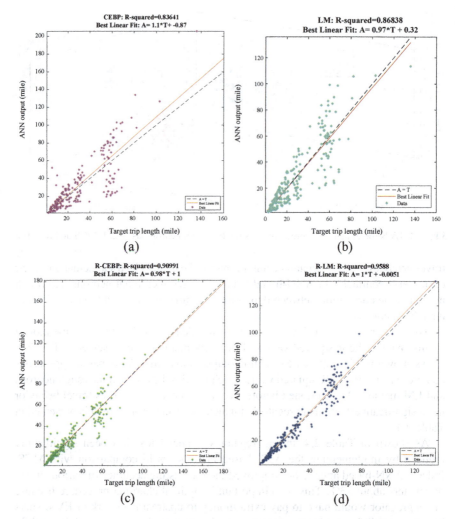

Fig. 2.9 Regression result of Trip length in different ANN approaches: (**a**) CEBP, (**b**) LM, (**c**) R-CEBP, (**d**) R-LM

2.6 Conclusion

In this chapter, an artificial intelligence-based approach for modeling the EVs travel behavior is presented. Travel behavior of EVs has high stochastic and intermittent profiles, and to handle the uncertainty of this input data, rough neuron-based ANN, which is a robust form of ANNs, is implemented. Indeed, rough neurons, by employing the interval-based neurons, have high ability in feature extraction of input data. As shown in the numerical results of this study, rough ANN outperforms the Monte Carlo Simulation method which is the benchmark approach in this field.

Table 2.4 Aggregator's financial loss by different forecasting approaches for 210 EVs in a day

Scenarios		Price ($)	Aggregator financial loss ($)	R-squared for trip length forecasting	Average forecasting error (MAPE%)
Real data		3825.032	___	___	___
Conventional ANN	CEBP	3849.809	24.777	0.83641	34.636
	LM	3817.754	7.277	0.86838	27.263
Rough ANN	R-CEBP	3830.478	5.446	0.89690	24.790
	R-LM	3829.465	4.432	0.94479	17.926
Monte Carlo		3811.283	13.749	___	___

Accurate forecasting of travel behavior parameters has a significant effect on the aggregator's financial income. To illustrate this claim, we compared the forecasting results of different approaches with real data travels in the optimal charging procedure.

In this regard, by improving the precision of the forecasting results, we can increase the aggregators' income which is so important for aggregators. In the near future, by increasing the penetration of EVs in the transportation fleet, accurate forecasting of EVs behavior will be of the utmost importance in the optimal charging procedure. The findings of this chapter imply that for accurate modeling of EVs behavior, we must use modern data engineering approaches which are among the best solutions in large dimension problems.

Appendix A

The Nomenclature is presented as follows

Indices

i	Input data index
g	Output sample index
j	Hidden layer index
k	Iteration number index
l	Electric vehicle index
p	Node of power network index
q	Node of power network index
S	Layer index
t	Time sample index
u	Equipment index

Parameters

$B(p,q)$	Susceptance value between bus p and bus q
C_{inf}	Cost of charging infrastructure
C_{eff}	EV's efficiency factor
$Cap_{bat,\,l}$	l-th EV's battery capacity
$G(p,q)$	Conductance value between bus p and bus q
M	Number of input data variables for LM
n_0	Input data dimension
n	Number of power system nodes
N_T	Number of training neurons for LM
N_w	Weight dimension for LM
N_{w_L}	Weight dimension of lower bound neurons for LM
N_{w_U}	Weight dimension of upper bound neurons for LM
n_{pev}	Overall number of EVs
n_{eq}	Overall number of equipment
nd	Dimension of sample data
P^u_{max}	Maximum value of active power for each equipment
Q^u_{max}	Maximum value of reactive power for each equipment
SOC_{dep}	Minimum SOC value of every EV at departure time
V_{max}	Maximum value of node's voltage
V_{min}	Minimum value of node's voltage
ρ_{chr}	Coefficient of the charging efficiency
η_w	Training factor for weights
η_ψ	Training factor for activation function in flexible mode

Variables

$Ca(t)$	Active power cost at time t
$Cr(t)$	Reactive power cost at time t
$d(k)$	Target value in k-th iteration
$e^s_j(k)$	Internal error of neuron j in k-th iteration for layer S
$e_{jM}(k)$	Internal error of neuron j f in k-th iteration for input data M in LM
$e_{jM_L}(k)$	Internal error of lower bound neuron j in k-th iteration for input data M in LM
$e_{jM_u}(k)$	Internal error of upper bound neuron j in k-th iteration for input data M in LM
$e_{L-M}(k)$	Overall error vector in LM method for k-th iteration
E	Overall value of sum square error
$E(k)$	Overall value of sum square error in k-th iteration
$f^s_j(k)$	Actuation function of neuron j for layer S k-th iteration
$H(k)$	Hessian matrix for k-th iteration
$I(k)$	Identity matrix for k-th iteration

(continued)

$J(k)$	Jacobian matrix for k-th iteration
$La(t)$	Active load at time t
$Lr(t)$	Reactive load at time t
$net(k)$	Activation function input in iteration k
$net_j^s(k)$	Activation function input for neuron j in layer S in iteration k
$net_{Lj}^s(k)$	Lower bound of activation function input for neuron j in layer S in iteration k
$net_{Uj}^s(k)$	Upper bound of activation function input for neuron j in layer S in iteration k
$O^s(k)$	Output of layer S in iteration k
$O_j^s(k)$	Output of neuron j in layer S in iteration k
$O_U^s(k)$	Output of upper bound neuron for layer S in iteration k
$O_L^s(k)$	Output of lower bound neuron for layer S in iteration k
$O_{Uj}^s(k)$	Output of upper bound neuron j for layer S in iteration k
$O_{Lj}^s(k)$	Output of lower bound neuron j for layer S in iteration k
$P^u(t)$	Active power value for u-th equipment at time t
$P_l^{chr}(t)$	Charging rate value for l-th EV at time t
$PEVa(t)$	Active power of EV at time t
$PEVr(t)$	Reactive power of EV at time t
$Plossa(t)$	Active power loss value at time t
$Plossr(t)$	Reactive power loss value at time t
$Q^u(t)$	Reactive power value of u-th equipment at time t
$Ra(t)$	Purchased active power value at time t
$Rr(t)$	Purchased reactive power value at time t
$SOC_{init,l}(t)$	Initial value of SOC for l-th EV at time t
$SOC_{dep,l}(t)$	Departure value of SOC for l-th EV at time t
$SOC_l(t)$	Value of SOC for l-th EV at time t
Tl_l	Trip length value for l-th EV
$w_j^s(k)$	Vector of weights for j-th neuron in layer S for k-th iteration
$W_{Uj}^S(k)$	Vector of weights for j-th upper bound neuron for k-th iteration in layer S
$W_{Lj}^S(k)$	Vector of weights for j-th lower bound neuron for k-th iteration in layer S
$W_{ij}^S(k)$	Vector of weights between i-th input sample and j-th neuron in hidden layer S in k-th iteration
$V(p,q,t)$	Voltage of line between node p and q at time t
$V_p(t)$	Voltage of p-th node time t
X	Vector of input data
X_i	Vector of i-th input data sample
Y_g	Vector of g-th output data sample
X_C	
$\widehat{Y_g}$	Vector of g-th target data sample
λ_j^s	Upper bound factor for neuron j in layer S
γ_j^s	Lower bound factor for neuron j in layer S
$\mu(k)$	Decision value of iteration k in LM

(continued)

$\theta_{p,\,t}$	Angel of voltage for *p-th* node at time t
$\theta_{q,\,t}$	Angel of voltage for q-*th* node at time t
$\psi(k)$	Variable of actuation function at iteration k

References

1. J.C. Mukherjee, A. Gupta, Distributed charge scheduling of plug-in electric vehicles using inter-aggregator collaboration. IEEE Trans. Smart Grid **8**(1), 331–341 (2017)
2. G.E. Hinton, S. Osindero, Y.-W. Teh, A fast learning algorithm for deep belief nets. Neural Comput. **18**(7), 1527–1554 (2006)
3. H. Jahangir et al., A novel electricity price forecasting approach based on dimension reduction strategy and rough artificial neural networks. IEEE Trans. Ind. Inf, 1–1 (2019), https://doi.org/10.1109/TII.2019.2933009. Accessed 05 Aug 2019
4. M.B. Arias, S. Bae, Electric vehicle charging demand forecasting model based on big data technologies. Appl. Energy **183**, 327–339 (2016)
5. K. Qian, C. Zhou, M. Allan, Y. Yuan, Modeling of load demand due to EV battery charging in distribution systems. IEEE Trans. Power Syst. **26**(2), 802–810 (2011)
6. M. Neaimeh et al., A probabilistic approach to combining smart meter and electric vehicle charging data to investigate distribution network impacts. Appl. Energy **157**, 688–698 (2015)
7. L. Yu, T. Zhao, Q. Chen, J. Zhang, Centralized bi-level spatial-temporal coordination charging strategy for area electric vehicles. CSEE J. Power Energy Syst. **1**(4), 74–83 (2015)
8. S.M. Mohseni-Bonab, A. Rabiee, Optimal reactive power dispatch: A review, and a new stochastic voltage stability constrained multi-objective model at the presence of uncertain wind power generation. IET Gener. Transm. Distrib. **11**, 815–829 (2017)
9. P.P. Biswas, P.N. Suganthan, G.A.J. Amaratunga, Optimal power flow solutions incorporating stochastic wind and solar power. Energy Convers. Manag. **148**, 1194–1207 (2017)
10. M.J. Morshed, J. Ben Hmida, A. Fekih, A probabilistic multi-objective approach for power flow optimization in hybrid wind-PV-PEV systems. Appl. Energy **211**, 1136–1149 (2018)
11. H. Jahangir, A. Ahmadian, M.A. Golkar, Optimal design of stand-alone microgrid resources based on proposed Monte-Carlo simulation, in *Proceedings of the 2015 IEEE Innovative Smart Grid Technologies - Asia, ISGT ASIA 2015*, (2016)
12. A. Ahmadian, M. Sedghi, B. Mohammadi-Ivatloo, A. Elkamel, M. Aliakbar Golkar, M. Fowler, Cost-benefit analysis of V2G implementation in distribution networks considering PEVs battery degradation. IEEE Trans. Sustain. Energy **9**, 961–970 (2018)
13. Q. Yang, S. Sun, S. Deng, Q. Zhao, M. Zhou, Optimal sizing of PEV fast charging stations with Markovian demand characterization. IEEE Trans. Smart Grid **10**, 4457–4466 (2018)
14. S. Sun, Q. Yang, W. Yan, A novel Markov-based temporal-SoC analysis for characterizing PEV charging demand. IEEE Trans. Ind. Inf **14**(1), 156–166 (2018)
15. Y. Wang, D. Infield, Markov chain Monte Carlo simulation of electric vehicle use for network integration studies. Int. J. Electr. Power Energy Syst. **99**, 85–94 (2018)
16. S. Sun, Q. Yang, W. Yan, A novel statistical Markov-based approach for modeling charging demand of plug-in electric vehicles, in *2016 China International Conference on Electricity Distribution (CICED)*, (2016), pp. 1–6
17. J. Tang, A. Brouste, K.L. Tsui, Some improvements of wind speed Markov chain modeling. Renew. Energy **81**, 52–56 (2015)
18. Y. Zhou, B. Qi, B. Zhang, Online prediction of electrical load for distributed management of PEV based on Grey-Markov model, in *2017 29th Chinese Control And Decision Conference (CCDC)*, (2017), pp. 6911–6916

19. G. Li, X.-P. Zhang, Modeling of plug-in hybrid electric vehicle charging demand in probabilistic power flow calculations. IEEE Trans. Smart Grid **3**(1), 492–499 (2012)
20. Y. Kongjeen, K. Bhumkittipich, Modeling of electric vehicle loads for power flow analysis based on PSAT, in *2016 13th International Conference on Electrical Engineering/Electronics, Computer, Telecommunications and Information Technology (ECTI-CON)*, (2016), pp. 1–6
21. S. Bae, A. Kwasinski, Spatial and temporal model of electric vehicle charging demand. IEEE Trans. Smart Grid **3**(1), 394–403 (2012)
22. J.G. Vlachogiannis, Probabilistic constrained load flow considering integration of wind power generation and electric vehicles. IEEE Trans. Power Syst. **24**(4), 1808–1817 (2009)
23. L.E. Bremermann, M. Matos, J.A.P. Lopes, M. Rosa, Electric vehicle models for evaluating the security of supply. Electr. Power Syst. Res. **111**, 32–39 (2014)
24. Z. Zhou, T. Lin, Spatial and temporal model for electric vehicle rapid charging demand, in *2012 IEEE Vehicle Power and Propulsion Conference*, (2012), pp. 345–348
25. O. Hafez, K. Bhattacharya, Queuing analysis based PEV load modeling considering battery charging behavior and their impact on distribution system operation. IEEE Trans. Smart Grid **9**, 261–273 (2018)
26. D. Wang, X. Guan, J. Wu, J. Gao, Analysis of multi-location PEV charging behaviors based on trip chain generation, in *2014 IEEE International Conference on Automation Science and Engineering (CASE)*, (2014), pp. 151–156
27. W. Jianfeng, X. Xiangning, Z. Jian, L. Kunyu, Y. Yang, T. Shun, Charging demand for electric vehicle based on stochastic analysis of trip chain. IET Gener. Transm. Distrib. **10**(11), 2689–2698 (2016)
28. Y. Mu, J. Wu, N. Jenkins, H. Jia, C. Wang, A spatial–temporal model for grid impact analysis of plug-in electric vehicles. Appl. Energy **114**, 456–465 (2014)
29. D. Panahi, S. Deilami, M.A.S. Masoum, S.M. Islam, Forecasting plug-in electric vehicles load profile using artificial neural networks, in *2015 Australasian Universities Power Engineering Conference (AUPEC)*, (2015), pp. 1–6
30. Z. Shi, H. Liang, V. Dinavahi, Direct interval forecast of uncertain wind power based on recurrent neural networks. IEEE Trans. Sustain. Energy **9**(3), 1177–1187 (2018)
31. M. Elhoseny, A. Nabil, A.E. Hassanien, D. Oliva, Hybrid rough neural network model for signature recognition, in *Advances in Soft Computing and Machine Learning in Image Processing*, (Springer, 2018), pp. 295–318
32. T. Zhang, D. Liu, D. Yue, Rough neuron based RBF neural networks for short-term load forecasting, in *Energy Internet (ICEI), IEEE International Conference on*, (2017), pp. 291–295
33. H. Jahangir et al., Charging demand of plug-in electric vehicles: Forecasting travel behavior based on a novel rough artificial neural network approach. J. Clean. Prod. **229**, 1029–1044 (2019)
34. D.E. Rumelhart, G.E. Hinton, R.J. Williams, *Learning internal representations by error propagation* (California Univ San Diego La Jolla Inst for Cognitive Science, San Diego, 1985)
35. H. Hikawa, A new digital pulse-mode neuron with adjustable activation function. IEEE Trans. Neural Netw. **14**(1), 236–242 (2003)
36. B.M. Wilamowski, H. Yu, Improved computation for Levenberg–Marquardt training. IEEE Trans. Neural Netw. **21**(6), 930–937 (2010)
37. C. Lv et al., Levenberg–Marquardt backpropagation training of multilayer neural networks for state estimation of a safety-critical cyber-physical system. IEEE Trans. Ind. Inf **14**(8), 3436–3446 (2018)
38. A. Behnood, E.M. Golafshani, Predicting the compressive strength of silica fume concrete using hybrid artificial neural network with multi-objective grey wolves. J. Clean. Prod. **202**, 54–64 (2018)
39. G. Lera, M. Pinzolas, Neighborhood based Levenberg-Marquardt algorithm for neural network training. IEEE Trans. Neural Netw. **13**(5), 1200–1203 (2002)

40. P. Lingras, Rough neural networks, in *Proceedings of the 6th International Conference on Information Processing and Management of Uncertainty in Knowledgebased Systems*, (1996), pp. 1445–1450
41. Z. He, S. Lin, Y. Deng, X. Li, Q. Qian, A rough membership neural network approach for fault classification in transmission lines. Int. J. Electr. Power Energy Syst. **61**, 429–439 (2014)
42. G. Ahmadi, M. Teshnehlab, Designing and implementation of stable sinusoidal rough-neural identifier. IEEE Trans. Neural Netw Learn. Syst. **28**(8), 1774–1786 (2017)
43. H. Jahangir, A. Ahmadian, M.A. Golkar, Multi-objective sizing of grid-connected micro-grid using Pareto front solutions, in *Proceedings of the 2015 IEEE Innovative Smart Grid Technologies - Asia, ISGT ASIA 2015*, (2016)
44. S. Deilami, A.S. Masoum, P.S. Moses, M.A.S. Masoum, Real-time coordination of plug-in electric vehicle charging in smart grids to minimize power losses and improve voltage profile. IEEE Trans. Smart Grid **2**(3), 456–467 (2011)
45. M. Sedghi, A. Ahmadian, M. Aliakbar-Golkar, Optimal storage planning in active distribution network considering uncertainty of wind power distributed generation. IEEE Trans. Power Syst. **31**, 304–316 (2016)
46. H. Xing, M. Fu, Z. Lin, Y. Mou, Decentralized optimal scheduling for charging and discharging of plug-in electric vehicles in smart grids. IEEE Trans. Power Syst. **31**(5), 4118–4127 (2016)
47. National household travel survey (2018), https://nhts.ornl.gov/. Accessed 05 Dec 2018
48. Alternative fuels data center (2018), https://afdc.energy.gov/data/10567. Accessed 05 Dec 2018
49. Y.-S. Cheng, M.-T. Chuang, Y.-H. Liu, S.-C. Wang, Z.-Z. Yang, A particle swarm optimization based power dispatch algorithm with roulette wheel re-distribution mechanism for equality constraint. Renew. Energy **88**, 58–72 (Apr. 2016)
50. E.G. Carrano, F.G. Guimarães, R.H.C. Takahashi, O.M. Neto, F. Campelo, Electric distribution network expansion under load-evolution uncertainty using an immun system inspired algorithm. IEEE Trans. Power Syst. **22**, 851–861 (2007)
51. The independent electricity system operator (IESO) (2018), http://ieso.ca/. Accessed 05 Dec 2018
52. C. Yu, Y. Li, M. Zhang, Comparative study on three new hybrid models using Elman neural network and empirical mode decomposition based technologies improved by singular Spectrum analysis for hour-ahead wind speed forecasting. Energy Convers. Manag. **147**, 75–85 (2017)
53. E.M. Golafshani, A. Behnood, Application of soft computing methods for predicting the elastic modulus of recycled aggregate concrete. J. Clean. Prod. **176**, 1163–1176 (2018)
54. J. Yuan, C. Farnham, C. Azuma, K. Emura, Predictive artificial neural network models to forecast the seasonal hourly electricity consumption for a University Campus. Sustain. Cities Soc **42**, 82–92 (2018)
55. N. Srivastava, G. Hinton, A. Krizhevsky, I. Sutskever, R. Salakhutdinov, Dropout: A simple way to prevent neural networks from overfitting. J. Mach. Learn. Res. **15**, 1929–1958 (2014)

Chapter 3
The Role of Off-Board EV Battery Chargers in Smart Homes and Smart Grids: Operation with Renewables and Energy Storage Systems

Vitor Monteiro, Jose Afonso, Tiago Sousa, and Joao L. Afonso

3.1 Introduction

The spread of electric mobility is experiencing a steady growth worldwide, in special concerning the private-level contribution, where its participation in the transportation sector is identified as a key paradigm for substituting conventional vehicles based on internal combustion engines. In this context, varied technologies are available, demonstrating an appropriate contribution to sustainability. The issues, challenges, and opportunities of vehicle electrification are investigated in [1, 2], and a survey about this topic in the context of the smart grid is offered in [3]. Among the different technologies, the most representative is the pure plug-in battery electric vehicle (EV) and the hybrid plug-in EV. From the power grid point of view, both two types of EVs are plugged-in to absorb power through EV battery charging, where the main difference is the required time for the charging since the capacity of their energy storage system (ESS) is very different. Therefore, for simplicity, in the scope of this book chapter, the designation "EV" is used for both types of vehicles. In this context, it is important to note that the use of hybrid ESS is conceivable for both cases, mainly based on batteries and ultra-capacitors as support for sudden requirements of power [4]. The number of plug-in EVs available for purchase is growing, especially along the last two decades, all of them equipped with an on-board EV battery charging system (EV-BCS) and some also permitting external charging using equipment designated as off-board EV-BCS.

V. Monteiro (✉) · T. Sousa · J. L. Afonso
ALGORITMI Research Centre, Department of Industrial Electronics, University of Minho, Braga, Portugal
e-mail: vmonteiro@dei.uminho.pt; tsousa@dei.uminho.pt; jla@dei.uminho.pt

J. Afonso
CMEMS-UMinho Center, Department of Industrial Electronics, University of Minho, Braga, Portugal
e-mail: jose.afonso@dei.uminho.pt

© Springer Nature Switzerland AG 2020
A. Ahmadian et al. (eds.), *Electric Vehicles in Energy Systems*,
https://doi.org/10.1007/978-3-030-34448-1_3

Concerning the power transfer interaction between most of the commercially available EVs and the power grid, only the EV battery charging, i.e., the unidirectional power transfer from the power grid to the EV, is possible [5, 6]. This unidirectional operation is designated in the literature as grid-to-vehicle (G2V), and it is common for both on-board and off-board EV-BCS. Nevertheless, since the power flows from the power grid to the EV, the latter can be understood in two distinct functions: (1) As a normal electrical appliance for the power grid, consuming power randomly in terms of schedules and charging point (independently of the power quality matters, in terms of harmonic current and power factor); or (2) As an electrical appliance with the possibility of flexible control schedules.

In this scenario, the EV introduction into the power grid is of utmost importance, since the EV is identified not only as a key element to mitigate the emission of greenhouse gases, but may also enable a useful power transfer collaboration with the power grid. In this perspective, for the power grid, the presence of the EV becomes even more relevant when it is controlled in a flexible way, allowing accomplishing three main features: (1) The EV operation can be controlled for absorbing power from the power grid in specific schedules, controlled by the smart grid, according to the power limits of the EV-BCS, independently of the place where it is plugged-in (e.g., at a smart home); (2) The EV can be set up for storing energy in a specific place where it is plugged-in, and, due to its natural mobility, transport the stored energy for a different place in the power grid (e.g., it can be interesting for power management between different smart homes); (3) The EV can be controlled for injecting power into the power grid, conferring the necessities of the electrical installation of the place where it is plugged-in, or the necessities of the smart grid (e.g., the EV can be plugged-in to the smart home, but injecting power as a service for the smart grid, independently of the smart home power management). Within this context, as an example, an admission and scheduling mechanism for EV charging is proposed in [7], and a scheduling strategy for the EV, framed in residential demand response programs, is proposed in [8].

Based on the possibilities of the EV interaction with the power grid, together with the G2V mode arises the vehicle-to-grid (V2G) mode, which is a bidirectional mode, permitting a power flow from the power grid (smart grid or smart home) to the EV and vice-versa [9–11]. The G2V/V2G power interaction may be investigated also from the perspective of coordinated controllability [12–14]. Contextualizing this scenario, an on-board EV-BCS incorporated into a smart home is illustrated in Fig. 3.1, where the G2V/V2G modes are identified. As it can be seen, since a smart home is considered, the smart home power management can communicate with the EV-BCS, with the smart grid, with the electrical switch-board, and with the controlled smart electrical appliances. As represented in Fig. 3.1, the EV can either consume power from the grid (G2V) or deliver power (V2G) for the smart home, the smart grid, or both at the same time (i.e., a parcel of the power injected by the EV is consumed by the electrical appliances of the smart home and another parcel is injected into the power grid). Besides the G2V/V2G controllability, particularly along the last decade, novel paradigms of operation were proposed targeting power quality improved features (e.g., in the presence of unpredictable power

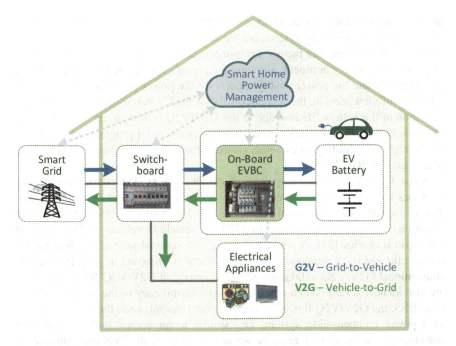

Fig. 3.1 On-board EV-BCS in a smart home scenario: G2V/V2G modes

outages, in islanded power grids, in situations of compensating harmonics, and in circumstances of producing reactive power).

Concerning all of the aforementioned aspects, the main contributions of this book chapter are: (1) A complete investigation about technologies of on-board and off-board EV-BCS and advanced operation modes outlined in smart homes and smart grids; (2) A comprehensive explanation about upcoming perspectives of operation for on-board and off-board EV-BCS, and their innovative association with renewable energy sources (RES) and ESS, when framed with smart grids and with ac, dc or hybrid smart homes; (3) A validation taking into account the upcoming perspectives, addressed in the previous point, for the off-board EV-BCS operation in smart homes and smart grids.

3.2 EV Operation Modes: An Overview

As described in the introductory section, there is a bidirectional opportunity associated with the EV interaction with the power grid through the G2V/V2G modes. In fact, the G2V/V2G modes are already a reality and an encouraging factor for enhancing the interaction with smart homes and smart grids. Nevertheless, these modes are only used for exchanging active power according to the necessities of the smart grid or smart home, i.e., the EV is controlled based on an on/off strategy. In this way, the schedules for the charging (G2V) or discharging (V2G) processes are defined by the management system, but the value of active power to be exchanged is defined neglecting relevant factors, for instance, other constraints of the smart home. This influence is very pertinent, allowing the use of the plugged-in EVs to improve the efficiency and the power quality aspects for the power grid side [15–18]. From this point of view, despite the increased wear of the EV battery and the consequent reduction of the battery lifetime due to the G2V/V2G operation, the EV driver can also benefit from this interaction, since interesting tariffs for programs of G2V/V2G are emerging, permitting to establish collective schedules for charging (G2V) and discharging (V2G) [19–21].

Since the EV can be plugged-in in distinct places, the control complexity increases for the smart grid, even with the flexibility in terms of operation modes. This is also valid for scenarios of microgrids [22, 23]. A specific case is related to the possibility of the EV to operate in V2G or vehicle-to-vehicle (V2V) mode in microgrid scenarios [24]. A smart microgrid with optimal joint scheduling for the EV and the electrical appliances of a smart home is proposed in [25]. Besides the conventional G2V/V2G validations [26, 27], emerging G2V/V2G future interfaces are also identified [28]. When considering the intermittency of the power produced from RES, the G2V/V2G flexibility is even more relevant, since the EV can be used as a power compensation system, i.e., similar to an energy buffer, capable of consuming, storing, or delivering power as a function of the RES intermittence. In this context, a strategy of accommodating the EV operation into the power grid, as a function of the power production from RES, is available in [29]. Considering the EV operation in G2V/V2G modes, a specific control algorithm based on RES production for demand-side management is presented in [30]. A specific cooperative combination between the EV and RES in terms of controllability, with the main objective of reducing emissions and costs, is offered in [31]. An optimal cost minimization about the EV charging with operation modes for the smart grid and smart home is presented in [32].

The associated operation of the EV with RES is not only limited to a smart grid perspective. In fact, this mixed operation is also very applicable for smart homes, as demonstrated in [33], since smart homes are a strategic contribution to smart grids. Therefore, technologies and foresight for the EV integration in smart homes are discussed in [34], and an EV optimization in a smart home from the customer point of view is analyzed in [35]. All the aforementioned technologies (in the context of smart grids and smart homes), as well as the cooperative operation with RES, are considered in the perspective of EV-BCS only in the G2V/V2G modes. Nevertheless, other opportunities are identified in the literature with relevant potential prospecting smart grids and smart homes.

The home-to-vehicle (H2V) is interesting in the smart home perspective, requiring a plugged-in EV. In fact, the H2V mode is similar to the G2V mode. Nevertheless, as a distinctive characteristic, the H2V mode offers the possibility of power controllability in opposite to the conventional on/off approach. This is predominantly relevant, since it involves the EV in the smart home management more effectively, offering more flexibility of controllability together with the controlled electrical appliances. In the H2V mode, the value of the operating power can range from zero to the nominal power. Moreover, for both the EV and the electrical appliances, strategic levels of precedence can be planned. For instance, through a mobile app, the EV user can outline their preferences. Considering the context of the H2V mode in the smart home, three main cases can be highlighted. In the first case (a), maximum priority is defined to the EV. In this way, the EV has more priority than the electrical appliances, and then it is charged with maximum power, where, in this circumstance, the electrical appliances are turned-off if necessary (this occurs when some electrical appliances can be turned-off and it is indispensable, as fast as possible, the EV charging). In a second case (b), the EV is defined to have priority only over some specific electrical appliances. In this way, the EV is charged with fixed power, but different from the maximum power that is permitted by the EV-BCS. In such circumstance, it is guaranteed a fixed value of power for the EV charging, and to avoid the circuit breaker tripping, the electrical appliances are programmed in different schedules. In a third case (c), a minimum priority is established to the EV, where the value of power results from the difference between the smart home nominal power and the instantaneous power consumed. If the electrical appliances are turned off, then the EV is charged with maximum power (similar to the case (a)). Instead, if the electrical appliances are turned on and turned off, then the EV is charged with variable power. Summarizing, the H2V mode is comparable to the G2V, but permitting the adjustable charging power. Instead, it is also important to note that the H2V strategy can also be used during the discharging process. In this way, the EV can inject power to the smart home or the smart grid, but as a function of the electrical appliances. A specific case occurs when the power consumed by the electrical appliances exceeds the nominal power of the smart home. In this circumstance, the EV can be controlled just to provide the difference of power (between the nominal value and the required by the electrical appliances).

3.3 Operation Modes: Future Perspectives

Future viewpoints for the EV are discussed in this section, highlighting the relation with the G2V/V2G/H2V modes, but establishing new opportunities for the EV-BCS, involving the requirements of the smart grids and the smart homes (with ac, dc, or hybrid electrical installations [36]).

3.3.1 On-Board EV BCS

Figure 3.2 illustrates an on-board EV-BCS in a smart home considering the G2V/V2G/H2V modes, as well as a new mode for its operation, which is related to power quality. Although the demonstration is for a smart home, this new operation mode can be also framed with smart grids, contributing to define new control strategies and energy policies under the smart grid scope.

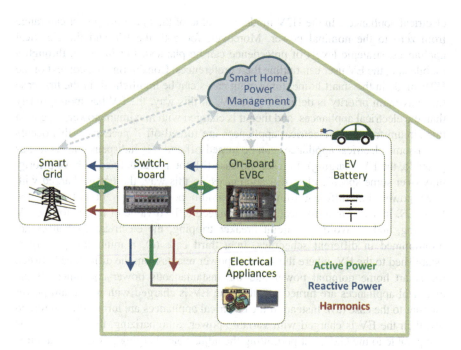

Fig. 3.2 On-board EV-BCS in a smart home scenario: Operation in G2V/V2G modes and compensating problems of power quality, both for smart home and smart grid (harmonic currents and low power factor)

Based on the analysis of Fig. 3.2, three different cases are identified: (a) the on-board EV-BCS can be controlled for exchanging power with the smart home, providing power only in accordance with the requirements of the smart home management system, i.e., the G2V/V2G/H2V operation is limited to the smart home scope; (b) the on-board EV-BCS can be controlled for exchanging power with the smart grid, providing power only in accordance with the requirements of the smart grid management system, i.e., the G2V/V2G operation is limited to the smart grid scope, and the H2V is not considered in this control strategy; (c) the on-board EV-BCS can be controlled, at the same time, with the smart home and with the smart grid, i.e., the three G2V/V2G/H2V modes are adjusted within the smart home and smart grid scope.

Besides the G2V/V2G/H2V modes, the on-board EV-BCS can also be controlled in a perspective of compensating power quality problems. This mode is identified as vehicle-for-grid (V4G) since the on-board EV-BCS is used to provide additional services for the grid. This mode is exceptionally relevant since it does not interfere with the G2V/V2G/H2V modes, i.e., the V4G mode can be combined with each of the G2V/V2G/H2V modes. Moreover, it does not require to use the EV battery, since only the front-end converter of the on-board EV-BCS is used. Using the EV in this mode, independently or not of the G2V/V2G/H2V modes, almost all the harmonic currents and the power factor of the smart home can be compensated. However, in the smart grid perspective, when the EV is plugged-in at the smart home, the on-board EV-BCS can be used only to produce specific harmonic currents and a specific value of reactive power for compensating the power factor (e.g., other EV plugged-in neighboring smart homes can operate with the same functionalities to compensate all the current harmonics and power factor of a specific part of the smart grid). This is particularly relevant, making emerge another perspective for the EV in smart homes, which is associated with the selective harmonic current compensation. In this strategy, each EV is controlled to produce a specific harmonic current for the smart grid. Notwithstanding the strong benefits of the on-board EV-BCS operating in G2V/V2G/H2V/V4G modes for the smart home and the smart grid, a crucial drawback is notorious: these modes are only conceivable if the EV is plugged-in at the smart home (since the on-board EV-BCS is used). However, from the power grid point of view, a new important benefit is recognized: the EV can operate in the G2V/V2G/H2V/V4G modes where it is plugged-in, representing a dynamic system conferring an important asset for smart grids. As previously demonstrated, the EV can be controlled in the G2V/V2G/H2V/V4G modes, where the specific V4G mode is linked to the compensation of harmonic currents and power factor.

This new opportunity is also relevant taking into account that the derived costs caused by power quality problems are substantially high around the world [37–39]. Therefore, the EV can be used as a dispersed and dynamic active power filter within the power grid, demonstrating that it can be an added value for supporting power quality. Equivalent opportunities are obtainable based on computer simulations in [40, 41], but in the perspective of the EV powertrain. The option just for producing reactive power, as a requirement of the power grid, is investigated in [42–45], and in [46], but neglecting the capability of harmonic current compensation. The option for compensating harmonics and reactive power only during the G2V mode is assumed in [47], which is a pertinent drawback since the EV manufacturers are presenting the V2G mode and this option is independent of the G2V/V2G mode. The exploitation of the V2G mode for power quality improvement is proposed in [48] for a smart grid perspective. The possibility to use an on-board EV-BCS in four quadrants is offered in [49], which is based on an experimental validation in G2V/V2G modes, but only considering the production of reactive power, i.e., neglecting the harmonic current compensation. In [50] is proposed an external system, only validated by computer simulations, where the necessary coupling passive filters are installed, limiting the option of linking the EV to any outlet. In [51] is presented a three-phase off-board EV-BCS offering the capability of harmonic current and reactive power compensation. Another three-phase off-board EV-BCS is proposed in [52], but only for harmonic current compensation, neglecting the power factor compensation, as well as the possibility of a combined operation in G2V/V2G modes.

Nevertheless, also within the V4G mode, the on-board EV-BCS can be controlled to compensate other power quality problem: power outages of short duration at the smart home level. Taking into account this scenario, besides the G2V/V2G modes, the operation of the EV as a power source for the electrical installation, when it is not plugged-in to the power grid, is considered in [53]. Correspondingly, the possibility of using the EV as a backup generator at the residential level is proposed in [54]. This particular concept is denominated as vehicle-to-home (V2H), since, when required, the EV can be used to feed the electrical appliances in the smart home. A more convenient situation is related to the possibility of using the EV as an off-line uninterruptible power supply (UPS). This possibility was initially considered in [55], based on preliminary results used to validate the operation mode when considering linear electrical appliances (representing an unrealistic condition). In this context, it must be highlighted that, sometimes, the designation of V2H is also considered for aggregating the G2V/V2G modes for the EV at residential level [34, 56], but without the possibility of operation as an off-line UPS. The possibility of operation in V2H mode is particularly committed to smart homes during the occurrence of power outages, representing a relevant contribution for improving reliability and security against failures at the power grid level. These methodologies comprise the operation of the EV in smart homes as an ESS [33], the flexibility for managing the power consumption and user comfort at the smart home [57, 58], and an optimal control scheduling considering the electrical appliances power consumption [25]. Nissan proposed the EV operation through the "LEAF to Home" [59], and Mitsubishi and Toyota have also similar platforms [60]. Nevertheless, in such platforms, the EV does not allow the operation as an off-line UPS.

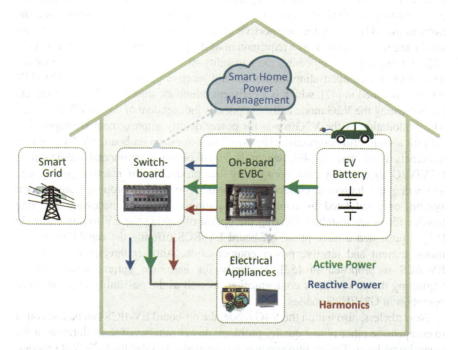

Fig. 3.3 On-board EV-BCS in a smart home scenario: Operation in G2V/V2G modes and compensating problems of power quality for the smart home (power outages using the EV as a power supply)

Figure 3.3 illustrates this case, where the EV battery is the power source of the smart home, the on-board EV-BCS is controlled as an inverter, and the smart home is not connected to the smart grid. In this particular case, the on-board EV-BCS is controlled with voltage feedback to guarantee a stable voltage, both in terms of amplitude and frequency, even with sudden variations of the electrical appliances (which will define the current waveform). Since the EV battery is the power source of the smart home, the management of the battery state-of-charge must be determined with maximum accuracy [61]. More specifically, when the EV is forced to operate in this mode, the power management of the smart home can control some of the non-priority electrical appliances to be turned-off, contributing to preserving the EV battery.

3.3.2 Off-Board EV BCS

Besides the application of the G2V/V2G/H2V/V4G modes for on-board EV-BCS, these modes can also be applied to off-board EV-BCS. Therefore, the previous descriptions of these modes are also valid when considering off-board EV-BCS. Figure 3.4 illustrates an off-board EV-BCS and an EV plugged-in at a smart home. Taking into account the use of an off-board EV-BCS at the smart home level, the

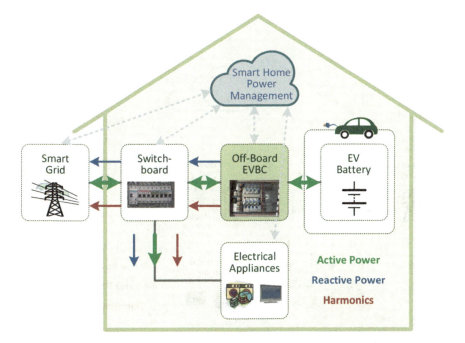

Fig. 3.4 Off-board EV-BCS in a smart home scenario with an EV plugged-in: Operation in G2V/V2G modes and compensating problems of power quality, both for smart home and smart grid (harmonic currents and low power factor)

existing opportunities are even more appropriate, since the off-board EV-BCS is permanently connected to the smart home. Consequently, independently of the EV presence at the smart home, some of the previous operation modes are also available, representing an added value for the smart home. As an example, the off-board EV-BCS can offer power quality functionalities, both for the smart home and for the smart grid, precisely as the on-board EV-BCS. Nevertheless, as a vital differencing factor, the functionalities offered by the V4G can be provided independently of the EV presence, while the G2V/V2G/H2V modes are only accessible when the EV is present (a situation that also occurs for on-board EV-BCS). Figure 3.5 illustrates a vision of an off-board EV-BCS in a smart home, but without an EV plugged-in. As shown, the V4G mode is possible in terms of compensation of harmonic currents and power factor, but, since the EV is not plugged-in, the possibility of operation during power outages is not possible.

Despite the relevance of the aforementioned modes, the foremost future opportunity associated with off-board EV-BCS is about the interfacing of other technologies for smart grids and smart homes, namely the technologies of ESS and RES [62]. In this way, the main objective consists of using the same off-board EV-BCS to interface a dc-dc converter for RES (unidirectional mode), as well as a dc-dc converter for ESS (bidirectional mode), where a shared dc-link is used for such purpose. This is distinct from the conventional solutions based on multiple power stages for encompassing the EV, the RES, and the ESS [63].

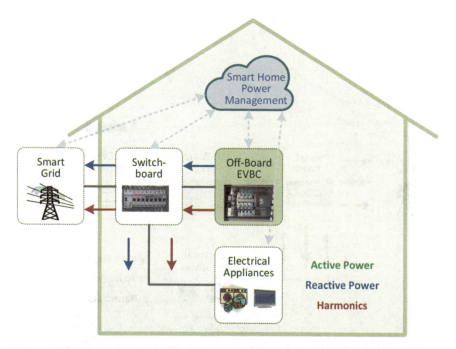

Fig. 3.5 Off-board EV-BCS in a smart home scenario without an EV plugged-in: Operation in G2V/V2G modes and compensating problems of power quality, both for smart home and smart grid (harmonic currents and low power factor)

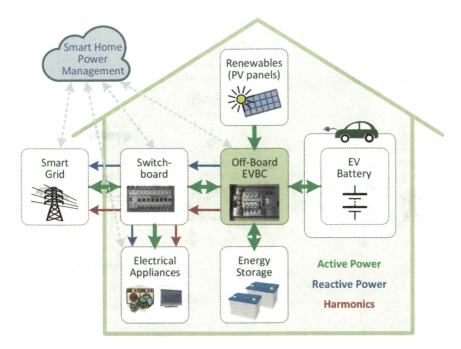

Fig. 3.6 Off-board EV-BCS in a hybrid smart home scenario with an EV plugged-in: Operation in G2V/V2G modes and compensating problems of power quality, both for smart home and smart grid (harmonic currents and low power factor). The RES (solar photovoltaic panels) and the ESS (batteries) are interfaced through a shared dc-link, while the electrical appliances are directly connected to the ac grid

It must be highlighted that the integration of an off-board EV-BCS encompassing this opportunity denotes a complete solution involving the smart home three key technologies: EV in bidirectional mode, ESS, and RES. This opportunity is different from the conventional cooperation between the EV and the ESS using independent systems [64]. Figure 3.6 illustrates this specific situation, where the single interface for the power grid is highlighted, avoiding the necessity of additional ac-dc converters to interface RES and ESS (i.e., this solution requires less two ac-dc converters). Furthermore, this opportunity is even more relevant when considering the migration from ac grids to dc grids, where the requirements of ac-dc converters are severely reduced. Moreover, since the majority of the electrical appliances include a front-end ac-dc converter that is only used to interface the ac grid, this opportunity gains new relevance to avoid the use of multiple ac-dc power converters (Figs. 3.7 and 3.8).

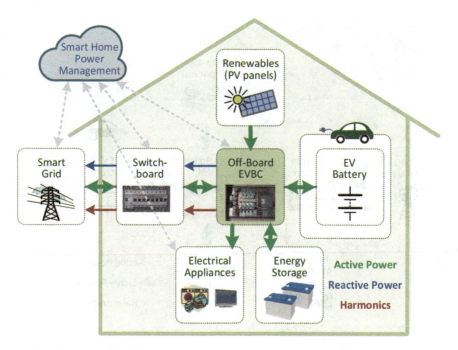

Fig. 3.7 Off-board EV-BCS in a dc smart home scenario with an EV plugged-in: Operation in G2V/V2G modes and compensating problems of power quality, both for smart home and smart grid (harmonic currents and low power factor). The RES (solar photovoltaic panels), the ESS (batteries), and the electrical appliances are interfaced through a shared dc-link

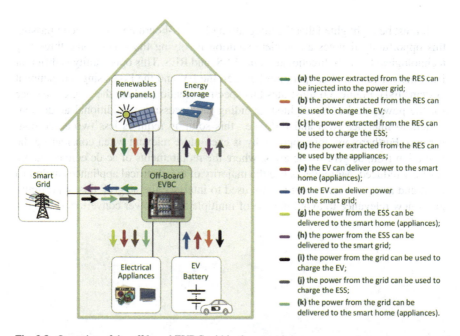

Fig. 3.8 Operation of the off-board EVBC within the smart home

3.4 Validation of Off-Board EV Battery Chargers when Contextualized in Smart Homes and Smart Grids

Taking into attention the possibilities offered by the integrated solutions based on the off-board EV-BCS for smart homes and smart grids, this section introduces an analysis that was performed considering three distinct cases of a smart home.

In the first case (a), a conventional ac smart home was considered, where the electrical appliances are coupled to the ac power grid, as well as an on-board EV-BCS, a power converter to interface RES, and a power converter to interface ESS. In this case, independent power converters for each technology are used, which are based on front-end (ac-dc) and back-end (dc-dc) power stages. In the second case (b), a hybrid ac/dc smart home was considered, where the electrical appliances are coupled to the ac power grid (as in the conventional case), but an off-board EV-BCS is used to interface RES and ESS (sharing a common dc-link and avoiding the necessity of ac-dc converters for interfacing these technologies). In the third case (c), a dc smart home was considered as a future perspective, including an off-board EV-BCS as main equipment. Therefore, for interfacing the technologies, dc-dc and dc-ac converters were considered. Figure 3.9 illustrates these three cases.

Dedicated simulation models were developed for each case based on the PSIM software. Considering the different technologies, the following situations were addressed: (a) a set of solar photovoltaic (PV) panels was considered as an example of RES, for a maximum power of 1.5 kW; (b) a set of lithium batteries was considered as an example of ESS, with nominal voltage of 200 V and capacity of 10 Ah; (c) a set of resistive loads was selected (dc electrical appliances); (d) an electric motor (induction) was selected (ac electrical appliances). On the other hand,

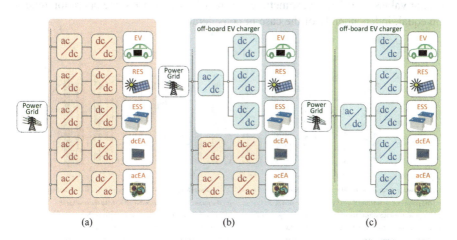

Fig. 3.9 Considered cases for analysis: (**a**) A conventional ac smart home (using ac-dc power converters for each technology); (**b**) A hybrid ac and dc smart home, where the electrical appliances are directly connected to the ac power grid, but an off-board EV-BCS interfaces RES and ESS; (**c**) A dc smart home, where an off-board EV-BCS is the only interface with the power grid equipment

for the power converters, the following conditions were addressed: (a) full-bridge converters as example of ac-dc converters to interface the power grid (voltage source converters controlled with current feedback); (b) half-bridge converters as example of dc-dc converters (also with current or voltage feedback); (c) full-bridge converters as example of dc-ac converters (also with voltage feedback).

3.4.1 Comparative Analysis: Efficiency of the Different Cases

Taking into account the innovative modes allowed by the future off-board EV-BCS (used in a dc smart home to interface the EV, RES, ESS, ac electrical appliances, and dc electrical appliances), as presented in Sect. 3.3 and based on the structures of Fig. 3.9, a comparative analysis was performed based on the efficiency of the different possibilities. Figure 3.10 presents the estimated efficiency for each mode of operation that was defined for the off-board EV-BCS and considering the three cases for the smart home (a conventional ac smart home, a hybrid ac/dc smart home, and a dc smart home). By analyzing the obtained results, it is clear that the dc smart home is the case that presents better results, independently of the operation mode defined for the off-board EV-BCS. These results makes sense, since a single ac interface is used, and, when it is necessary to exchange power between technologies, only the dc-dc converters are used (i.e., the quantity of power stages is considerably reduced). By analyzing the worst case in terms of efficiency, it is clear that the first case presents the worst results since ac-dc converters are always required for the power grid interface, even when it is necessary to exchange power between dc technologies, as, for instance, between the ESS and the EV. It must be noted that similar values of efficiency, sometimes, were achieved, since some operation modes are equal, independently of the case in consideration.

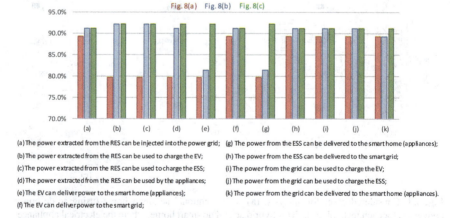

(a) The power extracted from the RES can be injected into the power grid;
(b) The power extracted from the RES can be used to charge the EV;
(c) The power extracted from the RES can be used to charge the ESS;
(d) The power extracted from the RES can be used by the appliances;
(e) The EV can deliver power to the smart home (appliances);
(f) The EV can deliver power to the smart grid;
(g) The power from the ESS can be delivered to the smart home (appliances);
(h) The power from the ESS can be delivered to the smart grid;
(i) The power from the grid can be used to charge the EV;
(j) The power from the grid can be used to charge the ESS;
(k) The power from the grid can be delivered to the smart home (appliances).

Fig. 3.10 Estimated efficiency for each case under analysis, where for each case were considered diverse perspectives of operation modes

3 The Role of Off-Board EV Battery Chargers in Smart Homes and Smart... 61

Aiming to obtain results as comprehensive as possible, the operation of the off-board EV-BCS was considered in the different modes according to the cases identified in Fig. 3.9: (a) a conventional ac smart home with ac-dc power converters for each technology; (b) a hybrid ac and dc smart home, where only a RES and an ESS are interfaced by the off-board EV-BCS; (c) a future dc smart home, where the off-board EV-BCS interfaces the RES and the ESS, as well as the dc or ac electrical appliances. The values of reference for the operation in each mode (i.e., the power for the EV charging, the power extracted from the RES, the power of the ESS, and the power of the electrical appliances) were selected as exemplification situations of a real scenario of the application. It is important to note that the development of a power management algorithm, for the smart home or the smart grid, is out of the scope of this book chapter, as well as the communication strategies.

3.4.2 Comparative Analysis: Operation of the Different Cases

Figure 3.11a shows the obtained results for the three cases when it is considered that only the electrical appliances are connected to the power grid, i.e., the power produced by the RES is zero, the EV is not plugged-in, and the ESS is not operating. For the electrical appliances, an active power of 1.2 kW and a reactive power of 225 VAr were measured. For the first and second cases, the current in the power grid ($i_{g\#1}$ and $i_{g\#2}$) is defined by the current consumption of the electrical appliances. Therefore, since the electrical appliances are categorized by linear and non-linear characteristics, the power grid presents harmonic distortion. In this case, the measured THD was 15.2% and the power factor was 0.97. In the third case, since an ac-dc converter is used to interface the power grid, the current in the power grid ($i_{g\#3}$)

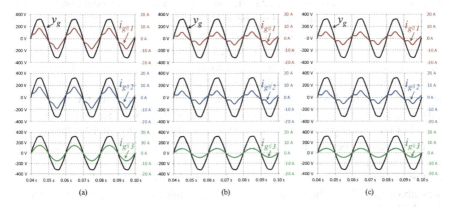

Fig. 3.11 Results concerning the three cases identified in Fig. 3.8: (**a**) When the power grid provides power exclusively to the electrical appliances; (**b**) When the power grid and the RES provide power to the electrical appliances; (**c**) When the power grid and the RES provide power to the electrical appliances and to the ESS

is sinusoidal, with a measured THD of 1.1% and with a unitary power factor. In this third case, the same electrical appliances were considered (i.e., the same active power), but removing the ac-dc converters to interface the power grid, since they are connected to the common dc-link through the dc-dc and dc-ac converters.

Figure 3.11b also shows the obtained results for the three cases considering that the same electrical appliances are connected to the power grid (an active power of 1.2 kW and a reactive power of 225 VAr were measured), but with a power production from the RES (with a power about of 550 W), while the EV is not plugged-in and the ESS is not operating. In this circumstance, for both cases #1 and #2, the power produced by the RES is injected into the power grid with a sinusoidal waveform. However, since the power required by the electrical appliances is greater than the power produced by the RES, a parcel of power is absorbed from the power grid, resulting in a non-sinusoidal current in the power grid side, as demonstrated by the waveforms of the currents $i_{g\#1}$ and $i_{g\#2}$. In these cases, when compared with the situation reported in Fig. 3.11a, the measured active power was 660 W and the reactive power was 225 VAr. For both cases, the measured THD of the current was 26.9%. In the third case, similar to the situation reported in Fig. 3.11a, the current in the power grid ($i_{g\#3}$) is sinusoidal, with a measured THD of 1.2% and with a unitary power factor. However, since the power produced by the RES is consumed by the electrical appliance, the power absorbed from the power grid is reduced, also meaning that the amplitude of the power grid current ($i_{g\#3}$) is reduced. In this case, the power from the RES is directly used by the electrical appliance through the dc-dc converters.

Finally, Fig. 3.11c shows the obtained results for the three cases considering that the same electrical appliances are connected to the power grid (an active power of 1.2 kW and a reactive power of 225 VAr were measured), with a power production from the RES (with a power about of 1 kW) and the ESS storing energy (with a power about of 450 W), while the EV is not plugged-in. Also, in this case, a parcel of power is absorbed from the power grid, resulting in a non-sinusoidal current in the power grid side, as demonstrated by the waveforms of the currents $i_{g\#1}$ and $i_{g\#2}$. For the first case, when compared with the situation reported in Fig. 3.11b, the increased power production from the RES was stored in the ESS, therefore, from the power grid point of view, the waveform of the current is the same. This is valid since the current from the RES is sinusoidal (injected into the power grid) and the current absorbed from the ESS is also sinusoidal. For the second case, the situation is different, since a unified topology is considered for RES and ESS. In this case, the parcel of the power produced by the RES is directly stored by the ESS through the dc-dc converters. For the first case, the measured active power was 660 W and the reactive power was 225 VAr, and, for the second case, the measured active power and reactive power were similar. For both cases, the measured THD of the current was about 26.9%. In the third case, which is similar to the situation reported in Fig. 3.11b, the current in the power grid ($i_{g\#3}$) is sinusoidal, with a measured THD of 1.2% and with a unitary power factor. When compared with the situation reported in Fig. 3.11b, the increment of the power production from the RES does not influence the current in the power grid side since it was directly stored by the ESS through the dc-dc converters.

3 The Role of Off-Board EV Battery Chargers in Smart Homes and Smart... 63

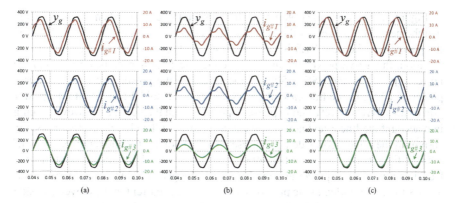

Fig. 3.12 Results concerning the three cases identified in Fig. 3.8: (**a**) When the power grid and the RES provide power to the electrical appliances and to the EV; (**b**) When the power grid and the ESS provide power to the electrical appliances; (**c**) When the power grid and the ESS provide power to the electrical appliances and to the EV

Similarly to Fig. 3.11, Fig. 3.12 shows some results where the off-board EV-BCS operates in different modes according to the different cases identified in Fig. 3.9. Figure 3.12a shows the obtained results for the three cases when it is considered that the necessary power for the electrical appliances and for the EV charging is provided by the RES and by the power grid, i.e., analyzing all the technologies, only the ESS is not operating. For the electrical appliances, an active power of 1.2 kW and a reactive power of 225 VAr were measured. To perform the EV charging, a power of 2 kW was established, while the power production from RES was about 1 kW. Therefore, the power extracted from the RES is used for the electrical appliances, but a parcel of power is required from the power grid, both for the electrical appliances and for the EV charging. In the first case, since a parcel of power absorbed from the power grid is used for the electrical appliances, the waveform of the power grid current ($i_{g\#1}$) is distorted, even with the presence of the EV consuming a sinusoidal current. In this case, the measured THD of the power grid current ($i_{g\#1}$) was 8.3% and the power factor was 0.91. The on-board EV-BCS presents a sinusoidal current with a THD of 1.1% and a unitary power factor. In the second case, since a unified topology is considered, the power extracted from the RES is directly used by the EV through the dc-dc converters. In this circumstance, the power required by the EV is greater than the power extracted from the RES, meaning that a parcel of power must be absorbed from the power grid for the EV charging, added by the necessary power for the electrical appliances. Therefore, the measured THD of the power grid current ($i_{g\#2}$) was 8.3% and the power factor was 0.91. In the third case, due to the presence of a unified topology with a single ac-dc converter used to interface the power grid, the current in the power grid ($i_{g\#3}$) is sinusoidal, with a measured THD of 1.1% and with a unitary power factor. Again, it must be noted that the same electrical appliances were considered (i.e., the same active power), but removing the ac-dc converters to interface the power grid, since they are connected to the common dc-link through the dc-dc and dc-ac converters.

Figure 3.12b shows the obtained results for the three cases when it is considered that the necessary power for the electrical appliances is provided by the ESS and by the power grid, i.e., the RES is not operating and the EV is not plugged-in. In this situation, the active power required by the electrical appliances is 1.5 kW and the power injected by the ESS is 500 W, meaning that the power grid must provide a power of 1 kW. In the first case, the ESS injects a sinusoidal current into the power grid (with a THD of 1.2% and in phase opposition with the voltage); however, taking into account the linear and non-linear characteristics of the electrical appliances, the current in the power grid side ($i_{g\#1}$) presents a THD of 18.4% and a power factor of 0.93. In the second case, a unified topology is used, but taking into account that the RES is not producing power and the EV is not plugged-in. This case is very similar to the situation reported in the previous case, where the current in the power grid side ($i_{g\#2}$) presents a THD of 18.4% and a power factor of 0.93. In the third case, the unified topology is considered, but the situation is very different from the situations reported in the previous case since the power from the ESS is directly used by the electrical appliances through the dc-dc and dc-ac converters. In this circumstance, power must be also absorbed from the power grid, but the current presents a sinusoidal waveform with a THD of 1% and a unitary power factor.

Figure 3.12c shows the obtained results for the three cases when it is considered that the necessary power for the electrical appliances and the EV charging is provided by the ESS and by the power grid, i.e., analyzing all the technologies, only the RES is not operating. In this situation, the EV requires a power of 2 kW, the ESS provides a power of 1 kW, and the electrical appliances require a power of 1.5 kW. Therefore, the power grid must provide a power of 2.5 kW. In the first case, a parcel of power is absorbed from the power grid, since the power injected by the ESS is not enough for the requirements of the EV and the electrical appliances. Therefore, the current waveform presents harmonic distortion, with a measured THD of 7.3%, and a power factor of 0.9. In this case, the EV operates with a sinusoidal current and with a unitary power factor, similar to the ESS (but with a current in phase opposition with the voltage), but the current of the electrical appliances presents harmonic distortion. In the second case, a unified topology is considered, where the power from the ESS is directly used for the EV charging, meaning that only the dc-dc converters are used for such purpose. However, in this situation, the power required by the EV is greater than the power provided by the ESS, meaning that a parcel of power must be absorbed from the power grid, added by the necessary power for the electrical appliances. In this case, the current in the power grid side presents a THD of 7.3% and a power factor of 0.9. In the third case, it is also necessary to absorb a parcel of power from the power grid, since the power provided by the ESS is not enough for the requirements of the EV and the electrical appliances. In this case, due to the presence of a unified topology, the current in the power grid ($i_{g\#3}$) is sinusoidal, with a measured THD of 1.1% and with a unitary power factor. Also, in this case, the same electrical appliances were considered.

Similarly to the previous situation, Fig. 3.13 shows some results where the off-board EV-BCS operates in different modes according to the different cases identified in Fig. 3.9. Figure 3.13a shows the obtained results for the three cases when it is considered that the necessary power for the electrical appliances is provided only by the ESS, meaning that the production from RES is zero and that the EV is not plugged-in. Moreover, as the ESS provides the necessary power for the electrical appliances, it is not necessary to absorb power from the power grid. In the first case, the necessary active power for electrical appliances is injected into the power grid by the ESS. However, the injected current is sinusoidal (in phase opposition with the voltage), but the consumed current by the electrical appliances is non-sinusoidal, meaning that a parcel of non-sinusoidal current is absorbed from the power grid. This situation is shown in the obtained waveforms of the currents. For the electrical appliances, an active power of 980 W and a reactive power of 172 VAr were measured, where the current has a THD of 18.6%. In this case, it is important to reinforce that the active power necessary for the electrical appliances is exclusively provided by the ESS, therefore, the current in the power grid ($i_{g\#1}$) is only responsible for providing the necessary harmonic currents. In the second case, a similar operation occurs. Despite the unified topology, in this case, when only the ESS is used, the power stages are the same. Therefore, the current in the power grid ($i_{g\#2}$) is very similar to the current in the power grid ($i_{g\#1}$) during the first case. In the third case, the operation is completely different, since the power is directly provided by the ESS to the electrical appliances through the dc-dc and dc-ac converters. In this way, it is not necessary to use the power grid, meaning that the current in the power grid ($i_{g\#3}$) is zero.

Figure 3.13b shows the obtained results for the three cases when it is considered that the off-board EV-BCS is used to provide services for the smart grid. In this case, the power extracted by the RES is injected into the power grid and the EV is also injecting power into the power grid. However, the power consumption from the electrical appliances is inferior to the injected power, meaning that the difference is used by the smart grid for power management control. In the first case, the necessary

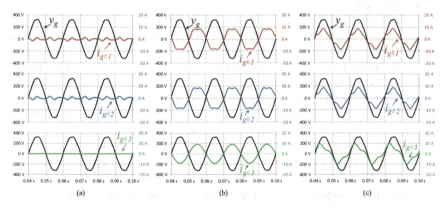

Fig. 3.13 Results concerning the three cases identified in Fig. 3.8: (**a**) When the ESS provides power to the electrical appliances; (**b**) When the RES and the EV provides power to the electrical appliances and to the smart grid; (**c**) When the power grid provides power to the electrical appliances and the off-board EV-BCS provides power quality services to the smart grid

active power for the electrical appliances is a parcel of the power injected by the RES and by the EV. However, the electrical appliances are categorized by linear and non-linear characteristics, meaning that the current in the power grid side ($i_{g\#1}$) has harmonic distortion with a THD of 11.8%, as well as a power factor of 0.98. This situation occurs because the converters of the RES and EV are controlled only to inject active power into the power grid. In a second case, a quite similar operation is observed, since the unified topology is controlled to inject a sinusoidal current into the power grid, but the current consumption of the electrical appliances has harmonic distortion. Therefore, harmonic currents are observed in the current of the power grid side ($i_{g\#2}$). The main difference from the previous case is the reduced number of necessary power converters, which only influence in the global efficiency. In the third case, the operation is completely different, since the necessary power for the electrical appliances is provided by the RES through the dc-dc and dc-ac converters. Therefore, a current with a sinusoidal waveform is injected into the power grid with a THD of 1.1% and in phase opposition with the voltage.

Figure 3.13c shows the obtained results for the three cases when considering that only the electrical appliances are connected to the power grid (i.e., the power produced by the RES is zero, the EV is not plugged-in) and that the ESS is not operating. For the electrical appliances, an active power of 1.2 kW and a reactive power of 225 VAr were measured. This situation was considered only to highlight the opportunities offered by the off-board EV-BCS for the smart grid. Therefore, the first two cases are equal to the reported cases in Fig. 3.11a. Therefore, in the third case, the current in the power grid side ($i_{g\#3}$) is composed of two parts: a fundamental component (corresponding to the active power necessary for the electrical appliances) and a selected harmonic current that is injected into the power grid for compensating the harmonic currents as requested by the smart grid. In this case, a third-order harmonic current was considered with an amplitude of 2 A and a phase of 143°. Moreover, the off-board EV-BCS is also controlled to produce reactive power for the smart grid, where a measured value of 600 VAr was considered. These values (harmonic order, amplitude, phase, and reactive power) were selected as exemplification since other values can be selected without jeopardizing the operation of the off-board EV-BCS. In these circumstances, the smart grid is responsible to establish a selective harmonic compensation control algorithm and to inform the different off-board EV-BCS about the required values. Moreover, the same off-board EV-BCS can be controlled for compensating more than one harmonic current (e.g., an off-board EV-BCS can produce the third-order and fifth-order harmonics and other off-board EV-BCS can produce the seventh-order harmonic).

Differently from the previous case, Fig. 3.14 shows a case when the RES (PV panels) deliver power to the power grid, i.e., it is directly injected into the power grid. In the figure, it is possible to visualize that the grid current ($i_{g\#3}$) is sinusoidal, but in opposition with the waveform of the voltage, meaning that the power grid is absorbing power. Since the control of the dc-dc converter, used to interface the RES, is based on a maximum power point tracking (MPPT) algorithm, it is possible to extract the maximum power from the RES at each instant. Due to this control algorithm, the extracted power can change, also forcing to change the

3 The Role of Off-Board EV Battery Chargers in Smart Homes and Smart... 67

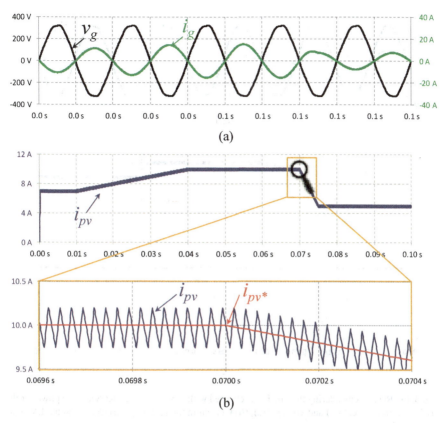

Fig. 3.14 Results considering that the power from the RES is injected into the power grid: (**a**) Power grid current (i_g) and voltage (v_g); (**b**) Current in the RES and its reference (i_{pv} and $i_{pv}*$)

reference current for the power grid side (assuming that the extracted power is injected into the power grid). Figure 3.14 shows this case for different levels of extracted power.

Besides the previous cases, Fig. 3.15 shows the results where the power for the EV charging is delivered by the RES and by the power grid. This is a combined situation that can be necessary if the power from the RES is not enough for charging the EV battery. Therefore, a parcel of power was absorbed from the power grid. In this figure are also shown the current (i_g) and the voltage (v_g) in the power grid side, as well as the currents in the dc-side, namely, the EV battery current (i_{ev}) and the RES current (i_{pv}). Since the EV battery is charged with constant current and the current in the RES is changing according to the MPPT algorithm, the current (i_g) in the power grid side is also changing, accordingly. Despite the observed variations, the current (i_g) in the power grid side does not present sudden variations capable of causing power quality problems.

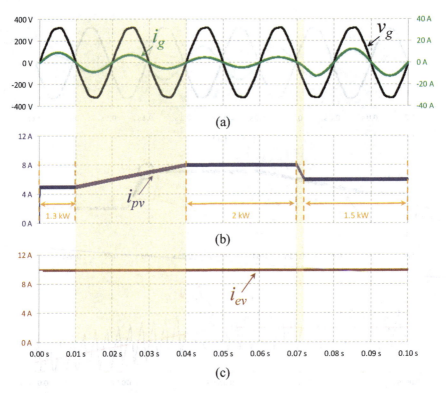

Fig. 3.15 Results considering that the EV is charged with power from the RES and the power grid: (**a**) Power grid current (i_g) and voltage (v_g); (**b**) Current in the RES (i_{pv}); (**c**) Current in the EV (i_{ev})

3.5 Conclusion

The dissemination of electric mobility has encouraged the appearance of new technologies and opportunities in terms of power management for smart homes and for smart grids, where the electric vehicle (EV) has emerged with a set of relevant valences for such purposes. Based on this background, this book chapter deals with the role of off-board EV battery chargers in terms of operation modes and new opportunities for smart homes and smart grids. Therefore, an analysis of the state-of-the-art is presented and used as a support for launching a relation with future perspectives. On-board and off-board EV battery charging systems (EV-BCS) are analyzed in the scope of this book chapter, but special focus is given to the off-board EV-BCS, particularly when interfacing renewable energy sources (RES) and energy storage systems (ESS). Moreover, as demonstrated throughout the chapter, an off-board EV-BCS can also be a central element in a future dc smart home, allowing to interface with electrical appliances. Based on this perspective, three distinct cases were considered: (1) A conventional ac smart home (using ac-dc power converters for each technology); (2) A hybrid ac and dc smart home, where the electrical appliances are directly connected to the ac power grid, but an off-board EV-BCS interfaces the RES and the ESS; (3) A dc smart home, where an off-board EV-BCS is the only interface with the electrical power grid. The results were obtained focusing on the efficiency and the power quality for each case when the off-board EV-BCS is operating in different modes. In terms of efficiency, the results show that the off-board EV-BCS in dc smart homes have better results, mainly due to the reduced number of required power stages. In terms of power quality, the off-board EV-BCS in dc smart homes also presents better results than the other cases, since the interface with the power grid is performed by a front-end ac-dc converter operating with a sinusoidal current, in phase or phase opposition with the power grid voltage. Moreover, for this case it was also demonstrated that the off-board EV-BCS can be used for providing power quality services for the smart grid, i.e., producing reactive power and operating in strategies of control based on selective harmonic compensation. By analyzing the obtained results, it is possible to infer that unified structures of off-board EV-BCS have several advantages when framed with hybrid or dc smart homes, as well as with smart grids.

References

1. A.G. Boulanger, A.C. Chu, S. Maxx, D.L. Waltz, Vehicle electrification: Status and issues. Proc. IEEE **99**(6), 1116–1138 (2011)
2. V. Monteiro, J.A. Afonso, J.C. Ferreira, J.L. Afonso, Vehicle electrification: New challenges and opportunities for smart grids. MDPI Energies **12**(1), 1–20 (2018)
3. S. Wencong, H. Rahimi-Eichi, W. Zeng, M.-Y. Chow, A survey on the electrification of transportation in a smart grid environment. IEEE Trans. Ind. Informat. **8**(1), 1–10 (2012)
4. L. Zhang, X. Hu, Z. Wang, F. Sun, J. Deng, D.G. Dorrell, Multiobjective optimal sizing of hybrid energy storage system for electric vehicles. IEEE Trans. Veh. Technol. **67**(2), 1027–1035 (2018)
5. D.S. Gautam, F. Musavi, M. Edington, W. Eberle, W.G. Dunford, An automotive onboard 3.3-kW battery charger for PHEV application. IEEE Trans. Veh. Technol. **61**(8), 3466–3474 (2012)
6. C.C. Chan, A. Bouscayrol, K. Chen, Electric, hybrid, and fuel-cell vehicles: Architectures and modeling. IEEE Trans. Veh. Technol. **59**(2), 589–598 (2010)
7. Y. Wang, J.S. Thompson, Two-stage admission and scheduling mechanism for electric vehicle charging. IEEE Trans. Smart Grid **10**(3), 2650–2660 (2019)
8. S. Pal, R. Kumar, Electric vehicle scheduling strategy in residential demand response programs with neighbor connection. IEEE Trans. Ind. Informat. **14**(3), 980–988 (2018)
9. J.C. Ferreira, V. Monteiro, J.L. Afonso, Vehicle-to-anything application (V2Anything app) for electric vehicles. IEEE Trans. Ind. Informat. **10**(3), 1927–1937 (2014)
10. V. Monteiro, B. Exposto, J.C. Ferreira, J.L. Afonso, Improved vehicle-to-home (iV2H) operation mode: Experimental analysis of the electric vehicle as off-line UPS. IEEE Trans Smart Grid **8**(6), 2702–2711 (2017)
11. M. Multin, F. Allerding, H. Schmeck, Integration of electric vehicles in smart homes - an ICT-based solution for V2G scenarios, in *IEEE ISGT PES Innovative Smart Grid Technologies*, (2012), pp. 1–8
12. Y. Ota, H. Taniguchi, T. Nakajima, K.M. Liyanage, J. Baba, A. Yokoyama, Autonomous distributed V2G (vehicle-to-grid) satisfying scheduled charging. IEEE Trans. Smart Grids **3**(1), 559–564 (2012)

13. M. Yilmaz, P.T. Krein, Review of the impact of vehicle-to-grid technologies on distribution systems and utility interfaces. IEEE Trans. Power Electron. **28**(12), 5673–5689 (2013)
14. R. Yu, W. Zhong, S. Xie, C. Yuen, S. Gjessing, Y. Zhang, Balancing power demand through EV mobility in vehicle-to-grid Mobile energy networks. IEEE Trans. Ind. Informat. **12**(1), 79–90 (2016)
15. J.A. Pecas Lopes, F. Soares, M. Pedro, R. Almeida, Integration of electric vehicles in the electric power systems. Proc. IEEE **99**(1), 168–183 (2011)
16. P. Richardson, D. Flynn, A. Keane, Optimal charging of electric vehicles in low-voltage distribution systems. IEEE Trans. Power Syst. **27**(1), 268–279 (2012)
17. V. Monteiro, A.A. Nogueiras Melendez, C. Couto, J.L. Afonso, Model predictive current control of a proposed single-switch three-level active rectifier applied to EV battery chargers, in *IEEE IECON Industrial Electronics Conference*, (Florence, 2016, Oct), pp. 1365–1370
18. A. Luo, Q. Xu, F. Ma, Y. Chen, Overview of power quality analysis and control Technology for the Smart Grid. J Mod Power Syst Clean Energy **4**(1), 1–9 (2016)
19. R.-C. Leou, Optimal charging/discharging control for electric vehicles considering power system constraints and operation costs. IEEE Trans. Power Syst. **31**(3), 1854–1860 (2016)
20. J.C. Ferreira, V. Monteiro, J.L. Afonso, Electric vehicle assistant based on driver profile. Int. J. Electr Hybrid Veh **6**(4), 335–349 (2014)
21. M. Zhang, J. Chen, The energy management and optimized operation of electric vehicles based on microgrid. IEEE Trans. Power Del. **29**(3), 1427–1435 (2014)
22. C. Gouveia, D. Rua, F. Ribeiro, L. Miranda, J.M. Rodrigues, C.L. Moreira, J.A. Peças Lopes, Experimental validation of smart distribution grids: Development of a microgrid and electric mobility laboratory. Electr. Power Energy Syst. **78**, 765–775 (2016)
23. M.D. Galus, M.G. Vaya, T. Krause, G. Andersson, The role of electric vehicles in smart grids. WIREs Energy Environ **2**, 384–400 (2013)
24. M.A. Masrur, A.G. Skowronska, J. Hancock, S.W. Kolhoff, D.Z. McGrew, J.C. Vandiver, J. Gatherer, Military-based vehicle-to-grid and vehicle-to-vehicle microgrid—System architecture and implementation. IEEE Trans. Trans Electr **4**(1), 157–171 (2018)
25. M.H.K. Tushar, C. Assi, M. Maier, M.F. Uddin, Smart microgrids: Optimal joint scheduling for electric vehicles and home appliances. IEEE Trans. Smart Grid **5**(1), 239–250 (2014)
26. V. Monteiro, J.C. Ferreira, J.L. Afonso, Operation modes of battery chargers for electric vehicles in the future smart grids, Chapter 44, in *Technological Innovation for Collective Awareness Systems*, ed. by L. M. Camarinha-Matos, L. M. Barreto, N. S. Mendonça, 1st edn., (Springer, 2014), pp. 401–408
27. V. Monteiro, J.C. Ferreira, A.A. Nogueiras Melendez, J.L. Afonso, Electric vehicles on-board battery charger for the future smart grids, Chapter 38, in *Technological Innovation for the Internet of Things*, ed. by L. M. Camarinha-Matos, S. Tomic, P. Graça, 1st edn., (Springer, 2013), pp. 351–358
28. D.P. Tuttle, R. Baldick, The evolution of plug-in electric vehicle-grid interactions. IEEE Trans. Smart Grid **3**(1), 500–505 (2012)
29. J.E. Hernandez, F. Kreikebaum, D. Divan, Flexible electric vehicle (EV) charging to meet renewable portfolio standard (RPS) mandates and minimize green house gas emissions, in *IEEE ECCE Energy Conversion Congress and Exposition*, (Atlanta, 2010, Sept), pp. 4270–4277
30. M.H.K. Tushar, A.W. Zeineddine, C. Assi, Demand-side management by regulating charging and discharging of the EV, ESS, and utilizing renewable energy. IEEE Trans. Ind. Informat. **14** (1), 117–126 (2018)
31. A.Y. Saber, G.K. Venayagamoorthy, Plug-in vehicles and renewable energy sources for cost and emission reductions. IEEE Trans. Ind. Electron. **58**(4), 1229–1238 (2011)
32. H. Turker, S. Bacha, Optimal minimization of plug-in electric vehicle charging cost with vehicle-to-home and vehicle-to-grid concepts. IEEE Trans. Veh. Technol. **67**(11), 10281–10292 (2018)
33. V.C. Gungor, D. Sahin, T. Kocak, S. Ergut, C. Buccella, C. Cecati, G.P. Hancke, Smart grid and smart homes - key players and pilot projects. IEEE Ind. Electron. Mag. **6**, 18–34 (2012)

34. C. Liu, K.T. Chau, W. Diyun, S. Gao, Opportunities and challenges of vehicle-to-home, vehicle-to-vehicle, and vehicle-to-grid technologies. Proc. IEEE **101**(11), 2409–2427 (2013)
35. C. Jin, J. Tang, P. Ghosh, Optimizing electric vehicle charging: A customer's perspective. IEEE Trans. Veh. Technol. **62**(7), 2919–2927 (2013)
36. M. Sanduleac, M. Albu, L. Toma, J. Martins, A.G. Pronto, V. Delgado-Gomes, Hybrid AC and DC smart home resilient architecture - transforming prosumers in UniRCons, in *IEEE ICE/ITMC International Conference on Engineering, Technology and Innovation*, (2017), pp. 1572–1577
37. Leonardo Energy, Poor power quality costs European business more than €150 billion a year, European Power Quality Survey, 2008
38. R. Targosz, D. Chapman, The cost of poor power quality, European Copper Institute, Leonardo Energy, Application Note, Oct 2015
39. Q. Zhong, W. Huang, S. Tao, X. Xiao, Survey on assessment of power quality cost in Shanghai China, in *IEEE PES General Meeting Conference and Exposition*, (2014, July), pp. 1–5
40. M.C.B.P. Rodrigues, H.J. Schettino, A.A. Ferreira, P.G. Barbosa, H.A.C. Braga, Active power filter operation of an electric vehicle applied to single-phase networks, in *IEEE/IAS INDUSCON International Conference on Industry Applications*, (Fortaleza, 2012, Nov), pp. 1–8
41. M.C.B.P. Rodrigues, I.D.N. Souza, A.A. Ferreira, P.G. Barbosa, H.A.C. Braga, Simultaneous active power filter and G2V (or V2G) operation of EV on-board power electronics, in *IEEE IECON Industrial Electronics Conference*, (Vienna, 2013, Nov), pp. 4684–4689
42. M.C. Kisacikoglu, B. Ozpineci, L.M. Tolbert, Examination of a PHEV bidirectional charger system for V2G reactive power compensation, in *IEEE APEC Applied Power Electronics Conference and Exposition*, (2010, Feb), pp. 458–465
43. M.C. Kisacikoglu, M. Kesler, L.M. Tolbert, Single-phase on-board bidirectional PEV charger for V2G reactive power operation. IEEE Trans. Smart Grid **6**(2), 767–775 (2015)
44. Y. Sun, W. Liu, M. Su, X. Li, H. Wang, J. Yang, A unified modeling and control of a multi-functional current source-typed converter for V2G application. Electr. Power Syst. Res. **106**, 12–20 (2014)
45. G. Buja, M. Bertoluzzo, C. Fontana, Reactive power compensation capabilities of V2G-enabled electric vehicles. IEEE Trans. Power Electon. **32**(12), 9447–9459 (2017)
46. V. Monteiro, J.G. Pinto, B. Exposto, J.C. Ferreira, C. Couto, J.L. Afonso, Assessment of a battery charger for electric vehicles with reactive power control, in *IEEE IECON Industrial Electronics Conference*, (2012, Oct), pp. 5124–5129
47. V. Monteiro, J.G. Pinto, J.L. Afonso, Operation modes for the electric vehicle in smart grids and smart homes: Present and proposed modes. IEEE Trans. Veh. Tech. **65**(3), 1007–1020 (2016)
48. M. Brenna, F. Foiadelli, M. Longo, The exploitation of vehicle-to-grid function for power quality improvement in a smart grid. IEEE Trans. Intell. Transp. Syst. **15**(5), 2169–2177 (2014)
49. A.R. Boynuegri, M. Uzunoglu, O. Erdinc, E. Gokalp, A new perspective in grid connection of electric vehicles: Different operating modes for elimination of energy quality problems. Appl. Energy **132**, 435–451 (2014)
50. M.C.B.P. Rodrigues, I. Souza, A.A. Ferreira, P.G. Barbosa, H.A.C. Braga, Integrated bidirectional single-phase vehicle-to-grid interface with active power filter capability, in *COBEP Power Electronics Conference (COBEP)*, (2013, Oct), pp. 993–1000
51. L. Rauchfuß, J. Foulquier, R. Werner, Charging station as an active filter for harmonics compensation of smart grid, in *IEEE ICHQP International Conference on Harmonics and Quality of Power*, (2014, May), pp. 181–184
52. H. Han, C. Zhang, Z. Lv, D. Huang, Power control strategy of electric vehicle for active distribution network, in *IEEE IECON Industrial Electronics Conference*, (2017, Nov), pp. 3907–3911
53. J.G. Pinto, V. Monteiro, H. Goncalves, B. Exposto, D. Pedrosa, C. Couto, J.L. Afonso, Bidirectional battery charger with grid-to-vehicle, vehicle-to-grid and vehicle-to-home technologies, in *IEEE IECON Industrial Electronics Conference*, (Vienna, 2013, Nov), pp. 5934–5939

54. D.P. Tuttle, R.L. Fares, R. Baldick, M.E. Webber, Plug-in vehicle to home (V2H) duration and power output capability, in *IEEE ITEC Transportation Electrification Conference and Expo*, (2013, June), pp. 1–7
55. V. Monteiro, B. Exposto, J.G. Pinto, R. Almeida, J.C. Ferreira, A.A.N. Melendez, J.L. Afonso, On-board electric vehicle battery charger with enhanced V2H operation mode, in *IEEE IECON Industrial Electronics Conference*, (2014, Oct), pp. 1636–1642
56. F. Berthold, A. Ravey, B. Blunier, D. Bouquain, S. Williamson, A. Miraoui, Design and development of a smart control strategy for plug-in hybrid vehicles including vehicle-to-home functionality. IEEE Trans. Trans Electr **1**(2), 168–177 (2015)
57. M. Multin, F. Allerding, H. Schmeck, Integration of electric vehicles in smart homes - an ICT-based solution for V2G scenarios, in *IEEE ISGT PES Innovative Smart Grid Technologies*, (2012, Jan), pp. 1–8
58. D.T. Nguyen, L.B. Le, Joint optimization of electric vehicle and home energy scheduling considering user comfort preference. IEEE Trans. Smart Grid **5**(1), 188–199 (2014)
59. Green Car Congress, Nissan to launch the 'LEAF to Home' V2Hpower supply system with Nichicon 'EV Power Station' in June (2012), http://www.greencarcongress.com/2012/05/leafvsh-20120530.html. Available 30 May 2012
60. N. Gordon-Bloomfield, Nissan, Mitsubishi, Toyota turn electric cars into backup batteries (2014). http://www.greencarreports.com/news/1063565_nissan-mitsubishi-toyota-turn-electric-cars-into-backup-batteries
61. R. Xiong, Y. Jiavyi Quanqing, H. He, F. Sun, Critical review on the battery state of charge estimation methods for electric vehicles. IEEE Access **6**, 1832–1843 (2017)
62. M. Stanley Whittingham, History, evolution, and future status of energy storage. Proc. IEEE **100**(Special Centennial Issue), 1518–1534 (2012)
63. L. Chandra, S. Chanana, Energy management of smart homes with energy storage, rooftop PV and electric vehicle, in *IEEE International Students' Conference on Electrical, Electronics and Computer Science*, (2018, Feb), pp. 1–6
64. V. Calderaro, V. Galdi, G. Graber, G. Graditi, F. Lamberti, Impact assessment of energy storage and electric vehicles on smart grids, in *IEEE PQ Electric Power Quality and Supply Reliability Conference*, (2014, July), pp. 15–18

Chapter 4
Optimal Charge Scheduling of Electric Vehicles in Solar Energy Integrated Power Systems Considering the Uncertainties

S. Muhammad Bagher Sadati, Jamal Moshtagh, Miadreza Shafie-Khah, Abdollah Rastgou, and João P. S. Catalão

4.1 Introduction

Nowadays, air pollution and dependence on fossil fuel resources are worldwide concerns. These issues are most taken into account in the transportation sectors and electricity generation system as the main consumers of fossil fuels. Electric vehicles (EVs) with the capability of Vehicle-to-Grid (V2G) are a solution to answer these concerns. Of course, most of the EVs, which will be added in the distribution system in the future, would highly consume energy, which leads to more energy production and consequently, increased the greenhouse gas emissions. However, this problem can be solved by charging/discharging schedule of the EVs as well as the usage of renewable-energy resources (RERs) such as the solar system.

Because of uncontrolled charging, controlled charging and charging/discharging schedule of the EVs, the planning and operation of the smart distribution network (SDN) have been intricated. Uncontrolled charging of the EVs has inappropriate

---

S. M. B. Sadati (✉)
National Iranian Oil Company (NIOC), Iranian Central Oil Fields Company (ICOFC), West Oil and Gas Production Company (WOGPC), Kermanshah, Iran

J. Moshtagh
Department of Electrical Engineering, Faculty of Engineering, University of Kurdistan, Sanandaj, Kurdistan, Iran
e-mail: j.moshtagh@uok.ac.ir

M. Shafie-Khah
School of Technology and Innovations, University of Vaasa, Vaasa, Finland

A. Rastgou
Department of Electrical Engineering, Kermanshah Branch, Islamic Azad University, Kermanshah, Iran
e-mail: a.rastgou@iauksh.ac.ir

J. P. S. Catalão
Faculty of Engineering of the University of Porto and INESC TEC, Porto, Portugal
e-mail: catalao@fe.up.pt

© Springer Nature Switzerland AG 2020
A. Ahmadian et al. (eds.), *Electric Vehicles in Energy Systems*,
https://doi.org/10.1007/978-3-030-34448-1_4

results such as increasing power losses and demand [1–4], imbalanced demand [5, 6], voltage drop [7], increasing of total harmonic distortion [8, 9], decreasing of cable and transformer life [10, 11], etc.. However, by using the controlled charging and charging/discharging schedule, as well as V2G capability of the EVs; the performance of the SDN is improved and is obtained some benefit such as ancillary service [12], peak load shaving [13, 14], emission's reduction [15], support for the integration of RERs [16, 17], losses reduction [18], improving voltage profile [19] and maximizing the profit [20, 21].

In addition, in [3, 22] are proved that charging of the EVs with only traditional power plants leads to unfit environmental impact. So, using of RERs along with traditional power plants is unavoidable. For this reason, charging of the EVs is explored with RERs i.e. solar system, wind turbine and both of them [23–28].

In addition, due to the uncertainties of the EVs, especially their availability and ensuring of the discharging power as well as the uncertainty of output power of the solar system, the SDN faces uncertainties. Therefore, it is necessary to introduce the risk-based model. Usually, risk control is done by using the risk measures. Value-at-risk (VaR) and conditional value-at-risk (CVaR) are the most important examples of risk measures. Due to the linear form of CVaR, this index is widely applied in the power system problems [29].

Although, the optimal operation of the SDN has been evaluated in different studies over the past few years; however, in this chapter, the operational scheduling of the SDN in the presence of solar-based EV PLs, within the bi-level framework has been investigated. The most important questions that are answered in this chapter, as follows:

1. What is the main aim of the optimal operation of the SDN?
2. What is the appropriate model with the PL owners as a new decision-maker?
3. What time the EVs will be charged and discharged?
4. How much is the total charging/discharging power of the EVs?
5. What is the amount of purchasing power from the wholesale market (WM) for the EVs and customers with regard to V2G capability?
6. what is the effect of the uncertainties on the SDN?
7. How does the risk effect on operational scheduling of the SDN?
8. What are the most important affecting factors on the SDN?
9. What is the proper method for solving the offered model?

The modeling of the EVs and the solar system are explained in Sects. 4.2 and 4.3, respectively. Section 4.4 gives modeling of operational scheduling of the SDN, i.e., bi-level model and single-level model. In Sect. 4.5 simulation results are presented. At last, conclusions are reported in Sect. 4.6.

4.2 Modeling of the EVs

The EVs can be categorized into three groups of battery-electric vehicles, hybrid-electric vehicles, and fuel cell electric vehicles. All these EVs have a battery as well as the V2G capability. Therefore, in the near future, EVs are widely used. With increasing the EVs, the batteries of them can provide a high-availability storage system for the SDN. In this way, the EVs can act as an active element during the parked times. So, the power stored in the batteries, particularly at the on-peak hours sells to the SDNO. The initial state of energy (SOE), arrival time/departure time of the EVs to/from the PLs, are the main uncertainties of each EV. Some studies are shown that the behavior of the EVs can be modeled with appropriate probability distribution function (PDF) such as a truncated Gaussian distribution [21]. Thus, the modeling of EVs is shown by Eqs. (4.1), (4.2), and (4.3).

$$SOE_{EV}^{ini} = f_{TG}\left(X; \mu_{SOE}; \sigma_{SOE}^2; \left(SOE_{EV}^{ini,\min}; SOE_{EV}^{ini,\max}\right)\right) \quad \forall EV \quad (4.1)$$

$$t_{EV}^{arv} = f_{TG}\left(X; \mu_{arv}; \sigma_{arv}^2; \left(t_{EV}^{arv,\min}; t_{EV}^{arv,\max}\right)\right) \quad \forall EV \quad (4.2)$$

$$t_{EV}^{dep} = f_{TG}\left(X; \mu_{dep}; \sigma_{dep}^2; \left(\max\left(t_{EV}^{dep,\min}, t_{EV}^{arv}\right); t_{EV}^{dep,\max}\right)\right) \quad \forall EV \quad (4.3)$$

Due to the large number of the EVs are in the PLs every day, the more energy is needed for charging of the EVs. Furthermore, due to the V2G capability, the performance of the SDN can be improved. Since the EVs are considered a load/source at the off-peak and mid-peak hours/during the on-peak hours, a complexity is created in the operation and planning of the SDN. Accordingly, proper PL's operation will only be possible if there is an energy management system (EMS) that be capable of controlling the process of charging and discharging of the EVs. Figure 4.1 illustrates the flowchart of charging or charging/discharging schedule of the EVs, and the power exchanged between the PLs and the SDNO. Based on this flowchart, after the entrance of the EVs to the PL, required data such as initial and desired SOE

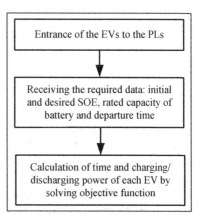

Fig. 4.1 The Flowchart of each EV's operation

of the EVs, the battery specifications and departure time are obtained from the EV owners. By computing the energy needed for each EV, the EMS determines the time and charging/discharging power of the EVs.

4.3 Modeling of the Solar System

Several cells create the solar system. This system transforms solar irradiance energy into electrical energy. The number of cells, the weather conditions, the direction of cells and the temperature are the main affecting factor of the power generated of the solar system. Of course, this power is an uncertain value due to the uncertainty of solar irradiance. The most usable PDF for modeling of solar irradiance is the Beta function that is explained in Eqs. (4.4), (4.5), and (4.6). In these equations, θ is the solar irradiance (kW/m^2). Also, by using the mean (μ) and variance (σ) of solar irradiance, α and β are computed [30].

$$f(\theta) = \begin{cases} \dfrac{\Gamma(\alpha+\beta)}{\Gamma(\alpha)+\Gamma(\beta)} \times \theta^{\alpha-1} \times (1-\theta)^{\beta-1} & 0 \leq \theta \leq 1, \alpha \geq 0, \beta \geq 0 \\ 0 & \text{otherwise} \end{cases} \tag{4.4}$$

$$\beta = (1-\mu) \times \left(\frac{\mu \times (1+\mu)}{\sigma^2} - 1 \right) \tag{4.5}$$

$$\alpha = \frac{\mu \times \beta}{1 - \mu} \tag{4.6}$$

The power generated of the solar system can be calculated by Eqs. (4.7), (4.8), (4.9), (4.10), and (4.11).

$$P_\theta = N \times FF \times V_y \times I_y \tag{4.7}$$

$$FF = \frac{V_{MPP} \times I_{MPP}}{V_{OC} \times I_{SC}} \tag{4.8}$$

$$V_y = V_{OC} - (K_v \times T_C) \tag{4.9}$$

$$I_y = \theta \times (I_{SC} + K_C \times (T_C - 25)) \tag{4.10}$$

$$T_C = T_a + \left(\theta \times \frac{T_N - 20}{0.8} \right) \tag{4.11}$$

Where voltage at the maximum power point and open circuit voltage are V_{MPP} and V_{oc}, respectively. I_{MPP} and I_{sc} are current at the maximum power point and short circuit current. The cell temperature is T_c in °C. The ambient and nominal operating temperatures are T_a and T_N in °C. k_v and k_c (in V/°C and A/°C) are the voltage temperature and the current temperature coefficient, respectively. N is the number of cells, P_θ is the power generated of the solar system, and FF is the fill factor [30].

4.4 Modeling of Operational Scheduling of the SDN

A bi-level model proposes when two decision-makers exist in the optimization problems. In this model, the upper-level and the lower-level are leader and follower, respectively. In this chapter, the SDNO as the leader and the PL owner as a follower are considered. The aims of the objective functions for leader and follower are maximizing the profit and minimizing the cost, respectively. The presented bi-level model investigates in two-parts. In the first part, the EVs only charge (controlled charging), and in the second part, the EVs participate in charging/charging schedule. The structure of the bi-level model shows in Fig. 4.2. Also, Fig. 4.3 shows how the decision-makers interact in this model. Based on Fig. 4.3, the power exchanged between the SDNO and the PL owners as well as the price of this power are considered as the decision variables of these two levels (in the controlled charging part, charging power and price, i.e. P^{ch} and Pr^{ch}, in the charging/discharging schedule part, charging/discharging power and price, i.e. P^{ch}, Pr^{ch} and P^{dch}, Pr^{dch}). The PL owner decides on the offered price for the power exchanged with the SDNO, which depends on the ability to charging or charging/discharging of the EVs. This decision affects the offered price, and the SDNO may change this price. The changing this price will also change the exchanging power. This action repeats several times in order to the problem reach the point of equilibrium.

4.4.1 Bi-Level Model with Controlled Charging

The proposed bi-level model with controlled charging of the EVs is defined in Eqs. (4.12), (4.13), (4.14), (4.15), (4.16), (4.17), (4.18), (4.19), (4.20), (4.21), (4.22), (4.23), (4.24), (4.25), (4.26), and (4.27). The goal of the upper-level is to maximize the profit of SDNO. Equations (4.12), (4.13), (4.14), (4.15), (4.16), (4.17), and (4.18) describe this level. The objective function is explained in Eq. (4.12). The decision variables of this level are the purchasing power from the WM, and the offered energy sold price to the PL owners. The parts of the objective function are as follows:

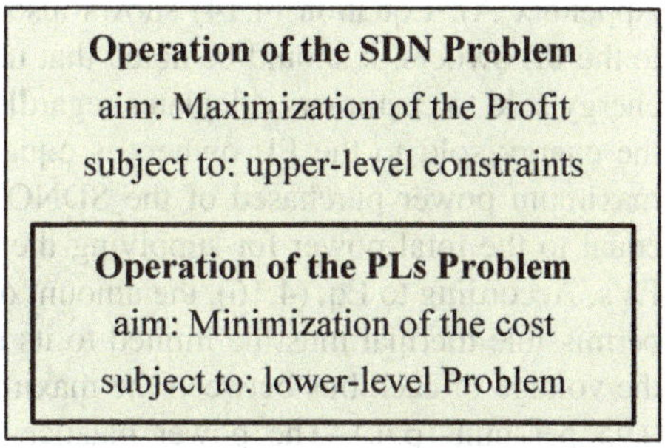

Fig. 4.2 Structure of the bi-level model

Operation of the SDN Problem
aim: Maximization of the Profit
subject to: upper-level constraints

Operation of the PLs Problem
aim: Minimization of the cost
subject to: lower-level Problem

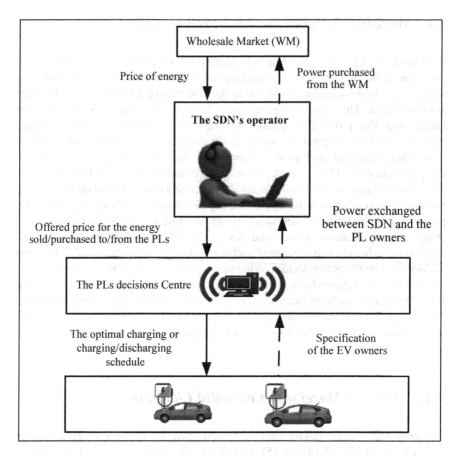

Fig. 4.3 Interaction with the SDNO and the PL owners in the bi-level model

Part 1. Selling energy to the customers (as an income term).
Part 2. Purchasing energy from the WM (as a cost term).
Part 3. The expected value of energy sold to the PL owner at off-peak/mid-peak hours (as an income term).

Equation (4.13) is the linear load flow, and is fully explained in [30] (see Appendix A). Equation (4.14) shows also the maximum price of the energy sold to the PL owners. It should be noted that in the next section, firstly, the price of the energy sold to customers calculates regardless of the EVs, so the maximum price of the energy sold to the PL owners is equal to this amount. The Eq. (4.15) is the maximum power purchased of the SDNO from the WM. This maximum limit is equal to the total power for supplying the customers' demand and charging of all EVs. According to Eq. (4.16), the amount of line current due to the capacity and the permissible thermal must be limited to its maximum value. Also, Eq. (4.17) limits the voltage of each bus between the maximum and minimum values, i.e., 1.05 and 0.95 per unit (p.u.). The power balance limit, i.e., equivalence the total power generated with the total power consumed, is shown in Eq. (4.18). The amount of loss in Eq. (4.18) is equal to multiply the value of the electrical resistance between the two lines and the squared of current between these lines, and is also linearized in [30].

Equations (4.19), (4.20), (4.21), (4.22), (4.23), (4.24), (4.25), (4.26), and (4.27) describe the lower-level. The cost minimization of the PL owners is the target of this level. At this level, the PL owners provide the optimal SOE of each EV at exiting time by charging the batteries of the EVs. The decision variables are the power purchased from the SDNO for charging of the EVs, the SOE of each EV, and the charging power of the EVs by the solar system. The objective function of this level is defined in Eq. (4.19), which minimizes the cost of the purchasing energy from the SDNO for EVs' charging during the off-peak and mid-peak hours.

To optimize the power purchased from the SDNO, it is necessary to be created proper scheduling for the charging power and charging time of the EVs. In fact, in the interval time between the arrival/departure time from/to the PLs, at the low energy prices, i.e. at the off-peak and mid-peak hours, the EVs should be charge so that the EVs leaves the PLs with the desired SOE. The time interval, i.e. charging/discharging time of the EVs and the customers' demand, is 1 hour ($\Delta t = 1$). Therefore, in these equations, Δt is neglected. The SOE of each EV, based on Eq. (4.20), should be less than its maximum value. Also, the total power purchased from the SDN and the power generated of the solar system for the EVs charging, according to Eq. (4.21) during the off-peak and mid-peak hours is limited to maximum and minimum values. According to Eq. (4.22), the EVs must not charge through the SDNO at the on-peak hours. Eq. (4.23) also shows that the EVs' charging power with the solar system at the on-peak hours should be limited to maximum and minimum values. Based on Eqs. (4.24) and (4.25), the SOE of each EV at each hour time is depended on to the remained SOE of the EV from the previous hour, the power purchased from the SDNO and the power generated by the solar system, charging efficiency, and the initial SOE of each EV. Based on Eq. (4.26), the SOE of the EVs reaches the desired SOE at the departure time. Equation (4.27) also shows that the power required for charging of the EVs through the solar system at each time is equal to the power generated of the solar system at the same time. Dual variables for the equal and unequal constraints of the lower-level problem are shown by λ. Figure 4.4 shows the proposed framework of this model.

Maximize

$$
\sum_{t=1}^{24} \left(\sum_{b=2}^{N_b} \left(P_{b,t}^L \times \mathrm{Pr}_t^L \right) - \sum_{sb=1}^{N_{sb}} \left(P_{sb,t}^{Wh2G} \times \mathrm{Pr}_t^{Wh2G} \right) \right) \\
+ \sum_{PL=1}^{N_{PL}} \sum_{EV=1}^{N_{EV}} \sum_{t=1}^{24} \left(\widehat{P}_{PL,EV,t^{mid/off-peak}}^{ch-grid} \times \mathrm{Pr}_{t^{mid/off-peak}}^{G2PL} \right)
$$

(4.12)

Subject to:

Inputs:
1. Specifications of the solar system as well as the EVs including arrival time, depurate time, initial and desired SOE, charging rate and battery capacity.
2. Real characteristics of the network such as the customers' demand, ohmic and inductive resistance, and power factor.

Upper-Level: Operational Scheduling of the SDN

Objective function: Maximizing the profit of the SDNO

Variables: The power purchased from the WM, the proposed energy sold price to the PL owners.

Limitations: Linear load flow, maximum and minimum the power purchased from the WM, maximum and minimum of the energy price, power balance, line capacity and bus voltage.

Lower-level: operation of the PLs

objective function: Minimizing the cost of the PLs

Variables: The power purchased from the SDNO for charging of the EVs, charging power of the EVs with the solar system, the SOE of the EVs.

Limitations: SOE, charging rate.

Outputs:
1- The energy sold price to the PL owners.
2- Charging scheduling of the EVs.
3- The power sold to the PLs.
4- Operational scheduling of the SDN.

Fig. 4.4 The proposed bi-level model framework with controlled charging

$$\text{Liner power flow} \tag{4.13}$$

$$0 < \Pr_{t,mid/off-peak}^{G2PL} \leq \Pr_{t,mid/off-peak}^{G2PL,\max} \tag{4.14}$$

$$0 < P_t^{Wh2G} \le P_t^{Wh2G,\,\max} \tag{4.15}$$

$$0 \le I_{b,t,s} \le I_{b,t}^{max} \tag{4.16}$$

$$V^{\min} \le V_{b,t,s} \le V^{\max} \tag{4.17}$$

$$P_{sb,t}^{Wh2G} \times \eta^{Trans} = P_{b,t}^{L} + P_{t,s}^{Loss} + \sum_{EV} \widehat{P}_{PL,EV,t^{mid/off-peak}}^{ch-grid} \tag{4.18}$$

Minimize

$$\sum_{s=1}^{N_s} \rho_s \sum_{PL=1}^{N_{PL}} \sum_{EV=1}^{N_{EV}} \sum_{t=1}^{24} \left(P_{PL,EV,t^{mid/off-peak},s}^{ch-grid} \times \Pr_{t^{mid/off-peak}}^{G2PL} \right) \tag{4.19}$$

Subject to

$$SOE_{PL,EV,t,s} \le SOE_{EV}^{\max} \quad \forall PL, EV, t, s \quad \lambda_{PL,EV,t,s}^{1} \tag{4.20}$$

$$0 \le P_{PL,EV,t^{mid/off-peak},s}^{ch-grid} + P_{PL,EV,t^{mid/off-peak},s}^{ch-Solar}$$
$$\le P^{\max} \quad \forall PL, EV, t^{mid/off-peak}, s \quad \lambda_{PL,EV,t^{mid/off-peak},s}^{2}, \lambda_{PL,EV,t^{mid/off-peak},s}^{3} \tag{4.21}$$

$$P_{PL,EV,t^{on-peak},s}^{ch-grid} = 0 \quad \forall PL, EV, t^{on-peak}, s \tag{4.22}$$

$$0 \le P_{PL,EV,t^{on-peak},s}^{ch-Solar}$$
$$\le P^{\max} \quad \forall PL, EV, t^{on-peak}, s \quad \lambda_{PL,EV,t^{on-peak},s}^{4}, \lambda_{PL,EV,t^{on-peak},s}^{5} \tag{4.23}$$

$$SOE_{PL,EV,t,s} = SOE_{PL,EV,t-1,s} + \left(P_{PL,EV,t,s}^{ch-grid} + P_{PL,EV,t,s}^{ch-Solar} \right) \times \eta_{ch} \quad \forall PL, EV, t$$
$$\succ t^{arv}, s \quad \lambda_{PL,EV,t\succ t^{arv},s}^{6} \tag{4.24}$$

$$SOE_{PL,EV,t,s} = SOE_{EV}^{arv} + \left(P_{PL,EV,t,s}^{ch-grid} + P_{PL,EV,t,s}^{ch-Solar} \right)$$
$$\times \eta_{ch} \quad \forall PL, EV, t^{arv}, s \quad \lambda_{PL,EV,t^{arv},s}^{7} \tag{4.25}$$

$$SOE_{PL,EV,t,s} = SOE_{EV}^{dep} \quad \forall PL, EV, t^{dep}, s \quad \lambda_{PL,EV,t^{dep},s}^{8} \tag{4.26}$$

$$\sum_{EV} P_{PL,EV,t,s}^{ch-Solar} = P_{PL,t,s}^{Solar} \quad \forall PL, EV, t, s \quad \lambda_{PL,EV,t,s}^{9} \tag{4.27}$$

4.4.2 Bi-Level Model with the Charging/Discharging Schedule

The presented bi-level model with the charging/discharging schedule of the EVs is described in Eqs. (4.28), (4.29), (4.30), (4.31), (4.32), (4.33), (4.34), (4.35), (4.36),

(4.37), (4.38), (4.39), (4.40), (4.41), (4.42), (4.43), (4.44), and (4.45). In this case, the SDNO at the on-peak hours uses the discharging power of the EVs as well as the power generated of the solar system for supplying the customers' demand. The goal of the upper-level is to maximize the profit of SDNO. This level is defined by Eqs. (4.28), (4.29), (4.30), (4.31), (4.32), (4.33), and (4.34). The objective function is presented in Eq. (4.28). The decision variables of this level are the power purchased from the WM, the energy purchased price from the PL owners. The energy sold price to the PL owners is calculated from the previous part and is considered as a parameter. The parts of this objective function are as follows:

Part 1. Selling energy to the customers (as an income term).
Part 2. Purchasing energy from the WM (as a cost term).
Part 3. The expected value of energy sold to the PL owners at off-peak/mid-peak hours (as an income term).
Part 4. The expected value of purchasing energy from the PL owners at the on-peak hours (as a cost term).
Part 5. The expected value of purchasing energy from the power generated of the solar system at the on-peak hours (as a cost term).

Equations (4.29), (4.30), (4.31), (4.32), (4.33), and (4.34) are the constraints of this level. Except Eq. (4.30), reminded equations are explained in Sect. 4.4.1. Equation (4.30) shows the maximum price of the energy purchased from the PL owners.

Equations (4.35), (4.36), (4.37), (4.38), (4.39), (4.40), (4.41), (4.42), (4.43), (4.44), and (4.45) describe the lower-level. The aim of this level is to minimization the cost of the PL owners. At this level, the PL owners provide the optimal SOE of each EV at the departure time by charging/discharging schedule of the EVs. The decision variables are the power exchanged between the SDNO and the PL owners, the SOE of each EV, and the charging power of the EVs by the solar system. The objective function of this level is described in Eq. (4.35). The parts of this objective function are as follows:

1. Purchasing energy from the SDNO for EVs' charging during the off-peak/mid-peak hours.
2. Purchasing energy from the EV owners at the on-peak hours for selling to the SDNO. In this case, it is supposed that half of this income is paid to the EV owners to encourage them to attend the V2G program.
3. The cost of battery depreciation that is paid to the EV owners due to many times discharging. This term is calculated by the exchanging power between each EV and the PL owner [21].

The constraints of this level explain in Eqs. (4.36), (4.37), (4.38), (4.39), (4.40), (4.41), (4.42), (4.43), (4.44), and (4.45). Based on the previous part, proper scheduling for the power and the time of the EVs charging/discharging is needed. In fact, in the interval time between the arrival/departure time from/to the PLs, at the low energy prices, i.e. the off-peak and mid-peak hours, the EVs should be charge and at the high energy prices, i.e. the on-peak hours, the EVs should be discharge. Also, the

4 Optimal Charge Scheduling of Electric Vehicles in Solar Energy...

EVs leaves the PLs with the desired SOE. The SOE of each EV, based on the Eq. (4.36), should be between the minimum and maximum value. Equations (4.37) and (4.38) are explained in the previous part. Equation (4.39) shows that the power generated of the solar system for charging of the EVs not used at the on-peak hours. In fact, at these hours, the discharging power of the EVs and the power generated of the solar system are applied in order to supply the customers' demand. The amount of discharging power of the EVs for selling to the SDNO at the on-peak hours is also limited between the maximum and minimum values, based on Eq. (4.40). According to Eq. (4.41), the discharging power must be zero during the off-peak/mid-peak hours. Equations (4.42), (4.43), (4.44), and (4.45) are also explained in the previous part. λ are dual variables for the equal and unequal constraints of the lower-level problem. Figure 4.5 shows the proposed framework for this model.

Maximize

$$
\begin{aligned}
&\sum_{t=1}^{24} \left(\sum_{b=2}^{N_b} \left(P_{b,t}^L \times \mathrm{Pr}_t^L \right) - \sum_{sb=1}^{N_{sb}} \left(P_{sb,t}^{Wh2G} \times \mathrm{Pr}_t^{Wh2G} \right) \right) \\
&+ \sum_{PL=1}^{N_{PL}} \sum_{EV=1}^{N_{EV}} \sum_{t=1}^{24} \left(\begin{array}{c} \left(\widehat{P}_{PL,EV,t^{mid/off-peak}}^{ch-grid} \times \mathrm{Pr}_{t^{mid/off-peak}}^{G2PL} \right) \\ - \left(\widehat{P}_{PL,EV,t^{on-peak}}^{dch} \times \mathrm{Pr}_{t^{on-peak}}^{PL2G} \right) \end{array} \right) \\
&- \sum_{PL=1}^{N_{PL}} \sum_{t=1}^{24} \left(\widehat{P}_{PL,t^{on-peak}}^{Solar} \times \mathrm{Pr}_{t^{on-peak}}^{PL2G} \right)
\end{aligned}
\tag{4.28}
$$

Subject to:

$$
\text{Liner power flow} \tag{4.29}
$$

$$
0 < \mathrm{Pr}_{t^{on-peak}}^{PL2G} \leq \mathrm{Pr}_{t^{on-peak}}^{PL2G,\,max} \tag{4.30}
$$

$$
0 < P_t^{Wh2G} \leq P_t^{Wh2G,\,max} \tag{4.31}
$$

$$
0 \leq I_{b,t,s} \leq I_{b,t}^{max} \tag{4.32}
$$

$$
V^{min} \leq V_{b,t,s} \leq V^{max} \tag{4.33}
$$

$$
P_{sb,t}^{Wh2G} \times \eta^{Trans} + \sum_{EV} \widehat{P}_{PL,EV,t^{on-peak}}^{dch} + \widehat{P}_{PL,t^{on-peak}}^{Solar}
$$

$$
= P_{b,t}^L + P_{t,s}^{Loss} + \sum_{EV} \widehat{P}_{PL,EV,t^{mid/off-peak}}^{ch-grid} \tag{4.34}
$$

Inputs:

1. Specifications of the solar system as well as the EVs including arrival time, depurate time, initial and final SOE, charging/discharging rate and battery capacity.

2. Real characteristics of the network such as the customers' demand, ohmic and inductive resistance, and power factor.

Upper -Level: Operational Scheduling of the SDN

Objective function: Maximizing the profit of the SDNO

Variables: The power purchased from the WM, the proposed energy purchased price from the PL owners.

Limitations: Linear load flow, maximum and minimum the power purchased from the WM, maximum and minimum energy price, power Balance, line capacity and bus voltage.

Lower-level: operation of the PLs

objective function: Minimizing the cost of the PLs

Variables: The power exchanged between the SDN and the PLs, charging power of the EVs with the solar system, the SOE of the EVs.

Limitations: SOE, charging/discharging rate.

Outputs:

1- The energy purchased price from the PL owners.

2- Charging/discharging scheduling of the EVs.

3- Power exchanged between the PLs and the SDN.

4- Operational scheduling of the SDN.

Fig. 4.5 The proposed bi-level model framework with charging/discharging schedule

4 Optimal Charge Scheduling of Electric Vehicles in Solar Energy...

Minimize

$$
\sum_{s=1}^{N_s} \rho_s \sum_{PL=1}^{N_{PL}} \sum_{EV=1}^{N_{EV}} \sum_{t=1}^{24} \left(\left(P^{ch-grid}_{PL,EV,t^{mid/off-peak},s} \times \mathrm{Pr}^{G2PL}_{t^{mid/off-peak}} \right) + \left(P^{dch}_{PL,EV,t^{on-peak},s} \times \left(0.5\mathrm{Pr}^{PL2G}_{t^{on-peak}} + C^{cd} \right) \right) \right)
\tag{4.35}
$$

Subject to:

$$
SOE^{min}_{EV} \leq SOE_{PL,EV,t,s} \leq SOE^{max}_{EV} \quad \forall PL, EV, t, s \quad \lambda^1_{PL,EV,t,s}, \lambda^2_{PL,EV,t,s}
\tag{4.36}
$$

$$
0 \leq P^{ch-grid}_{PL,EV,t^{mid/off-peak},s} + P^{ch-Solar}_{PL,EV,t^{mid/off-peak},s}
$$
$$
\leq P^{max} \quad \forall PL, EV, t^{mid/off-peak}, s \quad \lambda^3_{PL,EV,t^{mid/off-peak},s}, \lambda^4_{PL,EV,t^{mid/off-peak},s}
\tag{4.37}
$$

$$
P^{ch-grid}_{PL,EV,t^{on-peak},s} = 0 \quad \forall PL, EV, t^{on-peak}, s
\tag{4.38}
$$

$$
P^{ch-Solar}_{PL,EV,t^{on-peak},s} = 0 \quad \forall PL, EV, t^{on-peak}, s
\tag{4.39}
$$

$$
0 \leq P^{dch}_{PL,EV,t^{on-peak},s}
$$
$$
\leq P^{max} \quad \forall PL, EV, t^{on-peak}, s \quad \lambda^5_{PL,EV,t^{on-peak},s}, \lambda^6_{PL,EV,t^{on-peak},s}
\tag{4.40}
$$

$$
P^{dch}_{PL,EV,t^{mid/off-peak},s} = 0 \quad \forall PL, EV, t^{mid/off-peak}, s
\tag{4.41}
$$

$$
SOE_{PL,EV,t,s} = SOE_{PL,EV,t-1,s} - \left(\frac{P^{dch}_{PL,EV,t,s}}{\eta_{dch}} \right) + \left(P^{ch-grid}_{PL,EV,t,s} + P^{ch-Solar}_{PL,EV,t,s} \right)
$$
$$
\times \eta^{ch} \quad \forall PL, EV, t
$$
$$
\succ t^{arv}, s \quad \lambda^7_{PL,EV,t \succ t^{arv},s}
\tag{4.42}
$$

$$
SOE_{PL,EV,t,s} = SOE^{arv}_{EV} - \left(\frac{P^{dch}_{PL,EV,t,s}}{\eta_{dch}} \right) + \left(P^{ch-grid}_{PL,EV,t,s} + P^{ch-Solar}_{PL,EV,t,s} \right)
$$
$$
\times \eta_{ch} \quad \forall PL, EV, t^{arv}, s \quad \lambda^8_{PL,EV,t^{arv},s}
\tag{4.43}
$$

$$
SOE_{PL,EV,t,s} = SOE^{dep}_{EV} \quad \forall PL, EV, t^{dep}, s \quad \lambda^9_{PL,EV,t^{dep},s}
\tag{4.44}
$$

$$
\sum_{EV} P^{ch-Solar}_{PL,EV,t^{mid/off-peak},s} = P^{Solar}_{PL,EV,t^{mid/off-peak},s} \quad \forall PL, EV, t^{mid/off-peak}, s \quad \lambda^{10}_{PL,EV,t^{mid/off-peak},s}
\tag{4.45}
$$

4.4.3 A Bi-Level Problem Solving Method

The KKT conditions and the dual theory are applied to solve the non-linear bi-level model. The single-level steps and linearization of the bi-level model are as follows [21, 31]:

1. The energy sold price to PL owners in the controlled charging model as well as the energy purchased price from the PL owners in the charging/discharging schedule model; those are as variables in the upper-level, are considered as parameters in the lower-level. Therefore, the lower-level problem that is linear and continuous is replaced by KKT conditions.
2. With the using of the KKT conditions, the problem is still non-linear due to the multiplication of two variables. Therefore, by using the dual theory, the linear expressions of these non-linear parts are calculated and replaced.

The linear single-level model, whose steps are described in Appendix B, are expressed in Eqs. (4.46), (4.47), (4.48), (4.49), and (4.50) for controlled charging.

Maximize

$$
\begin{aligned}
\mathrm{OF}_1 + \sum_{s=1}^{N_s} \rho_s \times \mathrm{OF}_2 = {} & \sum_{t=1}^{24} \left(\sum_{b=2}^{N_b} \left(P_{b,t}^L \times \mathrm{Pr}_t^L \right) - \sum_{sb=1}^{N_{sb}} \left(P_{sb,t}^{Wh2G} \times \mathrm{Pr}_t^{Wh2G} \right) \right) \\
& + \sum_{s=1}^{N_s} \rho_s \sum_{PL=1}^{N_{PL}} \sum_{EV=1}^{N_{EV}} \sum_{t=1}^{24} \left(\begin{array}{l} -\left(SOE_{EV}^{\max} \times \lambda_{\mathrm{PL,EV},t,s}^1 \right) - \left(P^{\max} \times \lambda_{\mathrm{PL,EV},t^{mid/off-peak},s}^3 \right) \\ -\left(P^{\max} \times \lambda_{\mathrm{PL,EV},t^{on-peak},s}^5 \right) + \left(SOE_{EV}^{arv} \times \lambda_{\mathrm{PL,EV},t^{arv},s}^7 \right) \\ +\left(SOE_{EV}^{dep} \times \lambda_{\mathrm{PL,EV},t^{dep},s}^8 \right) + \left(P_{PL,t,s}^{Solar} \times \lambda_{\mathrm{PL,EV},t,s}^9 \right) \end{array} \right)
\end{aligned}
\tag{4.46}
$$

Subject to:

$$
\begin{array}{rr}
(4.13) \text{ to } (4.18) & (4.47) \\
(4.20) \text{ to } (4.27) & (4.48) \\
(4.I.11) \text{ to } (4.I.13) & (4.49) \\
(4.I.20) \text{ to } (4.I.24) & (4.50)
\end{array}
$$

Also, the charging/discharging schedule model is explained in Eqs. (4.51), (4.52), (4.53), (4.54), and (4.55).

4 Optimal Charge Scheduling of Electric Vehicles in Solar Energy...

$$\text{Maximize} \quad OF_3 + \sum_{s=1}^{N_S} \rho_s \times OF_4$$

$$= \sum_{t=1}^{24} \left(\sum_{b=2}^{N_b} \left(P_{b,t}^L \times \text{Pr}_t^L \right) - \sum_{sb=1}^{N_{sb}} \left(P_{sb,t}^{Wh2G} \times \text{Pr}_t^{Wh2G} \right) \right)$$

$$+ \sum_{PL=1}^{N_{PL}} \sum_{EV=1}^{N_{EV}} \sum_{t=1}^{24} \left(\widehat{P}_{PL,EV,t^{mid/off-peak}}^{ch-grid} \times \text{Pr}_{t^{mid/off-peak}}^{G2PL} \right)$$

$$- \sum_{PL=1}^{N_{PL}} \sum_{t=1}^{24} \left(\widehat{P}_{PL,t^{on-peak}}^{Solar} \times \text{Pr}_{t^{on-peak}}^{PL2G} \right)$$

$$- \sum_{s=1}^{Ns} \rho_s \sum_{PL=1}^{N_{PL}} \sum_{EV=1}^{N_{EV}} \sum_{t=1}^{24} 2 \times \begin{pmatrix} \left(SOE_{EV}^{min} \times \lambda_{PL,EV,t,s}^1 \right) - \left(SOE_{EV}^{max} \times \lambda_{PL,EV,t,s}^2 \right) \\ - \left(P^{max} \times \lambda_{PL,EV,t^{mid/off-peak},s}^4 \right) - \left(P^{max} \times \lambda_{PL,EV,t^{on-peak},s}^6 \right) \\ + \left(SOE_{EV}^{arv} \times \lambda_{PL,EV,t^{arv},s}^8 \right) + \left(SOE_{EV}^{dep} \times \lambda_{PL,EV,t^{dep},s}^9 \right) \\ + \left(P_{PL,t^{mid/off-peak},s}^{Solar} \times \lambda_{PL,EV,t^{mid/off-peak},s}^{10} \right) - \left(P_{PL,EV,t^{on-peak},s}^{dch} \times C^{cd} \right) \\ - \left(P_{PL,EV,t^{mid/off-peak},s}^{ch-grid} \times \text{Pr}_{t^{mid/off-peak}}^{G2PL} \right) \end{pmatrix}$$

$$(4.51)$$

Subject to:

(4.29) to (4.34)	(4.52)
(4.36) to (4.45)	(4.53)
(4.II.12) to (4.II.15)	(4.54)
(4.II.22) to (4.II.27)	(4.55)

4.4.4 Single-Level Model

In the single-level model, the SDNO also owns the PLs and the solar system; therefore, it must satisfy the owner of each EV in accordance with the limitations of the EVs. In fact, the constraints of the EVs that are described in the previous sections should be considered as the constraints of the SDNO.

4.4.4.1 Single-Level Model with Controlled Charging

In this case, the SDNO provides the total customers' demand and a part of the charging power of the EVs, from the WM. Also, the other part of the power needed for EVs' charging is provided through the power generated of the solar system. The

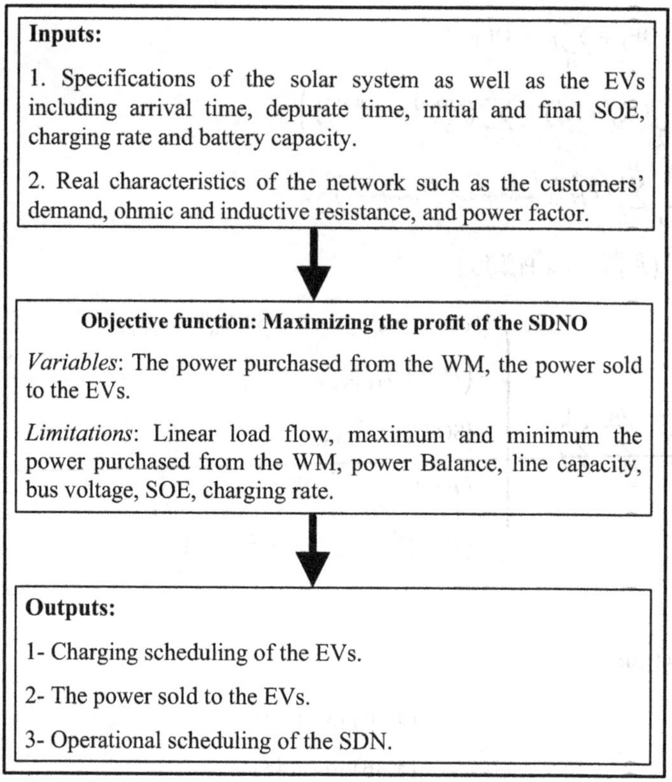

Inputs:

1. Specifications of the solar system as well as the EVs including arrival time, depurate time, initial and final SOE, charging rate and battery capacity.

2. Real characteristics of the network such as the customers' demand, ohmic and inductive resistance, and power factor.

Objective function: Maximizing the profit of the SDNO

Variables: The power purchased from the WM, the power sold to the EVs.

Limitations: Linear load flow, maximum and minimum the power purchased from the WM, power Balance, line capacity, bus voltage, SOE, charging rate.

Outputs:

1- Charging scheduling of the EVs.

2- The power sold to the EVs.

3- Operational scheduling of the SDN.

Fig. 4.6 The proposed single-level model framework with controlled charging

single-level model is defined in Eqs. (4.56), (4.57), (4.58), and (4.59). The objective function of the model is similar to the bi-level model, except for the last part, where the income from the selling energy to the EVs with the power generated of the solar system. Moreover, the energy sold price to the EVs, in this case, is equal to the energy sold price to the customer. The proposed framework of this model shows in Fig. 4.6.

$$\text{Maximize } \text{OF}_1 + \sum_{s=1}^{N_S} \rho_s \times \text{OF}_5$$

$$\sum_{t=1}^{24} \left(\sum_{b=2}^{N_b} \left(P_{b,t}^L \times \text{Pr}_t^L \right) - \sum_{sb=1}^{N_{sb}} \left(P_{sb,t}^{Wh2G} \times \text{Pr}_t^{Wh2G} \right) \right) \quad (4.56)$$

$$+ \sum_{s=1}^{N_s} \rho_s \sum_{PL=1}^{N_{PL}} \sum_{EV=1}^{N_{EV}} \sum_{t=1}^{24} \left(\begin{array}{c} \left(P_{PL,EV,t^{mid/off-peak},s}^{ch-grid} \times \text{Pr}_t^L \right) \\ + \left(P_{PL,EV,t,s}^{ch-Solar} \times \text{Pr}_t^L \right) \end{array} \right)$$

Subject to:

4 Optimal Charge Scheduling of Electric Vehicles in Solar Energy... 89

$$(4.13) \text{ and } (4.15) \text{ to } (4.18) \tag{4.57}$$

$$P_{sb,t}^{Wh2G} \times \eta^{Trans} + P_{PL,t,s}^{solar} = P_{b,t}^{L} + P_{t,s}^{Loss} + \sum_{EV} P_{PL,EV,t^{mid/off-peak},s}^{ch-grid}$$

$$+ \sum_{EV} P_{PL,EV,t,s}^{ch-solar} \tag{4.58}$$

$$(4.20) \text{ to } (4.27) \tag{4.59}$$

4.4.4.2 Single-Level Model with Charging/Discharging Schedule

In this case, the SDNO provides a part of the customers' demand and a part of the charging power of the EVs from the WM. Furthermore, a part of the customers' demand during the on-peak hours is provided by the power purchased from the EV owners, and the power generated by the solar system. A part of the charging power is being provided during the off-peak/mid-peak hours by the power generated of the solar system. The energy sold price to the EVs is equale to the energy sold price to the customer. It is also assumed that the energy purchased price from the EVs is equal to the minimum electricity price of the WM at the on-peak hours, i.e. 140 \$/MWh. The objective functions of this model are similar to the bi-level model, with two differences in the single-level model. The SDNO must pay the cost of depreciation of the battery to the EVs owners. Also, the SDNO gains the income from the selling energy to the EVs by the power generated of the solar system, so the single-level model is defined by the Eqs. (4.60), (4.61), (4.62), and (4.63). Figure 4.7 shows the proposed framework of the single-level model.

$$\text{Maximize } OF_1 + \sum_{s=1}^{N_S} \rho_s \times OF_6$$

$$\sum_{t=1}^{24} \left(\sum_{b=2}^{N_b} \left(P_{b,t}^{L} \times Pr_t^{L} \right) - \sum_{sb=1}^{N_{sb}} \left(P_{sb,t}^{Wh2G} \times Pr_t^{Wh2G} \right) \right)$$

$$+ \sum_{s=1}^{N_s} \rho_s \sum_{PL=1}^{N_{PL}} \sum_{EV=1}^{N_{EV}} \sum_{t=1}^{24} \left(\begin{array}{l} \left(P_{PL,EV,t^{mid/off-peak},s}^{ch-grid} \times Pr_{t^{mid/off-peak}}^{L} \right) + \left(P_{PL,EV,t^{mid/off-peak},s}^{ch-Solar} \times Pr_{t^{mid/off-peak}}^{L} \right) \\ - \left(P_{PL,EV,t^{on-peak},s}^{dch} \times Pr_{t^{on-peak}}^{min,Wh2G} \right) - \left(P_{PL,EV,t^{on-peak},s}^{dch} \times C^{cd} \right) \end{array} \right)$$

$$\tag{4.60}$$

Subject to:

$$(4.29), (4.31) \text{ to } (4.34) \tag{4.61}$$

> **Inputs:**
>
> 1. Specifications of the solar system as well as the EVs including arrival time, depurate time, initial and final SOE, charging rate and battery capacity.
>
> 2. Real characteristics of the network such as the customers' demand, ohmic and inductive resistance, and power factor.

> **Objective function: Maximizing the profit of the SDNO**
>
> *Variables*: The power purchased from the WM, the power exchanged between the SDN and the EVs.
>
> *Limitations*: Linear load flow, maximum and minimum the power purchased from the WM, power Balance, line capacity, bus voltage, SOE, charging/discharging rate.

> **Outputs:**
>
> 1- Charging/discharging scheduling of the EVs.
>
> 2- The power purchased/sold from/to the EVs.
>
> 3- Operational scheduling of the SDN.

Fig. 4.7 The proposed single-level model framework with charging/discharging schedule

$$P_{sb,t}^{Wh2G} \times \eta^{Trans} + \sum_{EV} P_{PL,EV,t^{on-peak},s}^{dch} + P_{PL,t,s}^{Solar} =$$
$$P_{b,t}^{L} + P_{t,s}^{Loss} + \sum_{EV} P_{PL,EV,t^{mid/off-peak},s}^{ch-grid} + \sum_{EV} P_{PL,EV,t^{mid/off-peak},s}^{ch-Solar} \qquad (4.62)$$

$$(4.36) \text{ to } (4.45) \qquad (4.63)$$

4.4.5 Risk Management

Due to uncertainties of the EVs and the solar system in the proposed model, the SDNO is faced to risk that a determined value is admissible. For controlling the risk level, three strategies, i.e. risk-seeker, risk-neutral, and risk-averse are offered [32].

4 Optimal Charge Scheduling of Electric Vehicles in Solar Energy...

1. By ignoring uncertainties, the SDNO has faced no risk. The model in this situation is solved with one scenario, i.e. $s = 1$.
2. By taking the several scenarios into account for uncertainties, i.e. Risk-neutral model, the optimal response is achieved by the expected value of scenarios.
3. If with considering scenarios, a term for controlling the risk of profit is added, the risk-averse model will be obtained. In this model, a non-suitable condition, e.g., a high probability of low profit is eliminated. Value-at-Risk (VaR) and Conditional Value-at-Risk (CVaR), are the most important of risk measures. In this chapter, CVaR is considered for risk measures because of the linear formulation. The CVaR at α confidence level is equal to the expected profit of the $(1 - \alpha)$ 100% scenarios with the worst value of profit. The confidence level of CVaR is set close to 1, so in this chapter is 0.95. The CVaR is explained by Eqs. (4.64), (4.65), and (4.66) [29]:

$$B_s = \zeta - \frac{1}{1 - \alpha} \sum_{s=1}^{N_s} \rho_s \eta_s \tag{4.64}$$

$$-B_s + \zeta - \eta_s \leq 0 \tag{4.65}$$

$$\eta_s \geq 0 \tag{4.66}$$

The risk-based models with CVaR index are introduced as follows.

4.4.5.1 Risk-Based Bi-Level Model

The risk-based bi-level model with CVaR index, for controlled charging model is defined in Eqs. (4.67), (4.68), (4.69), and (4.70).

Maximize

$$(1 - \beta) \times \left(OF_1 + \sum_{s=1}^{N_s} \rho_s \times OF_2 \right) + \beta \times \left(\zeta - \frac{1}{1 - \alpha} \sum_{s=1}^{N_s} \rho_s \eta_s \right) \tag{4.67}$$

Subject to:

$$(4.47) \text{ to } (4.50) \tag{4.68}$$

$$\eta_s \geq 0 \tag{4.69}$$

$$\zeta - \eta_s - (OF_1 + OF_2) \leq 0 \tag{4.70}$$

Also, Eqs. (4.71), (4.72), (4.73), and (4.74) explain the risk-based bi-level model in charging/discharging schedule.

Maximize

$$\text{OF}_3 + \sum_{s=1}^{N_s} \rho_s \times \text{OF}_4 + \beta \times \left(\zeta - \frac{1}{1-\alpha} \sum_{s=1}^{N_s} \rho_s \eta_s \right) \tag{4.71}$$

Subject to:

$$(4.52) \text{ to } (4.55) \tag{4.72}$$

$$\eta_s \geq 0 \tag{4.73}$$

$$\zeta - \eta_s - (\text{OF}_3 + \text{OF}_4) \leq 0 \tag{4.74}$$

4.4.5.2 Risk-Based Single-Level Model

The risk-based single-level model with CVaR index, for controlled charging model is described in Eqs. (4.75), (4.76), (4.77), and (4.78).

Maximize

$$(1 - \beta) \times \left(\text{OF}_1 + \sum_{s=1}^{N_s} \rho_s \times \text{OF}_5 \right) + \beta \times \left(\zeta - \frac{1}{1-\alpha} \sum_{s=1}^{N_s} \rho_s \eta_s \right) \tag{4.75}$$

Subject to:

$$(4.57) \text{ to } (4.59) \tag{4.76}$$

$$\eta_s \geq 0 \tag{4.77}$$

$$\zeta - \eta_s - (\text{OF}_1 + \text{OF}_5) \leq 0 \tag{4.78}$$

Also, Eqs. (4.79), (4.80), (4.81), and (4.82) explain the risk-based bi-level model in charging/discharging schedule.

Maximize

$$(1 - \beta) \times \left(\text{OF}_1 + \sum_{s=1}^{N_s} \rho_s \times \text{OF}_6 \right) + \beta \times \left(\zeta - \frac{1}{1-\alpha} \sum_{s=1}^{N_s} \rho_s \eta_s \right) \tag{4.79}$$

Subject to:

$$(4.61) \text{ to } (4.63) \tag{4.80}$$

$$\eta_s \geq 0 \tag{4.81}$$

$$\zeta - \eta_s - (\text{OF}_1 + \text{OF}_6) \leq 0 \tag{4.82}$$

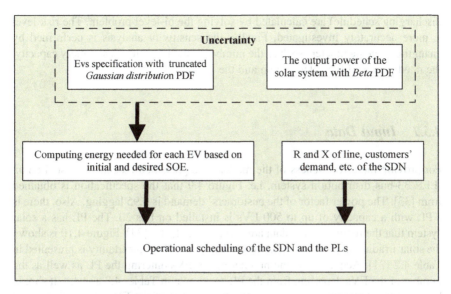

Fig. 4.8 The stochastic programming framework for optimal operation of the SDN

4.4.6 The Problem Solving Process

For solving the models, a flowchart based on stochastic programming is suggested, and is shown in Fig. 4.8. Forasmuch as the models are mixed-integer linear programming (MILP) problems, the simulation is performed through CPLEX solver of GAMS. By using the Kantorovich distance approach, the scenarios for modeling of uncertainty are decreased to 8. The simulation is carried out in a laptop with Corei7 up to 3.5 GHz CPU, 12 GB RAM (DDR4), and 4 MB Cash.

4.5 Simulation Results

In the following, based on the proposed models, the simulations are carried out on an IEEE 33-bus distribution system. At first, the maximum profit of the SDNO is calculated without the EV. In this program, the price of the energy sold to the customer is also obtained. Additionally, the customers' demand and the power purchased from the WM are investigated. Then, considering the EVs and controlled charging and charging/discharging schedule with and without the solar system, in the single-level model and the bi-level model, different parts of the objective functions such as charging/discharging power of the EVs are evaluated over a 24-hours. The price of the energy sold to the PL owner (controlled charging mode), and the price of the energy purchased from the PL owner (charging/

discharging schedule) are calculated by solving the bi-level problem. The risk level is more accurately investigated. Finally, the sensitivity analysis is performed by changing some parameters such as the number of the EVs, the EVs' battery capacity, the rated power of the solar system and the PL sitting.

4.5.1 Input Data

For proving the effectiveness of the models, the presented models are tested on an IEEE 33-bus distribution system, i.e. Figure 4.9 that the specification is obtained from [33]. The power factor of the customers' demand is 0.95 lagging. Also, there is a PL with a capacity of up to 500 EVs is installed on bus 20. The PL has a solar system that the requirements' data are given in Table 4.1 [34]. Figure 4.10 is shown the solar irradiation [35]. The data for modeling the EVs' uncertainty is presented in Table 4.2 [21]. Accordingly, the number of the EVs entering the PL as well as the number of the EVs departing from the PL are shown in Tables 4.3 and 4.4. It should be noted that between 10:00 and 18:00, the number of the EVs in the PL is fixed i.e. 500 EVs. In addition, Fig. 4.11 is illustrated the initial SOE of the EVs in scenario 1. The desired SOE at the departure time is considered 90% of battery capacity [21]. The minimum and maximum values of SOE are set to 15% and 90% battery capacity, respectively. The charging/discharging efficiency, EVs' battery capacity, and the maximum charging/discharging rate are 90%, 95%, 50 kWh and 10 kWh, respectively. The price of the battery depreciation of the EVs is 30 $/MWh [21]. Moreover, Fig. 4.12 shows the electricity price of WM. Hours (1:00–8:00), (23:00–24:00) and (9:00–12:00), (19:00–22:00) and (13,00–18:00) are the off-peak, mid-peak and on-peak hours, respectively [33]. The price of energy sold to customers' demand is 80,120 and 240 at the off-peak, mid-peak and on-peak hours, respectively.

4.5.2 The System Without the EVs and the Solar System

Initially, the model is solved for a situation in which the EVs do not exist, in order to determine the maximum profit of the SDNO along with the optimal price of the energy sold to the customer. In addition in this section, the customers' demand and the power purchased from the WM are calculated.

1. The maximum profit of the SDNO

    Table 4.5 shows the maximum profit of SDNO in the absence of the EVs. The solution time of each program is also given in Table 4.5.

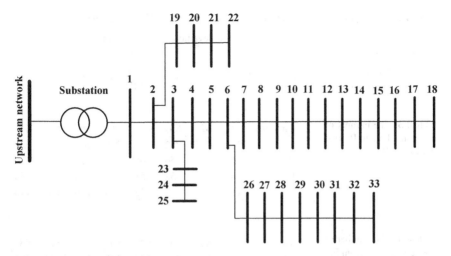

Fig. 4.9 The IEEE 33-bus distribution system

Table 4.1 The parameters of the solar system

Parameters	Value	Parameters	Value
Open circuit voltage (V)	21	Voltage temperature coefficient (V/c)	0.088
Short circuit current (A)	3.4	Current temperature coefficient (A/c)	0.0015
Voltage at maximum power (v)	17.4	Normal operating temperature (c)	34
Current at maximum power (v)	3.05	Ambient temperature (c)	25
Cell number	2000	Rated power (kW)	400

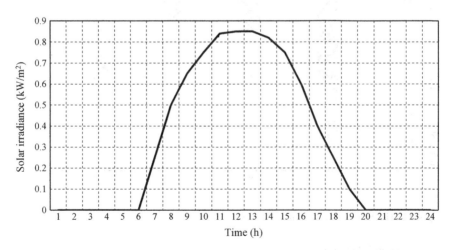

Fig. 4.10 The solar irradiance (kW/m^2)

Table 4.2 The required data for modeling of initial SOE, arrival/departure time of EVs

	Mean	Standard deviation	Minimum	Maximum
Initial SOE (%)	50	25	30	60
Arrival hours (h)	8	3	7	10
Departure hours (h)	20	3	19	24

Table 4.3 The number of entered the EVs to PL from 7:00 to 11:00

Time (h)	S1	S2	S3	S4	S5	S6	S7	S8
7	262	234	262	258	268	244	237	251
8	57	65	54	68	58	76	70	80
9	66	71	52	51	61	54	67	53
10	115	130	132	123	113	126	126	116

Table 4.4 The number of departed the EVs from PL from 19:00 to 24:00

Time (h)	S1	S2	S3	S4	S5	S6	S7	S8
19	235	233	268	245	243	255	259	233
20	67	66	80	62	66	71	60	72
21	73	63	50	54	68	49	56	59
22	49	54	31	52	45	42	50	59
23	33	34	30	27	34	38	35	30
24	43	50	41	60	44	45	40	47

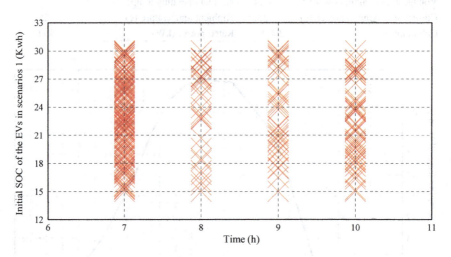

Fig. 4.11 The SOE of 500 EVs in scenario 1

2. Customers' demand

Figure 4.13 shows customers' demand. Table 4.6 shows the customers' demand at different time intervals. It is noted that the benefit of the SDNO from the energy sold to the customer is 24256.64 $.

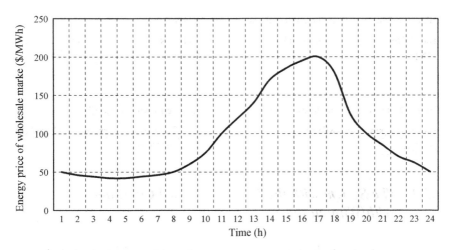

Fig. 4.12 The energy price of the WM

Table 4.5 The maximum profit of the SDNO

Profit of the SDNO ($)	Solution time (s)
5929.33	0.39

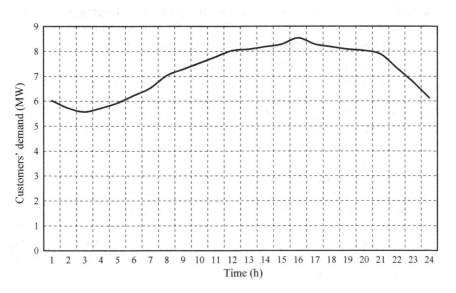

Fig. 4.13 The customers' demand in two models

3. Power purchased from the WM

The power purchased from the WM, i.e. the sum of the customers' demand and network losses and also its cost are shown in Table 4.7. Moreover, the SDN loss is 7.415 MWh.

Table 4.6 The customers' demand at the interval time (MWh)

Totald demand	Off-peak hours	Mid-peak hours	On-peak hours
173.139	61.646	61.946	49.547

Table 4.7 The energy purchased from the wholesale and its cost without the EVs

The energy purchased (MWh)	The cost of the energy purchased ($)
180.554	18327.31

4.5.3 The System with the EVs (Controlled Charging) With/Without the Solar System

In this section, the model is solved for a situation that the EVs are only charged i.e. controlled charging mode. The model is investigated in two parts: single-level and bi-level. Also, in each part, the effect of the solar system is evaluated. In the single-level model forasmuch as the SDNO is the owner of the PL and the solar system, the price of the energy sold to the EV owners is equal to the price of the energy sold to the customers. However, in the bi-level model, the price of the energy sold to the PL owner is calculated by solving the model. The charging power of the EVs and the power purchased from the WM are also examined in both models. It should be noted that the customers' demand, in this case, is the same as Fig. 4.13.

1. The maximum profit of SDNO

   Table 4.8 shows the maximum profit of the SDNO in the single-level and bi-level models with/without the solar system. The single-level model has more profit than the bi-level model. According to Tables 4.9 and 4.10, the main reason can be considered by the price of the energy sold to the EV owners and the PL owner. In the single-level model, this price is equal to the price of the energy sold to the customer; however, in the bi-level model, this price, due to the interaction between the two decision-makers, i.e. the SDNO and the PL owner, is lower than the price of the energy sold to the customer. The second reason is the revenue by the energy sold to the EV owners by the power generated of the solar system (in the single-level model, SDNO owns the solar system). The solution times are also presented in Table 4.8. With the presence of the EVs and the solar system, the solution time raise. Of course, in the bi-level model, due to the complexity of the problem, this time will be greatly increased.

2. Charging power of the EVs

   Due to the controlled charging of the EVs, at the off-peak and mid-peak hours, the EVs are charged. The maximum charging rate of the EVs is 10 kWh. Forasmuch as at some hours, there are 500 EVs in the PL, the maximum power that can be imposed on the system for charging of the EVs can be up to 5 MWh. In this regard, the charging power of the EVs by the SDNO and the solar system as well

4 Optimal Charge Scheduling of Electric Vehicles in Solar Energy... 99

Table 4.8 The maximum profit of the SDNO and solution time in all programs

Program	Profit of the SDNO ($)	Solution time (s)
1. Single-level model without the solar system	6430.646	22.859
2. Single-level model with the solar system	6600.369	44.766
3. Bi-level model without the solar system	6164.578	72.359
4. Bi-level model with the solar system	6225.330	288.266

Table 4.9 The selling energy price to the EV owners in the single-level model ($/MWh)

Hour	Energy price
7:00–8:00	80
9:00–12:00 and 19:00–22:00	120
23:00–24:00	80

Table 4.10 The selling energy price to the PL owner in the bi-level model ($/MWh)

	Energy price	
Hour	With solar	Without solar
7:00–8:00	71.6	72.2
9:00–12:00 and 19:00–22:00	114	118
23:00–24:00	71.6	72.2

Table 4.11 The power charged of the EVs in both models (MW)

Program	Total charging power of the EVs	Charging power of the EVs by the SDNO	Charging power of the EVs by the solar system
1	11.903	11.903	–
2	11.903	8.960	2.943
3	11.903	11.903	–
4	11.903	8.960	2.943

Table 4.12 The revenue of the energy sold to the EV owners or the PL owner ($)

Program	EV owners or PL owner
1	1383.34
2	1059.51
3	1123.16
4	937.547

as the benefit of its, in both models in different programs are presented in Tables 4.11 and 4.12. According to these tables, the EVs' charging power in each program is equal because of the condition of each EV, such as arrival time, departure time and the initial and desired SOE is the same. The price of the energy sold to the EVs, the price of the energy purchased from the WM, as well as the number of the EVs in the PL, are the main factors in the charging power of the EVs.

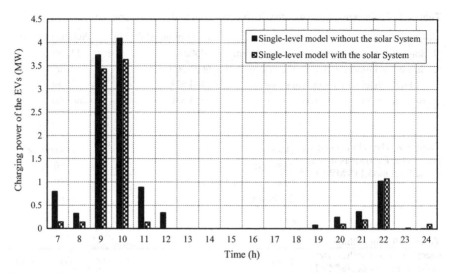

Fig. 4.14 The charging power of all EVs by the SDNO in the single-level model

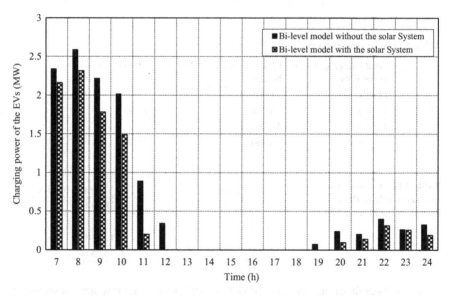

Fig. 4.15 The charging power of all EVs by the SDNO in the bi-level model

Also, Figs. 4.14 and 4.15 illustrate the total charging power of the EVs through the SDNO. Based on Fig. 4.14 in the single-level model, at the 7:00 and 8:00, the difference between the price of the energy purchased from the WM and the energy sold to EVs owner is low; therefore, at these times, the SDNO purchases less energy for charging of the EVs. However, at 9:00, 10:00, and 22:00, because of the higher difference between the two prices, the SDNO purchases high energy. Furthermore, in the bi-level model according to Fig. 4.15, the PL owner's decision is also effective, so SDNO purchases more energy for charging of the EVs at the 7:00 and 8:00 that the price of energy is low. In this situation, the SDNO gains more profit by the cheaper electricity price of the WM. Therefore, the purchasing behavior of the SDNO from the WM for charging of the EVs in this model is slightly different from the single-level model.

It is noted that the EVs will not be charged by the SDNO from 13:00 to 18:00 due to the on-peak hours. Additionally, at the 12:00 and 19:00, due to the difference between these two prices (purchasing from the WM and selling to the EV owners or the PL owner) is zero or negative and because of charging of the EVs by the solar system during the on-peak hours, no energy for charging of the EVs is purchased. Of course, in the system without the solar system, due to many EVs leave the PL at 19:00, and in accordance with the constraints, especially the EV owner's satisfaction (desired SOE), the energy is also purchased at 12:00 and 19:00. The power for charging of the EVs that is provided by the solar system is shown in Fig. 4.16.

3. Power purchased from the WM

   The power purchased from the WM with regard to the customers' demand, network losses, the power generated of the solar system and charging power of the EVs, along with its cost, are shown in Table 4.13. Figure 4.17 shows a comparison between the power purchased from the WM in the single-level and bi-level models. According to Fig. 4.17, until the arrival of the EVs, the purchasing power from the WM is the same. From 7:00, with the arrival of the EVs, this power will increase and will continue until 11:00. In these hours, purchasing the power from the WM in the single-level and bi-level models is slightly different. In

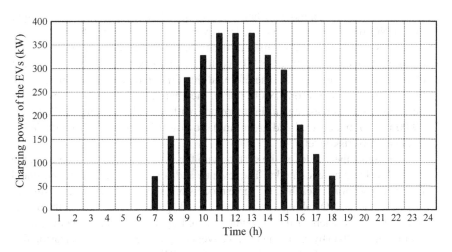

Fig. 4.16 The charging power of all EVs by the solar system in both models

Table 4.13 The energy purchased from the WM as well as its cost

Program	The energy purchased (MWh)	The cost of the energy purchased ($)
1	192.938	19209.334
2	191.767	19059.966
3	192.870	19113.120
4	191.714	18968.856

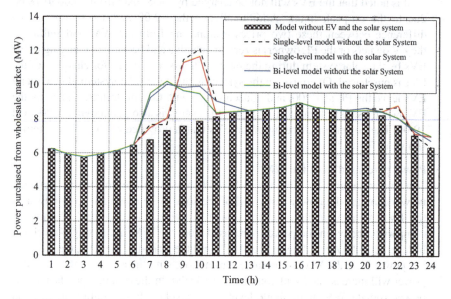

Fig. 4.17 The power purchased from the WM in both models

fact, in the single-level model, when the price of selling energy to the EVs is high, the SDNO purchases more power. However, in the bi-level model, the SDNO purchases more power when the electricity price of the WM is low, i.e. at the off-peak hours. At 13:00 to 18:00 due to the on-peak hours, the EVs are not charged through the SDNO. Therefore, at this time, the purchasing power from the WM is the same. From 19:00, due to the charging of some EVs, the power purchased will increase again. Furthermore, with the solar system, at 12:00 and 19:00 in two models (red and green line), no power is purchased from the WM.

4. Evaluation of risk level

To investigate the risk level, the system with the solar system is considered in the single-level and bi-level model. The revenue and cost of the SDNO are presented in separate sections in each of the three models of risk in Table 4.14. According to this table, the SDNO, taking into account the risk, gains less profit from the power sold to the EV owners or the PL owner. Also, Fig. 4.18 illustrates the maximum profit of the SDNO by changing the risk aversion parameter, i.e. β. Increasing this amount leads to a reduction in the profit of the SDNO.

4 Optimal Charge Scheduling of Electric Vehicles in Solar Energy... 103

Table 4.14 The revenue and cost of the SDNO in the three models of risk ($)

Income	Model	Bi-level model	Single-level model
Energy sold to the customer	Risk-seeker	24256.64	24256.64
	Risk-neutral	24256.64	24256.64
	Risk-averse	24256.64	24256.64
Energy sold to the EV owners by the solar system	Risk-seeker	–	344.185
	Risk-neutral	–	344.185
	Risk-averse	–	344.185
Energy sold to the EV owners or the PL owner by the SDNO	Risk-seeker	1040.057	1063.127
	Risk-neutral	937.547	1059.510
	Risk-averse	893.792	1048.646
Cost			
Energy purchased from the WM	Risk-seeker	18824.068	18951.786
	Risk-neutral	18968.856	19059.966
	Risk-averse	18967.328	19067.241
Profit			
Profit	Risk-seeker	6472.630	6712.166
	Risk-neutral	6225.330	6600.368
	Risk-averse	6183.103	6547.916

5. Sensitivity analysis

Finally, for investigation the affecting factors on the maximum profit of the SDNO in the risk-neutral model, sensitivity analysis is carried out according to Table 4.15 by changing some parameters such as the number of the EVs, the EVs' battery capacity and the rated power of the solar system in 6 modes for the single-level and bi-level model with the solar system. Based on Table 4.15, increasing the EVs' battery capacity, the number of the EVs as well as the rated power of the solar system will bring more profit to the SDNO due to increasing the energy sold to the EVs.

Additionally, for evaluating the effect of the PL sitting on the maximum profit of the SDNO, Table 4.16 is presented. In this regard, three buses are randomly

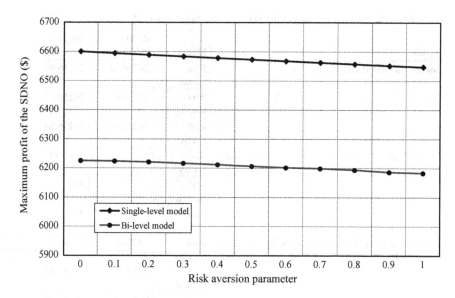

Fig. 4.18 The effect of risk aversion parameter on the maximum profit of the SDNO in both models

Table 4.15 Sensitivity analysis of the affecting factors on the maximum profit of the SDNO

				Maximum profit	
No	EVs no.	Battery capacity (kWh)	Rated power of the solar system (kW)	Single-level model	Bi-level model
1	500	50	400	6600.369	6225.330
2	500	24	400	6326.914	6185.083
3	500	50	500	6630.211	6367.970
4	1000	50	400	7114.364	6589.192
5	1000	24	400	6639.413	6373.397
6	1000	50	500	7150.733	6629.801

selected considering the situation of first and sixth sensitivity analysis. With the changing of the PL sitting, the difference between maximum profit occurs in the single-level model and bi-level model.

4.5.4 The System with the EVs (Charging/Discharging) With/Without the Solar System

In this section, the model is solved in the presence of the EVs with charging/discharging schedule as well as single-level and bi-level models. In the single-

4 Optimal Charge Scheduling of Electric Vehicles in Solar Energy...

Table 4.16 Evaluation of the effect of the PL sitting on the maximum profit of the SDNO

Sensitivity analysis No.	The bus of the PL	Maximum profit	
		Single-level model	Bi-level model
1	20	6600.369	6225.330
1	4	6586.420	6315.722
1	24	6565.793	6316.835
6	20	7150.733	6629.801
6	4	7110.169	6606.545
6	24	6943.132	6429.363

Table 4.17 The maximum profit of the SDNO and solution time in all programs

Program	Profit of the SDNO ($)	Solution time (s)
1. Single-level model without the solar system	6721.098	27.469
2. Single-level model with the solar system	6961.287	78.984
3. Bi-level model without the solar system	6645.461	243.67
4. Bi-level model with the solar system	6684.246	574.56

level model, the SDNO is the owner of the PL, so the price of the energy sold to the EV owners is equal to the price of the energy sold to the customers (see Table 4.9). Also, the maximum limit of the price of the energy purchased from the EV owners is 140 $/MWh, i.e. the minimum electricity price of the WM. In the bi-level model, the price of the energy sold to the PL owner is the same as Table 4.10. The price of the energy purchased from the PL owner (in the bi-level model) is calculated by solving the problem. The maximum profit of SDNO, the charging/discharging power of the EVs and the power purchased from the WM are examined in both models. It should be noted that the customers' demand is the same as Fig. 4.13.

1. The maximum profit of SDNO

   Table 4.17 shows the maximum profit of the SDNO in the single-level and bi-level models. In the single-level model, the SDNO gains more profit than the bi-level model. The reason can be seen in several factors. In the single-level model because of the power generated of the solar system, the SDNO purchases less power from the WM at the on-peak hours. Another reason is the price of the energy sold to the EV owners. In the single-level model, this price is equal to the price of the energy sold to the customer; however, in the bi-level model, this price is lower than the price of the energy sold to the customer. Moreover, in the bi-level model, the owner of the PL due to the minimization of cost purchases less power from the SDNO and therefore, has less power for selling to the SDNO during the on-peak hours. According to Table 4.18, the price of the energy purchased from the PL owner in the bi-level model is also lower than the single-level model. In addition, the solution times are presented in Table 4.17. With the presence of the EVs on the system, the solution time raise. Of course, in the bi-level model, due to the complexity of the problem, this time will be greatly increased.

Table 4.18 The price of the energy purchased from the EV owners and the PL owner ($/MWh)

Hour	EV owners (single-level model)	PL owner (bi-level model)
13:00–18:00	140	133

Table 4.19 The power charged of the EVs in the single-level model (MW)

Program	Total charging power of the EVs	Charging power of the EVs by the SDNO	Charging power of the EVs by the solar system
1	21.199	21.199	–
2	20.610	19.131	1.479
3	20.139	20.139	–
4	18.174	16.658	1.516

Table 4.20 The discharging power of the EVs in the single-level model (MW)

Program	Total discharging power of the EVs
1	7.948
2	7.444
3	7.001
4	6.505

2. Charging/discharging power of the EVs

Due to the charging/discharging schedule of the EVs, during the off-peak and mid-peak hours, the EVs are charged and at the on-peak hours are discharged. As previously mentioned, the maximum power that can be imposed on the SDN for charging of the EVs can be up to 5 MWh. The same amount of power during the on-peak hours is available due to discharging power of the EVs. In this regard, the charging/discharging power of the EVs, as well as its cost and benefit are presented in Tables 4.19, 4.20, 4.21, and 4.22. The power generated of the solar system is also used for charging the EVs and supplying the customers' demand. According to these tables, In the bi-level model, the aim of PL owner is influenced in the charging/discharging power, and therefore, less power is exchanged between the SDNO and the PL.

The Figs. 4.19 and 4.20 show the total charging/discharging power of the EVs through the SDNO in the single-level model. According to these figures, the charging/discharging schedule is properly done. At the off-peak and mid-peak hours, the EVs are charged and at the on-peak hours, the EVs are discharged. Since the discharging of the EVs occur at the on-peak hours, firstly, the EVs are fully charged, then they are discharged, and finally are again charged to achieve the desired SOE in the departure time. In accordance with Fig. 4.19, at 9:00 and 10:00 since the difference between the electricity price of the WM and the price of the energy sold to the EVs are high, so at these times, the SDNO sells more power. Also at the on-peak hours, the EVs do not charge. At 19:00, unlike the controlled charging mode, since most of the EVs participate in the discharging schedule and according to Table 4.4, about 50% of the EVs leave the PL, more power is sold for meeting the desired SOE. After that, considering the existing EVs, less power is sold for charging of the EVs.

Table 4.21 The revenue of the energy sold to the EV owners and the PL owner for charging of the EVs ($)

Program	EV owners or PL owner
1	2444.292
2	2233.900
3	1927.175
4	1826.865

Table 4.22 The cost of the energy purchased from the EV owners and the PL owner ($)

Program	EV owners or PL owner
1	1112.806
2	1042.281
3	931.012
4	865.213

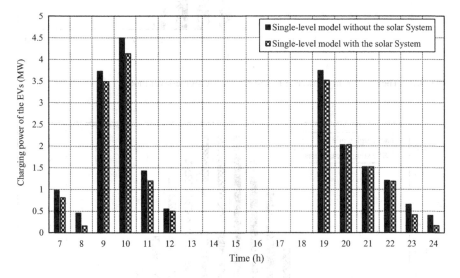

Fig. 4.19 The charging power of the all EVs by the SDNO in the single-level model

The discharging of the EVs occurs at the on-peak hours according to Fig. 4.20. Based on this figure, at 13:00, EVs do not discharge because at this time the discharging energy price is the same as the electricity price of the WM. In fact, the SDNO purchases the power discharged when the electricity price of the WM is very high, i.e. 17:00 and 16:00. At these times, the electricity price of the WM is 200 and 195 $/MWh, respectively.

Also, Fig. 4.21 shows the sharing of power generated by the solar system for charging of the EVs and feeding the customer in the single-level model. Based on Fig. 4.21, during the on-peak hours, the SDNO uses most of this power for feeding the customer due to the high electricity price of the WM.

Figure 4.22 shows the charging power of the EVs in the bi-level model. Because of the aim of the PL owner, i.e. cost of minimization, the PL owner

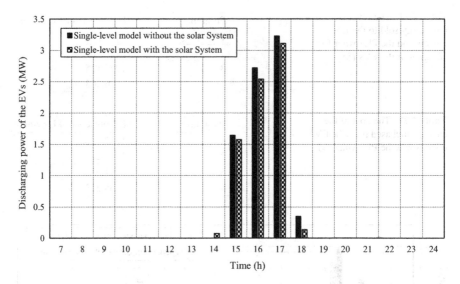

Fig. 4.20 The discharging power of the all EVs in the single-level model

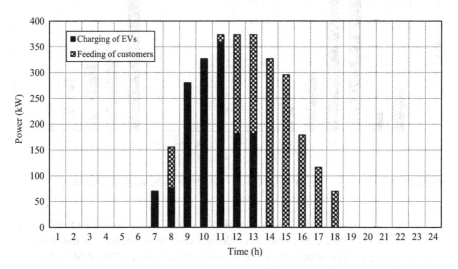

Fig. 4.21 Sharing of power generated by the solar system for charging of the EVs and feeding of the customer in the single-level model

purchases more power from the SDNO when the electricity price of the WM is low, i.e. at 7:00 and 8:00. Figure 4.23, also shows the discharging power of the EVs in the bi-level model that is the same as the single-level model. Also, Fig. 4.24 shows the sharing of power generated by the solar system for charging of the EVs and feeding the customer in the bi-level model.

4 Optimal Charge Scheduling of Electric Vehicles in Solar Energy...

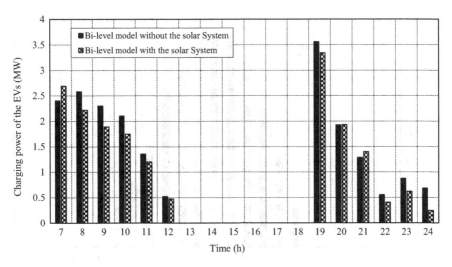

Fig. 4.22 The charging power of the all EVs by the SDNO in the bi-level model

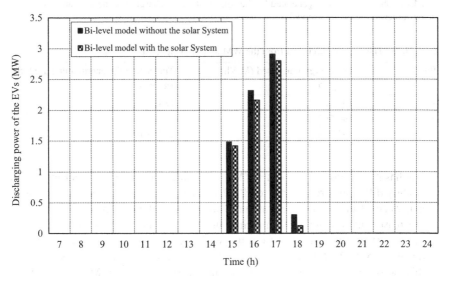

Fig. 4.23 The discharging power of the all EVs in the bi-level model

3. Power purchased from the WM (Table 4.23)

Table 4.25 shows the power purchased from the WM and its cost. Also, Figs. 4.24 shows a comparison between the power purchased from the WM in the single-level and bi-level models. Until the arrival of the EVs, i.e. 7:00, the power purchased from the WM is the same. Of course, this amount is slightly higher than the customers' demand due to network losses. From 7:00, with the arrival of the EVs, this power will increase and will continue until 11:00. In these hours,

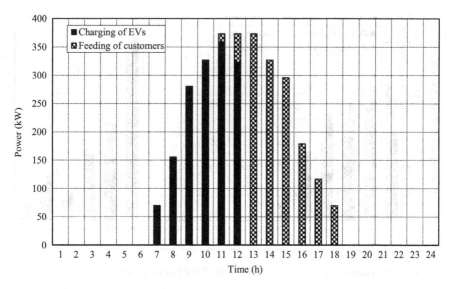

Fig. 4.24 Sharing of power generated by the solar system for charging of the EVs and feeding of the customer in the bi-level model

Table 4.23 The energy purchased from the WM as well as its cost

Program	The energy purchased (MWh)	The cost of the energy purchased ($)
1	194.572	18628.570
2	193.409	18435.302
3	194.503	18607.342
4	193.356	18344.192

purchasing the power from the WM in the single-level and bi-level models is slightly different. From 13:00 to 18:00, discharging power of the EVs or power generated of the solar system are used for meeting the customers' demand. For this reason, at these hours, the purchasing power from the WM is reduced, so that the lowest power purchased from the WM is at 17:00. From 19:00, due to the departure of 50% of the EVs from the PL and the satisfaction of the desired SOE, this power is increased. The power purchased from the WM after 19:00 is continued due to fewer numbers of the EVs in the PL and the customers' demand (Fig. 4.25).

4. Evaluation of risk level

   In order to investigate the risk level, the system with the solar system is considered in the single-level and bi-level model. The revenue and cost of the SDNO are presented in separate sections in each of the three models of risk in Table 4.24. In the risk-seeker model, the SDNO purchases more power for EVs' charging in order to get more profit, but in the risk-averse model, purchase less power for EVs' charging. Also, in the risk-seeker model, the SDNO by using discharging

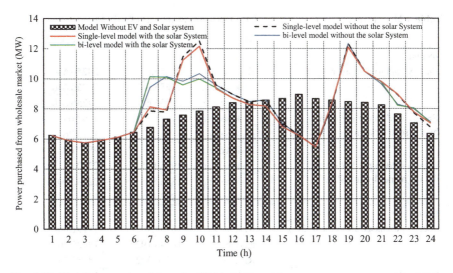

Fig. 4.25 The power purchased from the WM in both models

power to meeting the customers' demand at the on-peak hours, purchase less power from the WM. For this reason, in this model, the power purchased's cost of the EVs is the highest. So, in the risk-seeker model, the SDNO gains the most profit. Furthermore, Fig. 4.26 illustrates the maximum profit of the SDNO by changing the risk aversion parameter, i.e. β. Increasing this amount leads to a reduction in the profit of SDNO.

5. Sensitivity analysis

   Finally, for investigation the affecting factors on the maximum profit of the SDNO in the risk-neutral model, sensitivity analysis is carried out by changing some parameters such as the number of the EVs, the EVs' battery capacity and rated power of the solar system in 6 modes for both models, i.e. single-level and bi-level model with the solar system according to Table 4.25. Based on Table 4.15, increasing the EVs' battery capacity, number of the EVs as well as the rated power of the solar system will bring more profit to the SDNO due to increasing the energy sold to the EVs.

   Also, for evaluating the effect of the PL sitting on the maximum profit of the SDNO, Table 4.26 is presented. In this regard, three buses are randomly selected considering the situation of first and sixth sensitivity analysis. With the changing of the PL sitting, the difference between maximum profit occurs in the single-level model and bi-level model.

Table 4.24 The revenue and cost of the SDNO in the three models of risk ($)

	Model	Bi-level model	Single-level model
Income			
Energy sold to customer	Risk-seeker	24256.64	24256.640
	Risk-neutral	24256.64	24256.640
	Risk-averse	24256.64	24256.640
Energy sold to the EV owners by the solar system	Risk-seeker	–	149.803
	Risk-neutral	–	171.676
	Risk-averse	–	282.169
Energy sold to the EV owners or the PL owner by the SDNO	Risk-seeker	1906.865	2300.071
	Risk-neutral	1826.865	2233.900
	Risk-averse	1807.415	2223.475
Cost			
Energy purchased from the WM	Risk-seeker	18154.402	18191.885
	Risk-neutral	18344.192	18435.302
	Risk-averse	18326.522	18431.262
Battery depreciation	Risk-seeker	–	243.123
	Risk-neutral	–	223.345
	Risk-averse	–	249.266
Energy purchased from the EV owners or the PL owner (discharging power)	Risk-seeker	955.103	1134.576
	Risk-neutral	865.213	1042.280
	Risk-averse	936.124	1163.243
Energy purchased from the PL owner (power generated of the solar system)	Risk-seeker	215.413	–
	Risk-neutral	189.853	–
	Risk-averse	170.093	–
Profit			
Profit	Risk-seeker	6838.587	7136.930
	Risk-neutral	6684.246	6961.287
	Risk-averse	6631.316	6894.798

4.6 Conclusions

With modeling the EVs and the solar system and considering the private owner for the PLs (with two programs, i.e. controlled charging mode and smart charging/discharging mode), a new non-linear bi-level model was suggested for the operational scheduling of the SDN. The profit maximization of the SDNO and minimizing the cost of the PLs owner were the objective functions of each level. By using of KKT condition and the dual theory as well as the Fortuny-Amat and McCarl linearization method, the non-linear bi-level model was converted to single-level and linear models. Further, by supposing that the SDNO is the owner of the PLs, the single-level model was also proposed with the goal of profit maximization of the SDNO. Also, due to the uncertainties, three different strategies for risk management were introduced to evaluate the effect of the risk on the operational scheduling of the

4 Optimal Charge Scheduling of Electric Vehicles in Solar Energy...

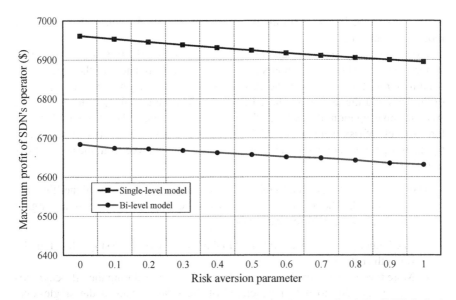

Fig. 4.26 The effect of risk aversion parameter on the maximum profit of the SDNO in both models

Table 4.25 Sensitivity analysis of the affecting factors on the maximum profit of the SDNO

				Maximum profit	
No	EVs no.	Battery capacity (kWh)	Rated power of the solar system (kW)	Single-level model	Bi-level model
1	500	50	400	6961.287	6684.246
2	500	24	400	6564.702	6303.117
3	500	50	500	7016.687	6751.966
4	1000	50	400	7662.344	7440.848
5	1000	24	400	6961.061	6693.670
6	1000	50	500	7721.598	7490.227

Table 4.26 Evaluation of the effect of the PL sitting on the maximum profit of the SDNO

		Maximum profit	
Sensitivity analysis No.	The bus of the PL	Single-level model	Bi-level model
1	20	6961.287	6684.246
1	4	6999.619	6707.128
1	24	6943.316	6645.155
6	20	7721.598	7490.227
6	4	7813.149	7549.778
6	24	7237.194	7150.823

SDN. By introducing a Conditional Value-at-Risk (CVaR) index, the risk-based model was defined.

After presenting these models, the simulations on the IEEE 33-bus distribution system were tested over a 24-hours for proving the effectiveness of the model. The maximum profit of the SDNO, the customers' demand, charging/discharging power of the EVs and the power purchased from the WM were evaluated in each mode. Also, for investigation of risk level, the amount of revenue and cost of the SDNO in three models of risk were presented. Finally, the sensitivity analysis was performed by changing some parameters. The main results were achieved from the case studies as follows:

1. The maximum profit of the SDNO in the single-level model was higher than the bi-level model. The reason in controlled charging and charging/discharging schedule can be seen in several factors:

    - The higher price of the energy sold to the EV owners in the single-level model (in both section)
    - More revenue from the energy sold to the EV owners during the off-peak/mid-peak hours due to power generated of the solar system in the single-level model (in both section)
    - More revenue from the less power purchased from the WM during the on-peak hours due to power generated of solar in the single-level model (in the charging/discharging schedule)
    - Less revenue from the energy sold to the PL owner in the bi-level model due to minimizing the cost (in the charging/discharging schedule)

2. The charging schedule and even charging/discharging schedule of the EVs were correctly done. So that the EVs' charging happened during the off-peak/mid-peak hours. Moreover, the EVs' discharging occurred during the on-peak hours. Of course, during the off-peak/mid-peak hours when the difference between the electricity price of the WM and the energy sold to the EV owners or the PL owner was negative or zero, discharging did not happen. Also, during the on-peak hours, the electricity price of the WM and the purchasing energy price from the EV owners or the PL owner were the main reason for the decision of the SDNO for purchasing energy. Therefore, most of the energy purchased from the EV owners or the PL owner was performed at 16:00 or 17:00. At this time, the energy purchased from the WM was the highest value.

3. By increasing the level of risk, the SDNO was more conservative done the charging/discharging schedule, so the SDNO was obtained the lowest profit in the risk-averse model. In fact, in the risk-averse model, since the EVs were less involved in charging/discharging schedule, the SDNO more power was purchased from the WM, and less profit was achieved.

4. Increasing the EV's battery capacity and increasing the number of EVs as well as the rated power of the solar system was brought more profit to the SDNO. Also, with the changing of sitting of PL, in some cases, there was a difference between the profit of the SDNO.

4 Optimal Charge Scheduling of Electric Vehicles in Solar Energy... 115

5. The results of the single-level and bi-level models were proved the effectiveness of these models. For solving the bi-level model, the dual theory, the KKT conditions, and the Fortuny-Amat and McCarl methods were applied. So, the non-linear bi-level model was transformed into a single-level and linear model that can be easily solved by the optimization solver.

Appendix A: Linear Power Flow

In this chapter, a linear power flow is used based on [20, 30]. This power flow is used only in radial distribution networks. For this purpose, a term is considered as a block to avoid nonlinearities. Note that the EVs in the PLs act as a source at the on-peak hours and as a load at the off-peak or mid-peak hours. The active and reactive power balance in this power flow is shown in Eqs. (4.A.1) and (4.A.2). Of course in the single-level model, instead of the expected value of the charging/discharging power and the output power of the solar system in Eq. (4.A.1), their scenario values are replaced.

$$
\begin{aligned}
P_{sb,t}^{Wh2G} \times \eta^{Trans} + \widehat{P}_{PL,t}^{Solar} + \sum_{EV} \widehat{P}_{PL,EV,t}^{dch} - \sum_{EV} \widehat{P}_{PL,EV,t}^{ch} - \sum_{b'} \left[\left(P_{b,b',t,s}^{+} - P_{b,b',t,s}^{-} \right) + R_{b,b'} I2_{b,b',t,s} \right] \\
+ \sum_{b'} \left(P_{b',b,t,s}^{+} - P_{b',b,t,s}^{-} \right) - P_{b,t}^{L} = 0 \qquad \forall t, s
\end{aligned}
$$

(4.A.1)

$$
\begin{aligned}
Q_{sb,t}^{Wh2G} - \sum_{b'} \left[\left(Q_{b,b',t,s}^{+} - Q_{b,b',t,s}^{-} \right) + X_{b,b'} I2_{b,b',t,s} \right] \\
+ \sum_{b'} \left(Q_{b',b,t,s}^{+} - Q_{b',b,t,s}^{-} \right) - Q_{b,t}^{L} \\
= 0 \qquad \forall t, s
\end{aligned}
$$

(4.A.2)

Note that I2 refers to an auxiliary variable linearly representing the squared current flow I2 in a given branch. At most one of these two positive auxiliary variables, i.e., $P_{b,b,t,s}$ and $Q_{b,b,t,s}$, can be different from zero at a time. This condition is again implicitly enforced by optimality. Moreover, Eqs. (4.A.3) and (4.A.4) limit these variables by the maximum apparent power for the sake of completeness.

$$
0 \leq \left(P_{b,b',t,s}^{+} + P_{b,b',t,s}^{-} \right) \leq V^{Rated} \times I^{\max,b,b}
$$

(4.A.3)

$$
0 \leq \left(Q_{b,b',t,s}^{+} + Q_{b,b',t,s}^{-} \right) \leq V^{Rated} \times I^{\max,b,b'}
$$

(4.A.4)

Equation (4.A.5) is presented for the balancing of voltage between two nodes. It should be noted that V2 in Eq. (4.A.5) is an auxiliary variable representing the squared voltage relation.

$$V2_{b,t,s} - V2_{b',t,s} - Z_{b,b'}^2 I2_{b,b',t,s} - 2R_{b,b'}\left(P_{b,b',t,s}^+ - P_{b,b',t,s}^-\right)$$

$$- 2X_{b,b'}\left(Q_{b,b',t,s}^+ - Q_{b,b',t,s}^-\right)$$

$$= 0 \tag{4.A.5}$$

Equation (4.A.6) is employed for linearizing the active and reactive power flows that appear in the apparent power expression.

$$V2_b^{Rated} I2_{b,b',t,s} = \sum_f \left[(2f-1)\Delta S_{b,b'}\Delta P_{b,b',f,t,s}\right]$$

$$+ \sum_f \left[(2f-1)\Delta S_{b,b'}\Delta Q_{b,b',f,t,s}\right] \tag{4.A.6}$$

For the piecewise linearization, Eqs. (4.A.7), (4.A.8), (4.A.9), (4.A.10), and (4.A.11) are represented. The number of blocks required to linearize the quadratic curve is set to 10 according to [20], which strikes the right balance between accuracy and computational requirements. Further descriptions, justifications, and derivations of the network model used in this chapter can be found in [30].

$$P_{b,b',t,s}^+ + P_{b,b',t,s}^- = \sum_f \Delta P_{b,b',f,t,s} \tag{4.A.7}$$

$$Q_{b,b',t,s}^+ + Q_{b,b',t,s}^- = \sum_f \Delta Q_{b,b',f,t,s} \tag{4.A.8}$$

$$0 \le \Delta P_{b,b',f,t,s} \le \Delta S_{b,b'} \tag{4.A.9}$$

$$0 \le \Delta Q_{b,b',f,t,s} \le \Delta S_{b,b'} \tag{4.A.10}$$

$$\Delta S_{b,b'} = \frac{V^{Rated} \times I^{max,b,b'}}{f} \tag{4.A.11}$$

Appendix B: Converting the Bi-Level Model to the Single-Level Model

The presented non-linear bi-level model by using the KKT conditions and the dual theory is converted into a linear single-level model. Firstly, by using of KKT optimization conditions (which a series of equal and unequal constraints that are inherently non-linear) a single-level model will be achieved. The presence of

4 Optimal Charge Scheduling of Electric Vehicles in Solar Energy...

complementary constraints is caused by the model to be non-linear. These series of constraints by the Fortuny-Amat and McCarl method which include binary variables, and a very large positive integer will be linear. Then, by using dual theory, the non-linear objective function becomes linear. When the bi-level model is converted to a single-level model, the main objective function of the final model is the linearly objective function of the upper level. Also the constraints of this model are the upper and lower level constraints, KKT's optimization constraints and linearly KKT's complementary constraints.

Converting Controlled Charging the Bi-Level Model to the Single-Level Model

At first, the constraints of the lower-level are described as Eqs. (4.I.1), (4.I.2), (4.I.3), (4.I.4), (4.I.5), (4.I.6), (4.I.7), (4.I.8), and (4.I.9):

$$C_1 = SOE_{EV}^{\max} - SOE_{PL,EV,t,s} \geq 0 \quad \forall PL, EV, t, s \quad \lambda_{PL,EV,t,s}^1 \tag{4.I.1}$$

$$C_2 = P_{PL,EV,t^{mid/off-peak},s}^{ch-grid} + P_{PL,EV,t^{mid/off-peak},s}^{ch-Solar}$$
$$\geq 0 \quad \forall PL, EV, t^{mid/off-peak}, s \quad \lambda_{PL,EV,t^{mid/off-peak},s}^2 \tag{4.I.2}$$

$$C_3 = P^{\max} - P_{PL,EV,t^{mid/off-peak},s}^{ch-grid} - P_{PL,EV,t^{mid/off-peak},s}^{ch-Solar}$$
$$\geq 0 \quad \forall PL, EV, t_{mid/off-peak}, s \quad \lambda_{PL,EV,t^{mid/off-peak},s}^3 \tag{4.I.3}$$

$$C_4 = P_{PL,EV,t^{on-peak},s}^{ch-solar} \geq 0 \quad \forall PL, EV, t^{on-peak}, s \quad \lambda_{PL,EV,t^{on-peak},s}^4 \tag{4.I.4}$$

$$C_5 = P^{\max} - P_{PL,EV,t^{on-peak},s}^{ch-solar} \geq 0 \quad \forall PL, EV, t^{on-peak}, s \quad \lambda_{PL,EV,t^{on-peak},s}^5 \tag{4.I.5}$$

$$SOE_{PL,EV,t,s} - SOE_{PL,EV,t-1,s} - \left(\left(P_{PL,EV,t,s}^{ch-grid} + P_{PL,EV,t,s}^{ch-Solar} \right) \times \eta^{ch} \right)$$
$$= 0 \quad \forall PL, EV, t \succ t^{arv}, s \quad \lambda_{PL,EV,t \succ t^{arv},s}^6 \tag{4.I.6}$$

$$SOE_{PL,EV,t,s} - SOE_{EV}^{arv} - \left(\left(P_{PL,EV,t,s}^{ch-grid} + P_{PL,EV,t,s}^{ch-Solar} \right) \times \eta_{ch} \right)$$
$$= 0 \quad \forall PL, EV, t^{arv}, s \quad \lambda_{PL,EV,t^{arv},s}^7 \tag{4.I.7}$$

$$SOE_{PL,EV,t,s} - SOE_{EV}^{dep} = 0 \quad \forall PL, EV, t^{dep}, s \quad \lambda_{PL,EV,t^{dep},s}^8 \tag{4.I.8}$$

$$\sum_{EV} P_{PL,EV,t,s}^{ch-Solar} - P_{PL,t,s}^{Solar} = 0 \quad \forall PL, EV, t, s \quad \lambda_{PL,EV,t,s}^9 \tag{4.I.9}$$

So, the Lagrangian function can be achieved by Eq. (4.I.10):

$$L = \sum_{s=1}^{Ns} \rho_s \sum_{PL=1}^{N_{PL}} \sum_{EV=1}^{N_{EV}} \sum_{t=1}^{24} P_{PL,EV,t^{mid/off-peak},s}^{ch-grid} \times \text{Pr}_{t^{mid/off-peak}}^{G2PL}$$

$$- \left(SOE_{EV}^{\max} - SOE_{PL,EV,t,s} \right) \lambda_{PL,EV,t,s}^1$$

$$- \left(P_{PL,EV,t^{mid/off-peak},s}^{ch-grid} + P_{PL,EV,t^{mid/off-peak},s}^{ch-solar} \right) \lambda_{PL,EV,t^{mid/off-peak},s}^2$$

$$- \left(P^{\max} - P_{PL,EV,t^{mid/off-peak},s}^{ch-grid} - P_{PL,EV,t^{mid/off-peak},s}^{ch-solar} \right) \lambda_{PL,EV,t^{mid/off-peak},s}^3$$

$$- \left(P_{PL,EV,t^{on-peak},s}^{ch-solar} \right) \lambda_{PL,EV,t^{on-peak},s}^4 - \left(P^{\max} - P_{PL,EV,t^{on-peak},s}^{ch-solar} \right) \lambda_{PL,EV,t^{on-peak},s}^5$$

$$- \left(SOE_{PL,EV,t,s} - SOE_{PL,EV,t-1,s} - \left(P_{PL,EV,t,s}^{ch-grid} \times \eta^{ch} \right) - \left(P_{PL,EV,t,s}^{ch-solar} \times \eta^{ch} \right) \right) \lambda_{PL,EV,t \succ t^{arv},s}^6$$

$$- \left(SOE_{PL,EV,t,s} - SOE_{EV}^{arv} - \left(P_{PL,EV,t,s}^{ch-grid} \times \eta^{ch} \right) - \left(P_{PL,EV,t,s}^{ch-solar} \times \eta^{ch} \right) \right) \lambda_{PL,EV,t^{arv},s}^7$$

$$- \left(SOE_{PL,EV,t,s} - SOE_{EV}^{dep} \right) \lambda_{PL,EV,t^{dep},s}^8 - \left(\sum_{EV} P_{PL,EV,t,s}^{ch-solar} - P_{PL,t,s}^{solar} \right) \lambda_{PL,EV,t,s}^9$$

$$(4.I.10)$$

Due to the decision variable in this model, KKT conditions are explained in Eqs. (4.I.11), (4.I.12), and (4.I.13):

$$\frac{\partial L}{\partial P_{PL,EV,t,s}^{ch-grid}} = \text{Pr}_{t^{mid/off-peak}}^{G2PL} - \lambda_{PL,EV,t^{mid/off-peak},s}^2 + \lambda_{PL,EV,t^{mid/off-peak},s}^3$$

$$+ \left(\eta_{ch} \times \lambda_{PL,EV,t,s}^6 \Big|_{t \succ t^{arv}} \right) + \left(\eta_{ch} \times \lambda_{PL,EV,t,s}^7 \Big|_{t=t^{arv}} \right) = 0$$

$$(4.I.11)$$

$$\frac{\partial L}{\partial P_{PL,EV,t,s}^{ch-solar}} = -\lambda_{PL,EV,t^{mid/off-peak},s}^2 + \lambda_{PL,EV,t^{mid/off-peak},s}^3 - \lambda_{PL,EV,t^{on-peak},s}^4 + \lambda_{PL,EV,t^{on-peak},s}^5$$

$$+ \left(\eta^{ch} \times \lambda_{PL,EV,t,s}^6 \Big|_{t \succ t_{arv}} \right) + \left(\eta^{ch} \times \lambda_{PL,EV,t,s}^7 \Big|_{t=t^{arv}} \right) - \lambda_{PL,EV,t,s}^9 = 0$$

$$(4.I.12)$$

$$\frac{\partial L}{SOC_{PL,EV,t,s}} = \lambda_{PL,EV,t,s}^1 + \lambda_{PL,EV,t+1,s}^6 \Big|_{t \succ t_{arv}} - \lambda_{PL,EV,t,s}^6 \Big|_{t \succ t_{arv}}$$

$$- \lambda_{PL,EV,t,s}^7 \Big|_{t=t_{arv}} - \lambda_{PL,EV,t,s}^8 \Big|_{t=t_{dep}} = 0$$

$$(4.I.13)$$

The dual variables of unequal constraints are equal or greater than zero, and the dual variables whose constraints are equal to zero are unrestricted in sign. For Eqs. (4.I.1), (4.I.2), (4.I.3), (4.I.4) and (4.I.5), the complementary constraints are as follows, i.e. Eqs. (4.I.14), (4.I.15), (4.I.16), (4.I.17), and (4.I.18).

$$0 \leq SOE_{EV}^{\max} \text{-} SOE_{PL,EV,t,s} \perp \lambda_{PL,EV,t,s}^1 \geq 0 \qquad (4.I.14)$$

4 Optimal Charge Scheduling of Electric Vehicles in Solar Energy...

$$0 \leq P^{ch-grid}_{PL,EV,t^{mid/off-peak},s} + P^{ch-solar}_{PL,EV,t^{mid/off-peak},s} \perp \lambda^2_{PL,EV,t^{mid/off-peak},s} \geq 0 \qquad (4.I.15)$$

$$0 \leq P^{max} - P^{ch-grid}_{PL,EV,t^{mid/off-peak},s} + P^{ch-Solar}_{PL,EV,t^{mid/off-peak},s} \perp \lambda^3_{PL,EV,t^{mid/off-peak},s} \geq 0 \qquad (4.I.16)$$

$$0 \leq P^{ch-solar}_{PL,EV,t^{on-peak},s} \perp \lambda^4_{PL,EV,t,s} \geq 0 \qquad (4.I.17)$$

$$0 \leq P^{max} - P^{ch-solar}_{PL,EV,t^{on-peak},s} \perp \lambda^5_{PL,EV,t,s} \geq 0 \qquad (4.I.18)$$

The linearization of complementary constraints is performed by Fortuny-Amat and McCarl linearization method by Eq. (4.I.19) [21]. Then, Eqs. (4.I.20), (4.I.21), (4.I.22), (4.I.23), and (4.I.24) are obtained.

$$
\begin{aligned}
&0 \leq F_1 \perp F_2 \geq 0 \\
&0 \leq F_1 \leq U \times M \\
&0 \leq F_2 \leq (1 - U) \times M \\
&U \varepsilon [0, 1]
\end{aligned}
\qquad (4.I.19)
$$

$$
\begin{aligned}
&0 \leq SOE^{max}_{EV} - SOE_{PL,EV,t,s} \leq U^1_{PL,EV,t,s} \times M^1 \\
&0 \leq \lambda^1_{PL,EV,t,s} \leq \left(1 - U^1_{PL,EV,t,s}\right) \times M^2
\end{aligned}
\qquad (4.I.20)
$$

$$
\begin{aligned}
&0 \leq P^{ch-grid}_{PL,EV,t^{mid/off-peak},s} + P^{ch-solar}_{PL,EV,t^{mid/off-peak},s} \leq U^2_{PL,EV,t^{mid/off-peak},s} \times M^1 \\
&0 \leq \lambda^2_{PL,EV,t^{mid/off-peak},s} \leq \left(1 - U^2_{PL,EV,t^{mid/off-peak},s}\right) \times M^2
\end{aligned}
\qquad (4.I.21)
$$

$$
\begin{aligned}
&0 \leq P^{max} - P^{ch-grid}_{PL,EV,t^{mid/off-peak},s} - P^{ch-solar}_{PL,EV,t^{mid/off-peak},s} \leq U^3_{PL,EV,t^{mid/off-peak},s} \times M^1 \\
&0 \leq \lambda^3_{PL,EV,t^{mid/off-peak},s} \leq \left(1 - U^3_{PL,EV,t^{mid/off-peak},s}\right) \times M^2
\end{aligned}
$$
$$\qquad (4.I.22)$$

$$
\begin{aligned}
&0 \leq P^{ch-solar}_{PL,EV,t^{on-peak},s} \leq U^4_{PL,EV,t^{on-peak},s} \times M^1 \\
&0 \leq \lambda^4_{PL,EV,t^{on-peak},s} \leq \left(1 - U^4_{PL,EV,t^{on-peak},s}\right) \times M^2
\end{aligned}
\qquad (4.I.23)
$$

$$
\begin{aligned}
&0 \leq P^{max} - P^{ch-solar}_{PL,EV,t^{on-peak},s} \leq U^5_{PL,EV,t^{on-peak},s} \times M^1 \\
&0 \leq \lambda^5_{PL,EV,t^{on-peak},s} \leq \left(1 - U^5_{PL,EV,t^{on-peak},s}\right) \times M^2
\end{aligned}
\qquad (4.I.24)
$$

The obtained model is a non-linear single-level model, which must be linearized using the dual theory. So, firstly, the dual objective function of the lower-level model is formed as Eq. (4.I.25):

Maximize

$$+\sum_{s=1}^{Ns}\rho_s\sum_{PL=1}^{N_{PL}}\sum_{EV=1}^{N_{EV}}\sum_{t=1}^{24}\left(\begin{array}{l} -\left(SOE_{EV}^{\max}\times\lambda_{PL,EV,t,s}^1\right)-\left(P^{\max}\times\lambda_{PL,EV,t^{mid/off-peak},s}^3\right)\\ -\left(P^{\max}\times\lambda_{PL,EV,t^{on-peak},s}^5\right)+\left(SOE_{EV}^{arv}\times\lambda_{PL,EV,t^{arv},s}^7\right)\\ +\left(SOE_{EV}^{dep}\times\lambda_{PL,EV,t^{dep},s}^8\right)+\left(P_{PL,t,s}^{Solar}\times\lambda_{PL,EV,t,s}^9\right) \end{array}\right)$$

$$(4.\text{I}.25)$$

According to the strong dual theory, the objective functions of the original and dual problems are equal at the optimal point of the decision variables of the two problems; therefore, the non-linear section of the objective function is linear according to Eq. (4.I.26).

$$\sum_{s=1}^{Ns}\rho_s\left(\sum_{PL=1}^{N_{PL}}\sum_{n=1}^{N}\sum_{t=1}^{24}P_{PL,n,t^{mid/off-peak},s}^{ch}\times\mathrm{Pr}_{t^{mid/off-peak}}^{G2PL}\right)=$$

$$\sum_{PL=1}^{N_{PL}}\sum_{EV=1}^{N_{EV}}\sum_{t=1}^{24}\left(\widehat{P}_{PL,EV,t^{mid/off-peak}}^{ch-grid}\times\mathrm{Pr}_{t^{mid/off-peak}}^{G2PL}\right)=$$

$$=\sum_{s=1}^{Ns}\rho_s\sum_{PL=1}^{N_{PL}}\sum_{EV=1}^{N_{EV}}\sum_{t=1}^{24}\left(\begin{array}{l} -\left(SOE_{EV}^{\max}\times\lambda_{PL,EV,t,s}^1\right)-\left(P^{\max}\times\lambda_{PL,EV,t^{mid/off-peak},s}^3\right)\\ -\left(P^{\max}\times\lambda_{PL,EV,t^{on-peak},s}^5\right)+\left(SOE_{EV}^{arv}\times\lambda_{PL,EV,t^{arv},s}^7\right)\\ +\left(SOE_{EV}^{dep}\times\lambda_{PL,EV,t^{dep},s}^8\right)+\left(P_{PL,t,s}^{Solar}\times\lambda_{PL,EV,t,s}^9\right) \end{array}\right)$$

$$(4.\text{I}.26)$$

Converting Charging/Discharging Schedule the Bi-Level Model to the Single-Level Model

At first, the constraints of the lower-level are described as Eqs. (4.II.1), (4.II.2), (4.II.3), (4.II.4), (4.II.5), (4.II.6), (4.II.7), (4.II.8), (4.II.9), and (4.II.10):

$$C_1 = SOE_{PL,EV,t,s} - SOE_{EV}^{\min} \geq 0 \quad \forall PL, EV, t, s \quad \lambda_{PL,EV,t,s}^1 \qquad (4.\text{II}.1)$$

$$C_2 = SOE_{EV}^{\max} - SOE_{PL,EV,t,s} \geq 0 \quad \forall PL, EV, t, s \quad \lambda_{PL,EV,t,s}^2 \qquad (4.\text{II}.2)$$

4 Optimal Charge Scheduling of Electric Vehicles in Solar Energy...

$$C_3 = P^{ch-grid}_{PL,EV,t^{mid/off-peak},s} + P^{ch-Solar}_{PL,EV,t^{mid/off-peak},s}$$

$$\geq 0 \quad \forall PL, EV, t^{mid/off-peak}, s \quad \lambda^3_{PL,EV,t^{mid/off-peak},s} \tag{4.II.3}$$

$$C_4 = P^{max} - P^{ch-grid}_{PL,EV,t_{mid/off-peak},s} - P^{ch-Solar}_{PL,EV,t^{mid/off-peak},s}$$

$$\geq 0 \quad \forall PL, EV, t^{mid/off-peak}, s \quad \lambda^4_{PL,EV,t^{mid/off-peak},s} \tag{4.II.4}$$

$$C_5 = P^{dch}_{PL,EV,t^{on-peak},s} \geq 0 \quad \forall PL, EV, t^{on-peak}, s \quad \lambda^5_{PL,EV,t^{on-peak},s} \tag{4.II.5}$$

$$C_6 = P^{max} - P^{dch}_{PL,EV,t^{on-peak},s} \geq 0 \quad \forall PL, EV, t^{on-peak}, s \quad \lambda^6_{PL,EV,t^{on-peak},s} \tag{4.II.6}$$

$$SOE_{PL,EV,t,s} - SOE_{PL,EV,t-1,s} + \left(\frac{P^{dch}_{PL,EV,t,s}}{\eta^{dch}} \right)$$

$$- \left(\left(P^{ch-grid}_{PL,EV,t,s} + P^{ch-Solar}_{PL,EV,t,s} \right) \times \eta^{dch} \right)$$

$$= 0 \quad \forall PL, EV, t \succ t^{arv}, s \quad \lambda^7_{PL,EV,t \succ t^{arv},s} \tag{4.II.7}$$

$$SOE_{PL,EV,t,s} - SOE^{arv}_{EV} + \left(\frac{P^{dch}_{PL,EV,t,s}}{\eta^{dch}} \right) - \left(\left(P^{ch-grid}_{PL,EV,t,s} + P^{ch-Solar}_{PL,EV,t,s} \right) \times \eta^{ch} \right)$$

$$= 0 \quad \forall PL, EV, t^{arv}, s \quad \lambda^8_{PL,EV,t^{arv},s} \tag{4.II.8}$$

$$SOE_{PL,EV,t,s} - SOE^{dep}_{EV} = 0 \quad \forall PL, EV, t^{dep}, s \quad \lambda^9_{PL,EV,t^{dep},s} \tag{4.II.9}$$

$$\sum_{EV} P^{ch-Solar}_{PL,EV,t^{mid/off-peak},s} - P^{Solar}_{PL,t^{mid/off-peak},s}$$

$$= 0 \quad \forall PL, EV, t^{mid/off-peak}, s \quad \lambda^{10}_{PL,EV,t^{mid/off-peak},s} \tag{4.II.10}$$

Based on the previous part, Eqs. (4.II.11), (4.II.12), (4.II.13), (4.II.14), (4.II.15), (4.II.16), (4.II.17), (4.II.18), (4.II.19), (4.II.20), (4.II.21), (4.II.22), (4.II.23), (4.II.24), (4.II.25), (4.II.26), and (4.II.27) is showing the single-level steps:

$$L =$$

$$\sum_{s=1}^{Ns} \rho_s \left(\begin{array}{c} \sum_{PL=1}^{N_{PL}} \sum_{EV=1}^{N_{EV}} \sum_{t=1}^{24} P^{ch-grid}_{PL,EV,t^{mid/off-peak},s} \times \mathrm{Pr}^{G2PL}_{t^{mid/off-peak}} \\ + \sum_{PL=1}^{N_{PL}} \sum_{EV=1}^{N_{EV}} \sum_{t=1}^{24} P^{dch}_{PL,EV,t^{on-peak},s} \times \left(0.5\mathrm{Pr}^{PL2G}_t + C^{cd}\right) \end{array} \right)$$

$$- \left(SOE_{PL,EV,t,s} - SOE^{\min}_{EV}\right) \lambda^1_{PL,EV,t,s}$$

$$- \left(SOE^{\max}_{EV} - SOE_{PL,EV,t,s}\right) \lambda^2_{PL,EV,t,s}$$

$$- \left(P^{ch-grid}_{PL,EV,t^{mid/off-peak},s} + P^{ch-Solar}_{PL,EV,t^{mid/off-peak},s}\right) \lambda^3_{PL,EV,t^{mid/off-peak},s}$$

$$- \left(P^{\max} - P^{ch-grid}_{PL,EV,t^{mid/off-peak},s} - P^{ch-Solar}_{PL,EV,t^{mid/off-peak},s}\right) \lambda^4_{PL,EV,t^{mid/off-peak},s}$$

$$- \left(P^{dch}_{PL,EV,t^{on-peak},s}\right) \lambda^5_{PL,EV,t^{on-peak},s} - \left(P^{\max} - P^{dch}_{PL,EV,t^{on-peak},s}\right) \lambda^6_{PL,EV,t^{on-peak},s}$$

$$- \left(\begin{array}{c} SOE_{PL,EV,t,s} - SOE_{PL,EV,t-1,s} - \left(P^{ch-grid}_{PL,EV,t,s} \times \eta^{ch}\right) \\ - \left(P^{ch-Solar}_{PL,EV,t,s} \times \eta^{ch}\right) + \left(\dfrac{P^{dch}_{PL,EV,t,s}}{\eta_{dch}}\right) \end{array} \right) \lambda^7_{PL,EV,t \succ t^{arv},s}$$

$$\quad (4.\mathrm{II}.11)$$

$$- \left(\begin{array}{c} SOE_{PL,EV,t,s} - SOE^{arv}_{EV} - \left(P^{ch-grid}_{PL,EV,t,s} \times \eta^{ch}\right) \\ - \left(P^{ch-Solar}_{PL,EV,t,s} \times \eta^{ch}\right) + \left(\dfrac{P^{dch}_{PL,EV,t,s}}{\eta^{dch}}\right) \end{array} \right) \lambda^8_{PL,EV,t^{arv},s}$$

$$- \left(SOE_{PL,EV,t,s} - SOE^{dep}_{EV}\right) \lambda^9_{PL,EV,t^{dep},s}$$

$$- \left(\sum_{EV} P^{ch-Solar}_{PL,EV,t^{mid/off-peak},s} - P^{Solar}_{PL,t^{mid/off-peak},s}\right) \lambda^{10}_{PL,EV,t^{mid/off-peak},s}$$

$$\frac{\partial L}{\partial P^{ch-grid}_{PL,EV,t,s}} = \mathrm{Pr}^{G2PL}_{t^{mid/off-peak}} - \lambda^3_{PL,EV,t^{mid/off-peak},s} + \lambda^4_{PL,EV,t^{mid/off-peak},s}$$

$$+ \left(\eta^{ch} \times \lambda^7_{PL,EV,t,s} \mid_{t \succ t^{arv}}\right) + \left(\eta^{ch} \times \lambda^8_{PL,EV,t,s} \mid_{t=t^{arv}}\right) = 0 \quad (4.\mathrm{II}.12)$$

$$\frac{\partial L}{\partial P^{ch-Solar}_{PL,EV,t,s}} = -\lambda^3_{PL,EV,t^{mid/off-peak},s} + \lambda^4_{PL,EV,t^{mid/off-peak},s} + \left(\eta_{ch} \times \lambda^7_{PL,EV,t,s} \mid_{t \succ t_{arv}}\right)$$

$$+ \left(\eta^{ch} \times \lambda^8_{PL,EV,t,s} \mid_{t=t^{arv}}\right) - \lambda^{10}_{PL,EV,t^{mid/off-peak},s} = 0$$

$$\quad (4.\mathrm{II}.13)$$

$$\frac{\partial L}{P^{dch}_{PL,EV,t,s}} = 0.5\mathrm{Pr}^{PL2G}_{t^{on-peak}} + C^{cd} - \left(\frac{\lambda^7_{PL,EV,t,s}}{\eta^{dch}} \mid_{t \succ t_{arv}}\right) - \left(\frac{\lambda^8_{PL,EV,t,s}}{\eta^{dch}} \mid_{t=t_{arv}}\right) \quad (4.\mathrm{II}.14)$$

$$- \lambda^5_{PL,EV,t_{on-peak},s} + \lambda^6_{PL,EV,t^{on-peak},s} = 0$$

4 Optimal Charge Scheduling of Electric Vehicles in Solar Energy...

$$\frac{\partial L}{SOC_{PL,EV,t,s}} = \lambda^7_{\text{PL,EV,t+1,s}}|_{t > t_{arv}} - \lambda^7_{\text{PL,EV,t,s}}|_{t > t_{arv}} - \lambda^8_{\text{PL,EV,t,s}}|_{t = t_{arv}}$$
$$-\lambda^9_{\text{PL,EV,t,s}}|_{t = t_{dep}} - \lambda^1_{\text{PL,EV,t,s}} + \lambda^2_{\text{PL,EV,t,s}} = 0 \tag{4.II.15}$$

$$0 \le SOE_{PL,EV,t,s} - \text{SOE}^{\min}_{EV} \perp \lambda^1_{\text{PL,EV,t,s}} \ge 0 \tag{4.II.16}$$

$$0 \le \text{SOE}^{\max}_{EV} - SOC_{PL,EV,t,s} \perp \lambda^2_{\text{PL,EV,t,s}} \ge 0 \tag{4.II.17}$$

$$0 \le P^{ch-grid}_{PL,EV,t_{mid/off-peak},s} + P^{ch-Solar}_{PL,EV,t_{mid/off-peak},s} \perp \lambda^3_{\text{PL,EV},t_{mid/off-peak},s} \ge 0 \tag{4.II.18}$$

$$0 \le P^{\max} - P^{ch-grid}_{PL,EV,t_{mid/off-peak},s} - P^{ch-Solar}_{PL,EV,t_{mid/off-peak},s} \perp \lambda^4_{\text{PL,EV},t_{mid/off-peak},s} \ge 0 \tag{4.II.19}$$

$$0 \le P^{dch}_{PL,EV,t_{on-peak},s} \perp \lambda^5_{\text{PL,EV},t_{on-peak},s} \ge 0 \tag{4.II.20}$$

$$0 \le P^{\max} - P^{dch}_{PL,EV,t_{on-peak},s} \perp \lambda^6_{\text{PL,EV},t_{on-peak},s} \ge 0 \tag{4.II.21}$$

$$0 \le SOE_{PL,EV,t,s} - \text{SOE}^{\min}_{EV} \le U^1_{PL,EV,t,s} \times M^1$$
$$0 \le \lambda^1_{PL,EV,t,s} \le \left(1 - U^1_{PL,n,t,s}\right) \times M^2 \tag{4.II.22}$$

$$0 \le \text{SOE}^{\max}_{EV} - SOE_{PL,EV,t,s} \le U^2_{PL,n,t,s} \times M^1$$
$$0 \le \lambda^2_{PL,EV,t,s} \le \left(1 - U^2_{PL,n,t,s}\right) \times M^2 \tag{4.II.23}$$

$$0 \le P^{ch-grid}_{PL,EV,t_{mid/off-peak},s} + P^{ch-Solar}_{PL,EV,t_{mid/off-peak},s} \le U^3_{PL,EV,t_{mid/off-peak},s} \times M^1$$
$$0 \le \lambda^3_{\text{PL,EV},t_{mid/off-peak},s} \le \left(1 - U^3_{PL,n,t_{mid/off-peak},s}\right) \times M^2 \tag{4.II.24}$$

$$0 \le P^{\max} - P^{ch-grid}_{PL,EV,t_{mid/off-peak},s} - P^{ch-Solar}_{PL,EV,t_{mid/off-peak},s} \le U^4_{PL,EV,t_{mid/off-peak},s} \times M^1$$
$$0 \le \lambda^4_{\text{PL,EV},t_{mid/off-peak},s} \le \left(1 - U^4_{PL,EV,t_{mid/off-peak},s}\right) \times M^2 \tag{4.II.25}$$

$$0 \le P^{dch}_{PL,EV,t_{on-peak},s} \le U^5_{PL,EV,t_{on-peak},s} \times M^1$$
$$0 \le \lambda^5_{\text{PL,EV},t_{on-peak},s} \le \left(1 - U^5_{PL,EV,t_{on-peak},s}\right) \times M^2 \tag{4.II.26}$$

$$0 \le P^{\max} - P^{dch}_{PL,EV,t_{on-peak},s} \le U^6_{PL,EV,t_{on-peak},s} \times M^1$$
$$0 \le \lambda^6_{\text{PL,EV},t_{on-peak},s} \le \left(1 - U^6_{PL,EV,t_{on-peak},s}\right) \times M^2 \tag{4.II.27}$$

The non-linear part of the objective function can be converted to a linear part with two equations i.e. (4.II.28) and (4.II.29).

Maximize

$$\sum_{s=1}^{Ns} \rho_s \sum_{PL=1}^{N_{PL}} \sum_{EV=1}^{N_{EV}} \sum_{t=1}^{24} \begin{pmatrix} \left(SOE_{EV}^{\min} \times \lambda_{PL,EV,t,s}^1\right) - \left(SOE_{EV}^{\max} \times \lambda_{PL,EV,t,s}^2\right) \\ -\left(P^{\max} \times \lambda_{PL,EV,t^{mid/off-peak},s}^4\right) - \left(P^{\max} \times \lambda_{PL,EV,t^{on-peak},s}^6\right) \\ +\left(SOE_{EV}^{arv} \times \lambda_{PL,EV,t^{arv},s}^8\right) + \left(SOE_{EV}^{dep} \times \lambda_{PL,EV,t^{dep},s}^9\right) \\ +\left(P_{PL,t^{mid/off-peak},s}^{Solar} \times \lambda_{PL,EV,t^{mid/off-peak},s}^{10}\right) \end{pmatrix}$$

$$(4.\text{II}.28)$$

$$\sum_{s=1}^{Ns} \rho_s \sum_{PL=1}^{N_{PL}} \sum_{n=1}^{N} \sum_{t=1}^{24} \begin{pmatrix} \left(P_{PL,n,t^{mid/off-peak},s}^{ch-grid} \times \text{Pr}_{t^{mid/off-peak}}^{G2PL}\right) \\ +\left(P_{PL,n,t^{on-peak},s}^{dch} \times \left(0.5\text{Pr}_{t^{on-peak}}^{PL2G} + C^{cd}\right)\right) \end{pmatrix}$$

$$= \sum_{s=1}^{Ns} \rho_s \sum_{PL=1}^{N_{PL}} \sum_{EV=1}^{N_{EV}} \sum_{t=1}^{24} \begin{pmatrix} \left(SOE_{EV}^{\min} \times \lambda_{PL,EV,t,s}^1\right) - \left(SOE_{EV}^{\max} \times \lambda_{PL,EV,t,s}^2\right) \\ -\left(P^{\max} \times \lambda_{PL,EV,t^{mid/off-peak},s}^4\right) - \left(P^{\max} \times \lambda_{PL,EV,t^{on-peak},s}^6\right) \\ +\left(SOE_{EV}^{arv} \times \lambda_{PL,EV,t^{arv},s}^8\right) + \left(SOE_{EV}^{dep} \times \lambda_{PL,EV,t^{dep},s}^9\right) \\ +\left(P_{PL,t^{mid/off-peak},s}^{Solar} \times \lambda_{PL,EV,t^{mid/off-peak},s}^{10}\right) \end{pmatrix}$$

So :

$$\sum_{s=1}^{Ns} \rho_s \sum_{PL=1}^{N_{PL}} \sum_{EV=1}^{N_{EV}} \sum_{t=1}^{24} P_{PL,EV,t^{on-peak},s}^{dch} \times \text{Pr}_{t^{on-peak}}^{PL2G} = \sum_{PL=1}^{N_{PL}} \sum_{EV=1}^{N_{EV}} \sum_{t=1}^{24} \left(\widehat{P}_{PL,EV,t^{mid/off-peak},s}^{dch} \times \text{Pr}_{t^{mid/off-peak}}^{G2PL}\right) =$$

$$= 2 \times \sum_{s=1}^{Ns} \rho_s \sum_{PL=1}^{N_{PL}} \sum_{EV=1}^{N_{EV}} \sum_{t=1}^{24} \begin{pmatrix} \left(SOE_{EV}^{\min} \times \lambda_{PL,EV,t,s}^1\right) - \left(SOE_{EV}^{\max} \times \lambda_{PL,EV,t,s}^2\right) \\ -\left(P^{\max} \times \lambda_{PL,EV,t^{mid/off-peak},s}^4\right) - \left(P^{\max} \times \lambda_{PL,EV,t^{on-peak},s}^6\right) \\ +\left(SOE_{EV}^{arv} \times \lambda_{PL,EV,t^{arv},s}^8\right) + \left(SOE_{EV}^{dep} \times \lambda_{PL,EV,t^{dep},s}^9\right) \\ +\left(P_{PL,t^{mid/off-peak},s}^{Solar} \times \lambda_{PL,EV,t^{mid/off-peak},s}^{10}\right) - \left(P_{PL,EV,t^{on-peak},s}^{dch} \times C^{cd}\right) \\ -\left(P_{PL,EV,t^{mid/off-peak},s}^{ch-grid} \times \text{Pr}_{t^{mid/off-peak}}^{G2PL}\right) \end{pmatrix}$$

$$(4.\text{II}.29)$$

Appendix C

The nomenclature is shown below.

Indices	
b, b'	Index for branch or bus
EV	Index for EV number
F	Index for linear partitions in linearization
s	Index for scenarios
sb	Index for slack bus
t, t'	Index for time (hour)
Parameters	
C^{cd}	Cost of equipment depreciation ($/kWh)
I^{max}	Upper limit of branches' current (A)
$I^{max, b, b'}$	Maximum current of branch b, b' (A)
M	Sufficiently large constants
P^L	The demand of customers (kW)
P^{max}	Nominal rate of charging/discharging of EVs (kWh)
P^{Solar}	Power generated of the solar system (kW)
Pr^L	Electricity price for the customer ($/kWh)
Pr^{Wh2G}	The wholesale market electricity price ($/kWh)
$R_{b, b'}$	Resistance between branch b, b' (Ω)
Q^L	Customer's reactive power (kVAR)
SOE^{arv}	Initial SOE of the EVs (kWh)
SOE^{dep}	Desired SOE of the EVs (kWh)
SOE^{max}	Upper limit of SOE (kWh)
SOE^{min}	Lower limit of SOE (kWh)
t^{arv}	Arrival time of the EVs to the PL
t^{dep}	Departure time of the EVs from the PL
V^{Rated}	Nominal voltage (V)
V^{max}	Maximum allowable voltage (V)
V^{min}	Minimum allowable voltage (V)
$X_{b, b'}$	Reactance between branch b, b' (Ω)
Z	Impedance (Ω)
η^{ch}	Charging efficiency (%)
η^{dch}	Discharging efficiency (%)
η^{Trans}	Transformer efficiency (%)
ρ	Probability of each scenario
α	Confidence level
β	Risk aversion parameter
ΔS	Upper limit in the discretization of quadratic flow terms (kVA)
Variables	
B	Profit in each scenario
I,I2	Current flow (A), squared current flow (A2)

(continued)

Indices	
$p^{ch\text{-}grid}$	Charging power of the EVs by the SDNO (kW)
$p^{ch\text{-}solar}$	Charging power of the EVs by the power generated of the solar system (kW)
$\hat{p}^{ch-grid}$	The expected value of charging power of the EVs by the SDNO (kW)
$\hat{p}^{solar}$	The expected value of the power generated of the solar system (kW)
p^{dch}	Discharging power of the EVs (kW)
$\hat{p}^{dch}$	The expected value of discharging power of the EVs (kW)
p^{Loss}	SDN's losses (kW)
p^{Wh2G}	Purchasing power from the wholesale by the SDNO (kW)
P^+	Active power flows in downstream directions (kW)
P^-	Active power flows in upstream directions (kW)
Pr^{G2PL}	Charging tariff of the EVs ($/kWh)
Pr^{PL2G}	Discharging tariff of the EVs ($/kWh)
Q^{Wh2G}	SDN's reactive power (kVAR)
Q^+	Reactive power flows in downstream directions (kVAR)
Q^-	Reactive power flows in upstream directions (kVAR)
SOE	State of energy (kWh)
U	Binary variable
λ	Dual variable ($/kWh)
η	Auxiliary variable for calculating CVaR
ξ	Value-at-risk

References

1. L.P. Fernandez, T.G. San Román, R. Cossent, C.M. Domingo, P. Frias, Assessment of the impact of plug-in electric vehicles on distribution networks. IEEE Trans. Power Syst. **26**, 206–213 (2011)
2. M.S. ElNozahy, M.M.A. Salama, A comprehensive study of the impacts of PHEVs on residential distribution networks. IEEE Trans. Sustain. Energy **5**, 332–342 (2014)
3. C. Weiller, Plug-in hybrid electric vehicle impacts on hourly electricity demand in the United States. Energy Policy **39**, 3766–3778 (2011)
4. J. Mullan, D. Harries, T. Bräunl, S. Whitely, Modelling the impacts of electric vehicle recharging on the Western Australian electricity supply system. Energy Policy **39**, 4349–4359 (2011)
5. J.Y. Yong, V.K. Ramachandaramurthy, K.M. Tan, N. Mithulananthan, A review on the state-of-the-art technologies of electric vehicle, its impacts and prospects. Renew. Sust. Energ. Rev. **49**, 365–385 (2015)
6. A. Jiménez, N. García, Voltage unbalance analysis of distribution systems using a three-phase power flow ans a genetic algorithm for PEV fleets scheduling, in *Proceedings of the Power and Energy Society General Meeting*, (San Diego, 2012, 22–26 July), pp. 1–8
7. H. Shareef, M.M. Islam, A. Mohamed, A review of the stage-of-the-art charging technologies, placement methodologies, and impacts of electric vehicles. Renew. Sust. Energ. Rev. **64**, 403–420 (2016)

8. H. Salman, K. Muhammad, R. Umar, Impact analysis of vehicle-to-grid technology and charging strategies of electric vehicles on distribution networks—A review. J. Power Sources **277**, 205–214 (2015)
9. P.S. Moses, S. Deilami, A.S. Masoum, M.A. Masoum, Power quality of smart grids with plug-in electric vehicles considering battery charging profile, in *Proceedings of the 2010 IEEE PES Innovative Smart Grid Technologies Conference Europe (ISGT Europe)*, (Chalmers Lindholmen, Gothenburg, 2010, 11–13 Oct), pp. 1–7
10. G. Razeghi, L. Zhang, T. Brown, S. Samuelsen, Impacts of plug-in hybrid electric vehicles on a residential transformer using stochastic and empirical analysis. J. Power Sources **252**, 277–285 (2014)
11. E. Akhavan-Rezai, M.F. Shaaban, E.F. El-Saadany, A. Zidan, Uncoordinated charging impacts of electric vehicles on electric distribution grids: Normal and fast charging comparison, in *Proceedings of the Power and Energy Society General Meeting*, (San Diego, 2012, 22–26 July), pp. 1–7
12. E. Sortomme, M.A. El-Sharkawi, Optimal scheduling of vehicle-to-grid energy and ancillary services. IEEE Trans. Smart Grid **3**(1), 351–359 (2012)
13. Z. Wang, S. Wang, Grid power peak shaving and valley filling using vehicle-to-grid systems. IEEE Trans. Power Del **28**(3), 1822–1829 (2013)
14. M.A. López, S. De la Torre, S. Martín, J.A. Aguado, Demand-side management in smart grid operation considering electric vehicles load shifting and vehicle-to-grid support. Int. J. Electr. Power Energy Syst. **64**, 689–698 (2015)
15. A. Zakariazadeh, S. Jadid, P. Siano, Multi-objective scheduling of electric vehicles in smart distribution system. Energy Convers. Manag. **79**, 43–53 (2014)
16. F. Fazelpour, M. Vafaeipour, O. Rahbari, M.A. Rosen, Intelligent optimization to integrate a plug-in hybrid electric vehicle smart parking lot with renewable energy resources and enhance grid characteristics. Energy Convers. Manag. **77**, 250–261 (2014)
17. F. Marra, G.Y. Yang, C. Træholt, E. Larsen, J. Østergaard, B. Blažič, W. Deprez, EV charging facilities and their application in LV feeders with photovoltaics. IEEE Trans. Smart Grid **4**(3), 1533–1540 (2013)
18. M.H. Amini, M.P. Moghaddam, O. Karabasoglu, Simultaneous allocation of electric vehicles' parking lots and distributed renewable resources in smart power distribution networks. Sustain. Cities Soc. **28**, 332–342 (2017)
19. S.M. Mohammadi-Hosseininejad, A. Fereidunian, H. Lesani, Reliability improvement considering plug-in hybrid electric vehicles parking lots ancillary services: A stochastic multi-criteria approach. IET Gener. Transm. Dis. **12**(4), 824–833 (2017)
20. S.M.B. Sadati, J. Moshtagh, M. Shafie-khah, J.P. Catalão, Smart distribution system operational scheduling considering electric vehicle parking lot and demand response programs. Electr. Power Syst. Res. **160**, 404–418 (2018)
21. S.M.B. Sadati, J. Moshtagh, M. Shafie-khah, A. Rastgou, J.P. Catalão, Operational scheduling of a smart distribution system considering electric vehicles parking lot: A bi-level approach. Int. J. Electr. Power Energy Syst. **105**, 159–178 (2019)
22. R. Sioshansi, J. Miller, Plug-in hybrid electric vehicles can be clean and economical in dirty power systems. Energy Policy **39**(10), 6151–6161 (2011)
23. M.S. ElNozahy, M.M.A. Salama, Studying the feasibility of charging plug-in hybrid electric vehicles using photovoltaic electricity in residential distribution systems. Electr. Power Syst. Res. **110**, 133–143 (2014)
24. M. Ghofrani, A. Arabali, M. Ghayekhloo, Optimal charging/discharging of grid-enabled electric vehicles for predictability enhancement of PV generation. Electr. Power Syst. Res. **117**, 134–142 (2014)
25. W. Hennings, S. Mischinger, J. Linssen, Utilization of excess wind power in electric vehicles. Energy Policy **62**, 139–144 (2013)

26. B.S.M. Borba, A. Szklo, R. Schaeffer, Plug-in hybrid electric vehicles as a way to maximize the integration of variable renewable energy in power systems: the case of wind generation in northeaster Brazil. Energy $37(1)$, 469–481 (2012)
27. D. Dallinger, S. Gerda, M. Wietschel, Integration of intermittent renewable power supply using grid-connected vehicles–a 2030 case study for California and Germany. Appl. Energy 104, 666–682 (2013)
28. C. Jin, X. Sheng, P. Ghosh, Optimized electric vehicle charging with intermittent renewable energy sources. IEEE J. Sel. Top. Signal Process. $8(6)$, 1063–1072 (2014)
29. A.J. Conejo, M. Carrión, J.M. Morales, *Decision making under uncertainty in electricity markets*, vol 1 (Springer, New York, 2010)
30. M. Shafie-khah, P. Siano, D.Z. Fitiwi, N. Mahmoudi, J.P. Catalão, An innovative two-level model for electric vehicle parking lots in distribution systems with renewable energy. IEEE Trans. Smart Grid 9, 1506–1520 (2017)
31. S. Bahramara, M.P. Moghaddam, M.R. Haghifam, Modelling hierarchical decision making framework for operation of active distribution grids. IET Gener. Transm. Dis. $9(16)$, 2555–2564 (2015)
32. M. Esmaeeli, A. Kazemi, H. Shayanfar, M.R. Haghifam, P. Siano, Risk-based planning of distribution substation considering technical and economic uncertainties. Electr. Power Syst. Res. 135, 18–26 (2016)
33. M. Mazidi, H. Monsef, P. Siano, Incorporating price-responsive customers in day-ahead scheduling of smart distribution networks. Energy Convers. Manag. 115, 103–116 (2016)
34. S. Talari, M. Yazdaninejad, M.R. Haghifam, Stochastic-based scheduling of the microgrid operation including wind turbines, photovoltaic cells, energy storages and responsive loads. IET Gener. Transm. Dis. $9(12)$, 1498–1509 (2015)
35. A. Zakariazadeh, S. Jadid, P. Siano, Stochastic multi-objective operational planning of smart distribution systems considering demand response programs. Electr. Power Syst. Res. 111, 156–168 (2014)

Chapter 5
Optimal Utilization of Solar Energy for Electric Vehicles Charging in a Typical Microgrid

Mohammad Saadatmandi and Seyed Mehdi Hakimi

Abbreviation

AER	All electric range
BCG	Binary conventional generation
CG	Conventional generation
ECPM	Energy consumption per mile
EV	Electric vehicle
HEV	Hybrid electric vehicle
HGC	Home general controller
PEN	Percentage of energy needed
PHEV	Plug-in hybrid electric vehicle
PNNL	Pacific Northwest National Laboratory
SOC	State of charge
TGC	Transformer general controller
V2G	Vehicle to grid

5.1 Introduction

The modernization of power grids has been the chief issue for power industry in different countries. As well as, the lack of energy and the growth of the electrical energy consumption are the major threats to the countries' economy. Thus

---

M. Saadatmandi · S. M. Hakimi (✉)
Department of Electrical Engineering and Renewable Energy Research Center, Damavand Branch, Islamic Azad University, Damavand, Iran
e-mail: sm_hakimi@damavandiau.ac.ir

© Springer Nature Switzerland AG 2020
A. Ahmadian et al. (eds.), *Electric Vehicles in Energy Systems*,
https://doi.org/10.1007/978-3-030-34448-1_5

increasing grid security during use of renewable energy, reduce the greenhouse gas emissions, and reduce losses in power transmission are the effective ways in order to be a smart power grid [1].

There are three main purposes of the smart grid: (i) improved reliability, (ii) optimal usage of distributed production resources, and (iii) energy efficiency. Demand response is one of the most important program in the energy efficiency issue, which is designed to change the pattern of consumers' consumption and load management. On the other hand, using demand response program and raise the generation efficiency by assuming creating smart grid will be reduced the load and the pollution caused by energy generation [2].

The used PHEV that is one of the electrical devices with charging feature through grid connection. The PHEV's batteries have high capacity, so needed to much energy to recharge them. It causes a peak load in the distribution grid and reduces the penetration of renewable resources into residential distribution grid. Demand response programs, PHEVs charging management can be solved such challenges. The aim of load managing is to change the PHEVs charging into other hours that decreases the received power from the conventional generation and increases the renewable resources consumption.

Additionally, the chapter is examined a residential grid model to optimize the use of renewable resources through of the plug-in vehicles charging & discharging management on 24 h. The residential grid consists of 5 smart houses which each of them has 14 photovoltaic panels to produce require power for houses. The loads of these houses are divided into static loads and Flexible loads. The smart home is equipped with the planned intelligent charging device for PHEV with able to exchange energy into a distribution grid through two ways [3].

Due to the static loads time of residential consumers the meteorological data are extracted of one-year period for renewable resources modeling. In addition, the cars features are derived from the manufactures cars as the flexible loads. so, the researcher used of Monte Carlo simulation algorithm with taking into account the uncertainty the mileage by the car owners during daytime, the initial capacity of the battery, and the time of the vehicle usage at different hours of the day; as well as used of the (PSO) algorithm to optimize the issue.

The aim of conduct this chapter is to increase the use of renewable energy resources by consumers. Hence, the objective function defined through minimized potential received from the conventional power grid. The results showed that, the received energy from conventional power grid is reduced significantly of control charging by use of the proposed method through transfer the time of the vehicle charging to hours with maximum renewable resource access. Moreover, the cost of consumers had decreased significantly by adding V2G and discharge control without reducing the comfort of the car owners, which indicates the effectiveness of this method.

5.2 Renewable Energy Resources

Since the beginning of the design and construction of power systems has always been raised the issue of uncertainty in the failure of units and equipment or the error in forecasting of demand. Nowadays with the advent of renewable energy especially solar energy, the uncertainty in the utilize of power systems has intensified. Hence the energy policy makers hope that the new energies will have a significant impact in supplying needed human energy. There has been proposed solar energy more than other types of energies, thus must take be considered an important role for planning and operating power systems. The uncertainty in output is one of the most important features of this energy so it must be considered properly in the planning of power systems generation; A smart home with solar panels, smart appliances and an electric vehicle shows as Fig. 5.1.

5.2.1 Solar Energy

In recent year, the process of using solar energy has been growing and evolving, in a way that technology has become more advanced in use of solar energy and it has an acceptant efficiency. The generation of solar systems depends on the amount of solar radiation. Some of the advantages of using solar panel as follows:

Fig. 5.1 Residential smart grid

1. Simple technology
2. The conversion of solar radiation into electrical energy directly
3. Low noise
4. Low cost and high efficiency through the advancement of science

5.2.2 Math Model of Solar Energy Output Power

The output power of solar panels is obtained by Eq. (5.1) [3]:

$$P_{pv} = N \times \frac{G_t}{1000} \times P_{mpv} \tag{5.1}$$

Where PPV is the output power of solar panels (w), Pmpv is the nominal power of each panel in terms of $G_t = 1000$, N is the number of solar panel, and Gt is the solar radiation intensity $(w/_{m^2})$.

5.3 Electrical Vehicle

Due to the reduction of fossil fuel resources and high environmental pollution by the high consumption of the vehicles as cars, motorcycles, and etc., that has attracted more attention to the use of equipment that use of other energy sources. The result of studies has shown that if the energy consumption continues to exist, the emission of carbon dioxide rate will be expected to double by 2050 than in 2005 which is not acceptable in terms of environmental perspective. According to global plans by 2050, this rate must be halved in the level of CO_2 in 2005.

To this end, the safest methods as the use of distributed generation of energy, simultaneous generation of electricity and warm, and use of motor vehicles that their driving forces supplied from electric power grid or batteries. The use of motor vehicles has been more important in all countries especially in developed countries such as united states and japan, or China and India that have made significant advances in this regard.

PHEVs are the electric vehicle whose use at least two types of energy resources to drive. The new generation of hybrid vehicles, two fuel and electric motors provide the required driving force to drive. But why used of two engines to drive these vehicles? The internal combustion engine (gasoline or gasoil) generates a bit power in the low engine rpm engine that has the higher efficiency, in contrast, the electric engine generates high torque with high efficiency in a low engine rpm. Therefore, simultaneous use of both engines can be reduced fuel consumption meanwhile achieving enough power to drive and accelerate. One of the most important features of an PHEV is that if the internal combustion engine generates

5 Optimal Utilization of Solar Energy for Electric Vehicles Charging in... 133

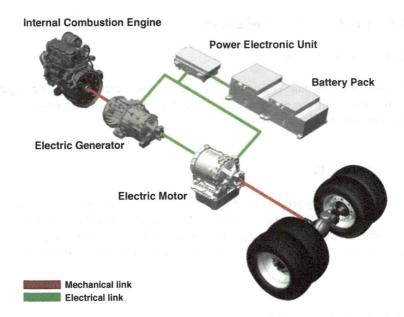

Fig. 5.2 The components of a hybrid vehicle

more force driving than needed, its battery will recharge by utilizing power engine directly during braking or moving downhill then provide the required power. All PHEVs have three factors: management, electronic and smart balance of engines connection. Hybrid vehicles are the progressed pure-electric vehicles that have modified somewhat defects of pure-electric vehicles and internal combustion engines vehicles. The main benefits of these vehicles than the internal combustion: (i) the function in the stable load and rpm and operating at the optimal point, which will be caused raise the engine efficiency, reduce pollution, and reduce fuel consumption; (ii) the energy store electrically in the battery during braking or negative accelerating which will be reduction of the combustion engine efficiency and thus reduce pollution and reduce fuel consumption, for instance, the Toyota Prius consumes 2.4 L fuel on per 100 km with the capacity of 1500 cc and four-cylinder engine; (iii) high mileage per battery charge. Although hybrid vehicles have different structures, but they must include of a power transmission system, a power generation unit, and an energy storage system. The initial choices for the energy storage system are the batteries, capacitors, and flywheels. Batteries are the first energy storage device due to their cheapness, being commercial, and lack of movable parts, but the major shortcoming is their short lifespan. However, the batteries are very expensive in new technology but today they are developing. So due to the fact that batteries are the major choice in this field, the research has started on other energy storage fields. The components of a hybrid vehicle are shown in Fig. 5.2.

5.3.1 The Components of a Hybrid Vehicle

5.3.1.1 Gasoline Engine

The nature of the gasoline engine in this car resembles to ordinary gasoline engines with two main differences: the higher technology and the smaller engine which reduces pollution and increases performance.

5.3.1.2 Electric Engine

Electric engine is very progressed in hybrid vehicles so it can operate as a generator and engine. For instance, whenever the engine is required, it will be able to provide the desired acceleration by using batteries and when the electric engine is not required (moving on a downhill) be able to restore the power to the battery as a generator.

5.3.1.3 Electric Generators

The engine and the generator are almost similar in the construction, but the generator just has the duty to provide the electricity required by the engine. The generator is used series in most hybrid vehicles.

Types of the hybrid systems were explained in more detail in the next sections.

5.3.1.4 Battery

The batteries are an energy storage device for the electric engine in the hybrid vehicle. Not only electric engine can be transferred fuel to the gasoline engine such as the gasoline in the tank (a one-way transfer from the fuel tank to the gasoline engine), but it can be restored energy to the battery which the gasoline engine cannot do that.

5.3.1.5 Comparing of the Efficiency of the Pure-Electric Vehicles and Hybrid Vehicles

Although electric vehicles have been introduced as the first way to reduce pollution, but due to the fact they did not succeed in long driving cycles thus was a failed product. The hybrid electric vehicles have such advantages than pure-electric vehicles. Due to the use of two sources of energy for the stimulus driving system generation has been minimized the pollution and fuel consumption issues and resolved the problems caused by pure-electric vehicles.

5.3.2 Performance of Hybrid Electric Vehicles

The gasoline engines of vehicles have a main problem. In order to drive at high speeds on highways, the cars have been designed to maximize their power. So the cars efficiency decreases when they barely move in heavy traffic. To this end, dual-engine hybrids vehicles are designed to use of the low-power electric engine in situations that low power is required, and use of the gasoline engine where needs to high power. The engines switching is automatically.

The gasoline engine simply turns off in the red lights and is often designed to operate on a limited range of power, thereby increasing its efficiency. The vehicle's battery is recharged by a gasoline engine. The generators are connected to the brake for convert some of the braking momentum (which is usually lost as heat) into electricity in order to recharge the battery.

5.3.3 Introducing Electric Vehicles

Electric vehicles are generally divided into three categories:

1. Electric vehicles (EV)
2. Hybrid electric vehicles (HEV)
3. Plug-in hybrid electric vehicles (PHEV)

5.3.3.1 Electric Vehicles (EV)

These vehicles have an electric engines and some batteries to provide electric energy that the energy of the batteries is used as the propulsion of the vehicles' electric engine and to provide the necessary energy for other equipment. The batteries could be recharged through power grid connection, the vehicle's braking power, and even of off grid electrical resources like solar panels.

The main advantages of electric vehicles:

- Completely free from greenhouse gas emissions
- very low-noise pollution
- Much higher efficiency than internal combustion engines
- Cheaper electrical engine
- The main defect of these vehicles is dependence on the battery (whose capacity and energy density is not comparable to fossil fuels)

5.3.3.2 Hybrid Electric Vehicles (HEV)

The vehicles have both fuel and electric engines with the proper battery capacity (1–3 kWh) that has the energy storage capability from the engine and the brakes. At the time of need, vehicles' batteries aid to generate the auxiliary force power or at low speeds by shutting off fuel engine provide vehicle propulsion.

In the past decade almost 1.5 million hybrid electric vehicles were sold. In developed countries including United States, the hybrid electric vehicles account for around 3% of all vehicles.

The disadvantages of these vehicles:

- Non-rechargeable batteries of the power grid
- Dependence on the fossil fuels engine (inability to move the vehicle by using only the electric engine)

5.3.3.3 Plug-in Hybrid Electric Vehicles (PHEV)

PHEVs are designed to eliminate the defects of hybrid electric vehicles so rechargeable from the grid thus need to more batteries than HEVs (about $5\times$). PHEVs have a complete fossil fuel engine system. The main difference between PHEV and HEV's batteries is that the PHEV batteries must have rapid dis/charging capability, while HEV batteries are usually keep fully charged and are rarely discharged.

The cost of PHEV batteries ranged between 1.3 to 1.5 times than EV batteries which due to fewer batteries, the total cost of PHEVs' batteries will be less than EVs.

The following can be mentioned for these vehicles:

- With the abundant battery production, the costs of battery may be reached to 750 dollars per kilowatt-hour so the total of its cost will be about 6000 dollars for each all electric range vehicle (40 k with 8 kWh battery capacity).
- If the car's life span is 200,000 km, the amount of the savings in the cost of fuel would be about 4000 dollars which is less than the cost of battery.
- Reducing the cost of the battery to 500 dollars per kilowatt-hour will be created the competition between plug-in hybrid electric vehicles and gasoline vehicles.

5.3.3.4 Comparing PHEV and HEV

The HEVs is a good start for changing the fuel of electric vehicles, but the mileage possible on one charge is limited. Hence, the PHEVs were raised which have the ability to connect to grid at the point where the electrical energy output is embedded. They have a battery and internal combustion engine. In general, most of PHEVs provide the required energy through electrical energy, but the internal combustion engine is also used when battery is not enough.

5.3.3.5 The Effect of PHEVs Performance on the Power Grid

Whenever the PHEVs are not used, the energy be received from the power grid to recharge them. The widespread use of PHEVs will bring challenges to the power grid because they have high electrical energy consumption and could be

5 Optimal Utilization of Solar Energy for Electric Vehicles Charging in...

Table 5.1 Consumption and battery capacity of various PHEVs

Vehicle type	ECPM (kWh/Mile)	PHEV (33-mile Battery) (kWh)
Compact Sedan	0.26	8.6
Mid-size Sedan	0.30	9.9
Mid-size SUV	0.38	12.5
Full-size SUV	0.46	15.2

connected to the distribution grid at any hours to charge. As well as, the distribution grid is usually used at their maximum capacity then the stability of the system will be compromised by the added load caused by improper use of electric vehicles.

In other words, the PHEVs should be charged at non-peak times or at other times such as energy storage devices.

5.3.4 Mathematical Model of the Electric Vehicles Battery

The battery capacity is the main factor to determining the mileage by a PHEV in the electrical mode. The PNNL[1] Institute has shown the power consumption per mile for all types of PHEV with a capacity of 33 miles and AER[2] = X in Table 5.1.

As it mentioned that AER determines the probable distance with the PHEV and the battery with fully charged. So the battery capacity is obtained of ECPM[3] multiplication in AER.

$$C = ECPM * AER \qquad (5.2)$$

Where C is the applied capacity of PHEV [4].

The Table 5.2 is shown the battery capacity for PHEVs with 30, 40, and 60 miles AERs based on the above-mentioned equation. Due to the table, there can be said that the PHEVs have the domain size of batteries from 7.8 to 27.6 kWh. Since the required energy to charge the PHEV depends on the capacity of the battery, so determining the types of AERs and PHEVs to analysing the impact of such vehicles is essential [5].

---

[1] Pacific Northwest National Laboratory

[2] All electric range

[3] Energy consumption per mile

Table 5.2 The size of PHEV batteries with various AERs [5]

Vehicle type	PHEV 30 mile AER (kWh)	PHEV 40 mile AER (kWh)	PHEV 60 mile AER (kWh)
Compact Sedan	7.8	10.4	15.6
Mid-size Sedan	9	12	18
Mid-size SUV	11.4	15.2	22.8
Full-size SUV	13.8	18.4	27.6

5.3.5 The Stored Energy in PHEVs' Batteries

SOC[4] is the amount of stored energy in the battery. Although it is assumed that PHEVs use of the charging system on the electric mode, but the combination of the function of the electric engine and the internal combustion engine is appropriate. Thus, PHEVs can be operated at any mode based on their required energy so they receive some part of energy from the battery at the lack of energy in the battery mode. In order to analyse all functions of PHEVs battery is used of the parameter λ for each PHEV. λ is the maximum mileage of a PHEV in the electric mode so $\lambda = $ AER. Then SOC o a PHEV is obtained through of the Eq. (5.3).

$$SOC = \begin{cases} \left(\dfrac{\lambda - d}{\lambda}\right) * 100 & d < \lambda \\ 0 & d \geq \lambda \end{cases} \tag{5.3}$$

Where λ is the AER and d is the total mileage by the car, hence the usable capacity of the battery is completely consumed in $d \geq \lambda$ mode and the SOC is zero [6].

Whenever the vehicle has travelled all the distance in electric mode the Eq. (5.3) be applied. In order to determine the battery capacity (SOC) in simultaneous use of both electrical and internal fuel consumption, a parameter should be introduced to determine the vehicle's mileage percentage in electric mode. There is assume that α represents this issue, in this case SOC is obtained of Eq. (5.4) [5].

$$SOC = \begin{cases} \left(1 - \dfrac{\alpha d}{AER}\right) * 100 & \alpha d < AER \\ 0 & \alpha d \geq AER \end{cases} \tag{5.4}$$

---

[4]State of charge

5 Optimal Utilization of Solar Energy for Electric Vehicles Charging in... 139

Table 5.3 Electrical specification of vehicles of various companies

Vehicle name	Type of battery	Battery capacity	Full charge time	Power charge	Maximum electrical distance	Electrical function
Chevy Volt	Lithium-ion	16 kWh	4 h-240 V	3.3 kW	64 km-40 miles	0.25 kWh/km
						0.4 kWh/mile
Renault Fluence Z.E	Lithium-ion	22 kWh	6–9 h-240 V	3.5 kW	185 km-115 miles	0.11 kWh/km
						0.19 kWh/mile
Toyota Prius	Lithium-ion	4.4 kWh	1.5 h-240 V	3.3 kW	23 km-14 miles	0.19 kWh/km
						0.31 kWh/mile
Ford C-max	Lithium-ion	7.6 kWh	3 h-240 V	3.3 kW	32 km-20 miles	0.23 kWh/km
						0.38 kWh/mile

5.3.6 The Amount of the Energy Required to PHEVs' Battery Charging

PEN[5] is the percentage of energy needed to fully charge the battery. The percentage of energy required to PHEVs' battery charging after the last time span is determined based on the SOC.

Such energy is obtained through the Eq. (5.5) to fully charge the battery.

$$PEN = 100 - SOC \tag{5.5}$$

The PEN is the energy needed to fully charge the battery.

As regards the efficiency of the converters to charge the battery is not 100% so the amount of energy that the grid should provide to charge the battery is achieved by follow equation.

$$PEN_R = \frac{PEN}{\eta} \tag{5.6}$$

The η is the efficiency of the battery converter charger and PENR is the actual energy needed to fully charge the battery which should be transferred from the grid to the battery. In References [7–9], the battery efficiency is assumed to be 88%, 90%, and 90%, respectively.

In other words, there supposed that a PHEV is a type of mid-size SUV with AER = 40 (mile). As Table 5.3 shown that a PHEV battery capacity must be equal to 2.15 kWh. If the vehicle mileage is 6.25 per day, then the SOC is calculated through the formula (5.4) 36% and the PEN should be equal to 73.9 kWh. According to Eq. (5.6), if assumed that the charging efficiency is 88%, PEN_R will be equal to 11.06.

---

[5]Percentage of Energy Needed

5.3.7 Required Time to Fully Charge

The required time to complete the vehicles battery charging depends on the amount of SOC battery at time t and the battery's efficiency. Due to the battery specification can be calculated required time through Eq. (5.7).

$$T - \text{charging} = \frac{\text{capacity} \times \text{PEN}_R}{\text{charging power}} \tag{5.7}$$

The capacity and the vehicle charging power is shown in the Table 5.3.

5.3.8 Technical Specifications of Electric Vehicle Battery

The private vehicles are often used to commuting miles between home, work, shopping, and entertainment. In Table 5.3 presented some of the vehicles' features, such as charging specifications and AER. However, it is possible to charging a vehicle in the home, work place, and large public places such as theatres or chain stores, but assumed that consumers charge their vehicles at home.

The energy demand of each PHEV depends on the specification of the vehicle's battery and the driving habits of each user in the vehicle charging process. Generally, the battery determines a maximum level "e_{max}" and a minimum level "e_{min}" of energy. The level of energy is a part of the full capacity of the battery at the initial charge mode and the vehicle connection to grid which depends on the vehicle mileage. Furthermore, the initial mode of vehicle charging can be provided as Eq. (5.8) which is the function of the vehicle mileage [3].

$$E(D) = \begin{cases} e_{max} & D = 0 \\ e_{max} - (D\xi) & 0 < D < d_{max} \\ e_{min} & D \geq d_{max} \end{cases} \tag{5.8}$$

d_{max} is the maximum distance that the PHEV can travel based on their electrical specification. D is the vehicle mileage before connecting to grid, ξ is the electrical performance of the vehicle (the energy consumption of the battery per unit of mileage).

In this research investigated Chevy volt with the specified features in the Table 5.3 [10].

Table 5.4 Voltage values and the authorized current for two level 1 and 2 of charge

Charging method	Nominal voltage power supply	Highest current rates
AC Level 1	V AC 120– single phase	A 16 – A 12
AC Level 2	V AC 240- 208- single phase	A 80 >

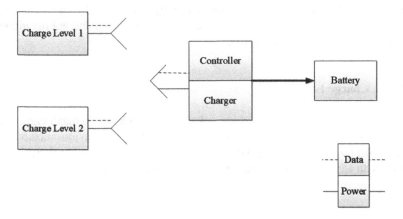

Fig. 5.3 Schematic of charge battery by two levels of charge

5.3.9 PHEV Charge Battery

The battery charging rate determines the speed of filling the battery that it depends on the charging place and the user. The various mode of the battery was discussed in the next sections. According to SAE J1772, in charging an electric vehicle connects to the power grid in order to energy transfer to charge the battery and for exchange the control information between vehicle and source. Generally, the charging process has three operations: one mechanical and the rest electrically. The electrical grid sends the electrical energy at different nominal voltage. The electric vehicle battery is a DC device that can operate based on the nominal voltage of the battery and the dis/charging rates at the variable voltages.

In the first electrical operation, AC is converted to DC, which is known as a rectifier. The second electrical operation, control and regulate of the source voltage in that the battery charge rate is compromised to the battery specification (voltage, capacity and other parameters). The mechanical operation is the physical coupler as it performs by the user.

Home charging system with charge rates shown in Table 5.4 which can be used in accordance Fig. 5.3.

5.3.9.1 Charge Level 1

The charge level 1 uses a standard single-phase-120 V, 12 A or 16 A. In fact, this is the lowest voltage level in residential areas in the United States. In the level 1 provides a less power (1.44 kW) to charge the battery so the batteries with a capacity greater than 10 kWh are charged at the over longer time which causes customers dissatisfaction. Therefore, Level 1 is just a basic voltage level and not the ultimate solution to the battery charging issue.

5.3.9.2 Charge Level 2

The charge level 2 uses a single-phase voltage 240–220 V with a maximum current 80 A and it is preferred due to the less charging time than the previous method. Due to the small size of the existing vehicles batteries, some chargers are limited to the level 2 with 15 A which the maximum charge power is 3.6 kW.

5.4 Related Study

In reference [3], the researchers presented a charging management program based on the availability of PHEVs in the distribution grid due to the renewable energy resources. The results show that the proposed program in addition to prevent of transformer overload, but also increase the dependence of subscribers on the use of renewable sources.

Paper [11], Modifying PEV parking behaviour due to exchange with solar energy sources. In this paper, a model has been proposed to illustrate the effects of different RERs on the profit and behaviour of PEV parking. Parking participation in different markets was modelled by considering energy and energy sources, as well as the uncertainty of PEV behaviour. Numerical results show that the parking behaviour changes in the participation in the electricity market by using different RERs [12].

In a study [13], the planning of a micro grid in the vision of EV was studied. The purpose is to decrease the electricity costs and optimize the charging of EV. The article [14] was aimed at intelligent grid optimization features that included renewable energy sources, which were carried out by a large number of EV parked in intelligent parking that has a V2G capability. In another study, the use of a Monte Carlo simulation to reduce the cost of consumers in managing the charging of PHEV is used to obtain charging patterns without vehicle management [15].

In the research [16], PHEV charging introduced as a way for Demand Side Management (DSM) problem and the researchers have proposed the methods to solve this problem: Multi-Agent Systems (MAS) and Quadratic Programming (QP). The second degree of planning has able to smooth the peak load optimization and appropriate charging vehicles, but not scalable; however, MAS solution is scalable adapted to complete and unanticipated information. The researchers are investigating the impact of the cars on micro grids with renewable resources [17].

5 Optimal Utilization of Solar Energy for Electric Vehicles Charging in... 143

The research [18] proposed that a home energy management system with the aim of reduce electric energy consumption, Peak to Average Ratio (PAR) and increase the consumers' welfare. The researchers used of an optimization Hybrid Bacterial Harmony algorithm (HBH) that consists of two algorithms: Bacteria Foraging Algorithm (BFA) and Harmony Algorithm Search (HAS). Their findings showed that it is possible to reduce the price of electric energy for consumers and increase the consumers' welfare by provide a coordinated program and in line with the generated electrical energy information of grid in order to use of appliances.

The researchers in article [19] proposed an Intelligent Residential Energy Management System (IREMS) for smart residential buildings that its benefits showed in a case study. The main object of IREMS was decrease in electricity, costs while less than the maximum power demand is limited to the various parameters such as the operation of residential loads and renewable energy resources. Moreover, the researchers used the battery as a suitable solution to reduce the loss of the power generation of renewable resources.

The research [20] provided an optimal overall framework for energy efficiency management and its components, which include a smart house with storage, the PHEV and photovoltaic. The aim of the researchers was to increase home profits provided that demand response and supply the energy required for PEV. in addition to that, the Battery Energy Storage System (BESS) is used in this model. In this project, according to different optimization times analysed the cost of the home battery energy storage system, the types and various modes of PEVs control, the parameters of the BESS and the electricity costs of in a systematic way. Their results showed that with the implementation of the CP (Convex programming) control program in V2H and H2V modes, the houses with a battery energy storage system will not purchase electricity at peak load hours.

In research [21], a home energy management system including photovoltaic panels and battery energy storage system was investigated. The researchers have examined the following items: (i) the effect of the electricity price mechanism using the Time-Of-Use pricing (TOU), the Real-Time Pricing (RTP), and the Stepwise Power Tariff (SPT); (ii) the impact of solar panels; and (iii) the variability of solar panels in different seasons. The management plan presented in this study was also programmable in the GAMS.

Research [22] examined the use of the Fuel Cell as an energy carrier to use in off grid mode. In this research, an Energy Management Algorithm (EMA) is used for Alternative Energy Sources (AES) in smart home systems. The fuzzy control logic is used to this purpose and is simulated in MATLAB.

5.5 Smart Charge

Normal charge is a mode where the vehicle is connected to the power to supply to fully charge the battery with the maximum power possible, or charge it until it is plugged in. The required time to charge a battery 10 kWh will be between 2 to 5 h based on the infrastructure of introduced charge. Since such vehicles don't need to much time to charge and most of them are charged at night so timing flexibility can be used to reduce the peak grid along with providing the vehicle owner's needs which is the smart charge. In fact, the smart charge not only is charge transfer to low-load times but is proper control charge per vehicle in order to provide the vehicle owner's needs (for example, prevent a high increase in peak distribution grid). There is a fact that a PHEV can be used of fossil fuels except the batteries so it has a high flexibility in the charging. Because it uses of fossil fuel when the vehicle is required and the battery is not fully charged.

Whenever the vehicle charging is carried out by the controller, the charge rate is unstable and the battery can be charged at various times and levels.

5.6 Determine the PHEV Specifications for Study

The PHEV specification is derived from reference information [23] and depends on the vehicle owner's usage which consists of mileage per day, start charging time, the number of appliances in each house and its type. The reference [24] assumes that PHEV owners charge the vehicle immediately upon arrival to home so the start charging time is the completion of driving and arrive to the destination. The data and arguments is needed to analyse this issue include the level of PHEV penetration, AER, daily load profile, home grown load. All data is needed to investigate the impact of PHEV on the distribution system in this model [5].

5.6.1 PHEV Charge Specification

5.6.1.1 Estimate the PHEVs Load Charge on a Large Scale

The charge and PHEVs capacity should be thoroughly investigated in order to coordination of the PHEVs large-scale. There is supposed that 1 day is evenly divided into T time and the number of considered PHEVs is N that each of them is independent. Thus, it is possible that the charged PHEV loads per day is equal to the total of each PHEV charge. As is shown in Eq. (5.9).

$$L_{PHEV}(t) = \sum_{i=1}^{N} P_i(t) \tag{5.9}$$

LPHEV(t) is the total charged loads (N) of PHEV at t time and Pi(t) is the power of each PHEV that is obtained through Eq. (5.10).

$$P_i^{CH}(t) = CR_i(t) \times P_i(t) \tag{5.10}$$

5 Optimal Utilization of Solar Energy for Electric Vehicles Charging in... 145

Pi(t) is the charging power that used to charge the PHEV, CRi(t) is the PHEV charging mode that can be zero or one (assumed that each PHEVs charge at their nominal power). The charging method is usually determined in accordance with the power grid of each country or region which includes predetermined voltage and current parameters. For instance, the authorized charging method in the Chinese government network has three main levels: level 1 (slow charge), level 2 (regular charge), and level 3 (fast charge). The period starts charging until its completion (t_{start} to t_{end}) is the charging process. There assumed that a PHEV requires to K times per day recharged, the PHEV arrival time to charging process is recognized by $t_{i,start}^k$ and the ending time of charge by $t_{i,end}^k$ which $K = 1, 2, 3, \ldots, k$ so charging mode of SOC at the $t_{i,start}^k$ and $t_{i,end}^k$ time are $SOC_{i,start}^k$ and $SOC_{i,end}^k$ respectively. The various charging modes including un/coordinated charge will be estimated in the following.

5.7 Random Charge

In the uncoordinated charge mode supposed that each PHEV enters the charging process at a desired time or immediately after entering the home so each PHEV charged based on $SOC_{i,start}^k$ of the battery $\left(t_{i,start}^k\right)$ and the plug of PHEV is disconnected of the power grid at their $SOC_{i,end}^k$ or at $\left(t_{i,end}^k\right)$ time. The time of PHEV dis/charging recognize by its nominal power. Each PHEV is connected to power grid for start charging process at $t_{i,start}^k \leq t \leq t_{i,end}^k$ time and disconnected at $t < t_{i,start}^k$, $t > t_{i,end}^k$ time.

$$CR_i(t) = \begin{cases} 1 & t_{i,start}^k \leq t \leq t_{i,end}^k \\ 0 & t < t_{i,start}^k, t > t_{i,end}^k \end{cases} \qquad K = 1, 2, \ldots, K \qquad (5.11)$$

The charging time shows as d_i^k that can be obtained by the Eq. (5.12):

$$d_i^k = \frac{\left(SOC_{i,end}^k - SOC_{i,start}^k\right) \times B_{i,c}}{P_i \times \eta_{ch}} \qquad k = 1, 2, \ldots, k, \quad i = 1, 2, \ldots, N \quad (5.12)$$

$B_{i,c}$ is the battery capacity of each PHEV and η_{ch} is the charge efficiency. The uncoordinated charging scenario has two modes: full charge which each PHEV is fully charged and the charging is constantly increasing.

$$SOC_{i,end}^k = SOC_{i,max}^k \qquad (5.13)$$

$SOC_{i,max}^k$ is for the time that the charge is completed. The ending time of charge is obtained through Eq. (5.14).

$$t_{i,end}^k = t_{i,start}^k + d_i^k \qquad (5.14)$$

In the continuous charge increase mode, the PHEV is charged at the predetermined constant time by the consumer. The start charging time should be predetermined and be available on the charge pattern chart. The charging process stops at the ending time of charge regardless of whether the PHEV is fully charged or not. The probability distribution functions of random variables in this field of the study can be derived from the official traffic reports.

5.8 Managed Charge

In order to investigate and control the status of PHEV loads, the coordinated charging scenario is provided that is based on the relationship between domestic load consumption and the amount of the renewable resources generation. The PHEVs is considered as the main factor of overloads and the penetration of renewable resources controller in this research. Hence, the coordinated charging scenario and time management of vehicle charging are presented to change in charging time to other times to the balance between load consumption and renewable resources generation. In this scenario, the priority of charging vehicles is possible in two ways that in accordance to their account and the load profile: (1) fast charge, (2) delayed charging.

The fast charge belongs to vehicles that when enter into the charging process the difference between renewable resources generation and load consumption is at the lowest level. Since, due to the fact that the urban power grid may be used to provide part of the required vehicles energy then the cost of consumers may be increased. This option is listed as the charging with priority time.

The second way delays charging time to hours with high renewable resources generation and to avoids of extra charge that can be lead to further financial savings; although, the vehicle will be charged in a longer time. This option is listed as the charging without priority time.

The choice of one of the two modes depends on whether consumer desire to save money through preventing of charging the vehicle at the times with a shortage of solar energy resources than consuming or trying to finish as soon as possible the process of charging battery and increasing the costs of energy.

The coordinated charging scenario relies on two basic subsystems:

- Home general controller (HGC)
- Transformer general controller (TGC)

Basically the HGC's performance is as follows: (i) increase demand energy of home loads to the demand energy of existing vehicle at a fixed time; (ii) checking the generated energy by renewable resources (solar), combination of two sources of generation, and estimate the amount of generated energy; (iii) calculating the differences in the total generation of the total consumption in which if the demanded energy exceeds of the amount of generation, the power grid will be compensated its shortages, otherwise their residue will be transferred to the power grid; (iv) after processing, all HGC's data will be presented as a report.

5 Optimal Utilization of Solar Energy for Electric Vehicles Charging in... 147

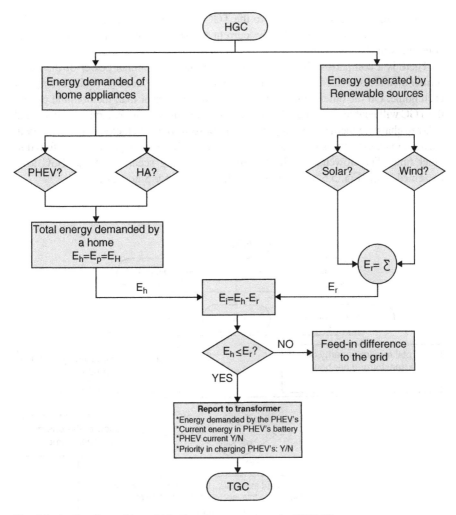

Fig. 5.4 the flowchart of the vehicle charging management by HGC [3]

This report contains some information on PHEV's terms:

- Whether connected to the grid or not?
- Demand energy by PHEV
- Energy in the PHEV battery
- Whether the PHEV has the priority for charging?

The HGC implements the coordinated charging scenario through this information.

The function of HGC is presented as flowchart in Fig. 5.4.

E_p represents the demanded energy by the vehicle and Eh is the energy of the home loads. E_r shows the total renewable resources generation (solar), Ed is the energy of vehicle discharge.

The HGC grid sends the report to the TGC, then TGC combines the report with the total available loads seen by the current transformer in order to determine the conditions of the residential grid.

There is a comparison between overall load and the nominal power of the transformer then if not exceed this capacity, the transformer provides the required energy of each home. On the other hands, if the overall load is greater than the nominal power, the TGC will enable coordinated charging algorithm in order to recognize vehicles and delay to charging them or discharging battery of some of the vehicles at a time interval. Charging vehicle causes overloads so this load should be managed and transferred to other time. The flowchart is the coordinated charging scenario on the TGC as it presented in the Fig. 5.5, (The target grid is supposed for 5 houses).

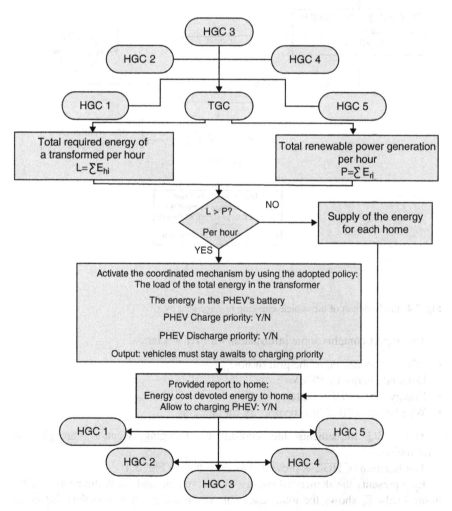

Fig. 5.5 the flowchart of the vehicle charging management by TGC [3]

5.9 The Coordinated V2G Mode

In managed V2G mode, the PHEV can be considered as a grid energy storage unit and a controller. Each PHEV has three modes: charge, deactivated and discharge that the optimal coordination PHEV dis/charging modes can be reduced the peak load as well as increases the penetration of renewable resources in the grid.

In any mode, the require to the charging and discharging PHEV should be considered.

The aim of this project is respond to consumer needs and minimize the total peak load through controlling $CR_i(t)$ and $DCR_i(t)$.

The subject has such limitations:

There is no charge and discharge in the outside of the dis/charging period. For example, dis/charging mode should be zero:

$$CR_i(t) = 0 \qquad t \notin \left[t^k_{i,start}, t^k_{i,end}\right] \qquad i = 1, 2, \ldots, N; \quad k = 1, 2, \ldots, K \quad (5.15)$$

$$DCR_i(t) = 0 \qquad t \notin \left[t^k_{i,start}, t^k_{i,end}\right] \qquad i = 1, 2, \ldots, N; \quad k = 1, 2, \ldots, K \quad (5.16)$$

Charge power should be as:

$$CR_i(t) \in \{0, 1\} \qquad t \in \left[t^k_{i,start}, t^k_{i,end}\right] \qquad i = 1, 2, \ldots, N; \quad k = 1, 2, \ldots, K \quad (5.17)$$

Charge power should be as:

$$DCR_i(t) \in \{0, -1\} \qquad t \in \left[t^k_{i,start}, t^k_{i,end}\right] \qquad (5.18)$$

The PHEV mode could be charging, discharging, or deactivated, as a result $CR_i(t)$ and $DCR_i(t)$ cannot be active at the same time.

$$|CR_i(t)| + |DCR_i(t)| \leq 1 \qquad t = 1, 2, \ldots, T; \quad i = 1, 2, \ldots, N \qquad (5.19)$$

And SOC should not be less than the specified value:

$$SOC_i(t) \geq B_l \qquad (5.20)$$

The B_l is the lower limit of the SOC which it is set to 20% in this program.

In the V2G mode, the car is not allowed to discharge when the battery capacity reached to this value.

5.10 Mathematical Model of the Issue

The aim of this project is to increase the renewable resources usage in the smart grid and to reduce the use of the conventional generation (CG). To this end, there should be obtained the power differences between the required energy for PHEVs charging, the amount of consumers' consumption, and above all the renewable energy resources. For doing this, the Eq. (20.3) is used.

So, the V2G process and discharge mode are considered to reduce the peak load in this equation.

$$E^j(t) = P_G^j(t) - \sum_{i=1}^{N} \left(P_i^j(t) \times (CR_i(t) + DCR_i(t)) \right)$$
$$- P_L^j(t) \qquad t = 1, 2, \ldots, T; \quad j = 1, 2, \ldots, m \tag{5.21}$$

The variables that used in the Eq. (5.21):

- The $E^j(t)$ is the energy differences between renewable resources generation and consumption at t time.
- "j" represent the considered house $\forall j \in (1, m)$
- The $P_G^j(t)$ is the total of the renewable resources generation, which equal the generation of the solar resources, it obtains through below equation:

$$P_G^j(t) = P_{pv} \tag{5.22}$$

- $P_i^j(t) \times (CR_i(t) + DCR_i(t))$ expresses the amount of consumption of PHEV charging power or V2G in discharging mode. Therefore, due to the Eqs. (5.17) and (5.18) equations, the value may be positive or negative.
- $P_L^j(t)$ is static load consumption by consumer that include all furniture, regardless vehicle energy.

According to Eq. (5.21), if $E^j(t)$ is obtained a positive value so the renewable resources more than of the consumer consumption which the load response is completed without any problems in the distribution grid. Otherwise, if $E^j(t)$ is obtained a negative value, the power should be transferred of conventional generation (CG) into the house in order to supply required energy of consumer which the received energy from CG is obtained by Eq. (5.23):

$$CG^j(t) = \begin{cases} P_L^j(t) + \sum_{i=1}^{N} \left(P_i^j(t) \times (CR_i(t) + DCR_i(t)) \right) - P_G^j(t) & E^j(t) < 0 \\ 0 & E^j(t) \geq 0 \end{cases} \tag{5.23}$$

The difference in consumer demand energy is equal to the renewable resources.

5.10.1 Optimization of the Issue

The minimizing the received power from conventional generation is the main factor of the objective function which contains such limitations:

$$\text{Min} \sum_{t=1}^{T} \sum_{j=1}^{m} CG_{t,j} \qquad (5.24)$$

1. The value of $E^j(t)$ must be smaller than zero, otherwise $CG_{t,j}$ is equal to zero.

$$E^j(t) < 0 \qquad (5.25)$$

2. In order to completing load response, the total resources generation that consist of renewable resources and conventional generation should be higher than or equal to the total consumer load.

$$P_G^j(t) + CG_{t,j} - \sum_{i=1}^{N} \left(P_i^j(t) \times (CR_i(t) + DCR_i(t)) \right) - P_L^j(t) \geq 0 \qquad (5.26)$$

3. The received energy of the power grid has the following limitation. BCG[6] is a binary number that indicating whether the energy is received of the conventional generation or not $CG \in \{0, 1\}$.

$$CG_{min}.BCG \leq CG \leq CG_{max}.BCG \qquad (5.27)$$

4. The vehicle capacity i at t time is equal to the vehicle capacity i at $(t - 1)$ time and the amount of received power for charge or discharge [6].

$$SOC_{i,t} = SOC_{i,t-1} + (P_{i,t} \times (CR_i(t) + DCR_i(t))) \qquad (5.28)$$

5. The limitation of the charge and discharge power are determined by the Eq. (5.29) [6].

$$DCR_{i,t}.P_{i,t_{min}} \leq P_{i,t} \leq CR_{i,t}.P_{i,t_{max}} \qquad (5.29)$$

6. Finally, the battery capacity has some limitations [6].

---

[6]Binary Conventional Generation

$$20\% \leq SOC_{i,t} \leq 100\% \tag{5.30}$$

The $SOC_{i_{max}}$ is considered 100% in the Eq. (5.30) and there is not allowed to discharge the vehicle with the capacity of less than 20%. This limitation is considered to increase battery life.

5.11 Multiple Results of Exploitation of the Studied Grid

The main purposes of this research chapter are discovering a suitable, automatically, and optimized way to managing PHEV charging, for minimize received energy from CG, and to increase energy consumption from renewable resources by using of PSO algorithm. In addition to Monte Carlo simulation used to obtain random variables and probability function. The result of PHEV charging program is generally seen on the grid for every home. As maintained above regarding smart homes, the houses equipped with the planned smart charging devices for their PHEVs and consumers declare the allowed time for their PHEVs charging. The allowed charging time for each of PHEV is assumed which is equal to time of return to home after the last travel and the start of first trip on the next day.

In the previous section provided an objective function which related to the operation of the proposed network, and all components were modelled separately. Then, the uncertainty mileage by cars during the day, the battery capacity of the vehicle at the first time, and also the absence of cars in the house is described. It is necessary to achieve optimum operation conditions of this grid by use of suitable tool. For do this, PSO algorithm is selected to optimize objective function. This algorithm optimized scenarios generated scenarios by Monte Carlo method.

It is simulated in the MATLAB program. A grid of solar and wind units with PHEVs as flexible loads also a transformer with the identified capacity to exchange houses energy with upstream grid is considered. Figure 5.1 Indicates sample grid of this work which consists of five smart houses with equipment for the renewable generation [3]. Next, is considered the effect of PHEVs charging in three modes: unmanaged charge, managed charge, and managed charge V2G, then the study will analyse the results of the simulation.

In order to managing energy is provided an initial scenario as well as used of Monte Carlo simulation to find a consistent driving pattern with current situation. For doing this, it is assumed that each house has a PHEV. The houses 1 and 2 use of their vehicle for commuting to workplace while the houses 3, 4, and 5 use of public transport vehicles and do not have a timing plan to use of their vehicles.

5 Optimal Utilization of Solar Energy for Electric Vehicles Charging in... 153

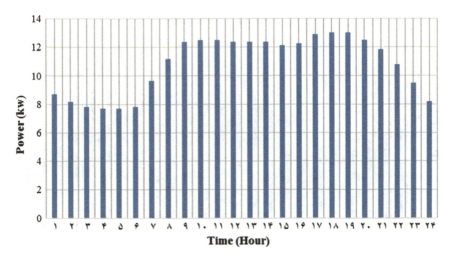

Fig. 5.6 The household consumption curve

5.11.1 Charging without Vehicle Management and V2G Capability

A distribution grid consists of five house with specified features. The loads of the houses are divided into two categories: static loads (such as lighting, cooling and heating systems; household appliances like washing machines, dishwashers, and electrical appliances) and flexible loads (such as electric vehicle battery). The Fig. 5.6 shows the static loads per house [25].

As it shown in the Fig. 5.6, the static load curve of the distribution grid represents the all of five houses, regardless of the loads of charging vehicle. The investigated grid is powered by a transformer that has the power 22.75 kW and is a ranged between 25 to 75 kVA per phase. (it is a standard power to supply between 4 to 7 houses) [3].

In Fig. 5.7 the daily generation of renewable resources with household consumption regardless of the load of charging PHEV were compared. In charging mode without management, vehicle owners are connected to the grid regardless of the grid status and the amount of the generation of renewable resources at the return time to home. This charging method has some problems: charging a large number of vehicle at the same time and creating a peak load in the distribution grid, which will be increased the load and consumers' cost. The lack of the amount of available renewable resources leads to a loss of electricity generated resources and increasing the conventional generation usage. The load curve without vehicle charging management shown in Fig. 5.8.

In Fig. 5.8, the amount of the generation of renewable resources with the curve of the household load were compared by take into account to PHEV charging. As it can be seen in the figure, there is no charging at the time with the maximum generation of

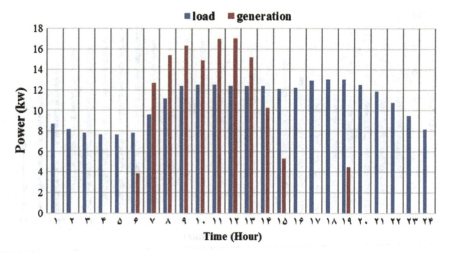

Fig. 5.7 The ratio of the generation of the renewable resources to static load consumption in the distribution grid

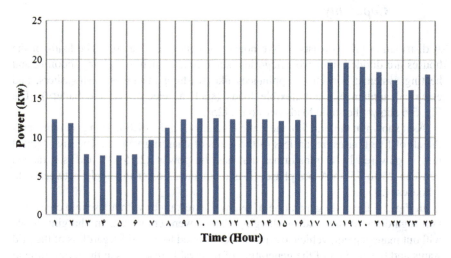

Fig. 5.8 The curve of household load without charging management

the renewable resources that increases demand energy of power grid and increases consumers' cost so the importance of PHEV charging management is revealed.

In Fig. 5.9, the use of photovoltaic panels and high-temperatures of the sun at 8–18 PM, the renewable resources generation is at its highest level which decrease at night. It is not taken into consideration in unmanaged electric vehicle charging so the vehicles enter to the charging process at the minimum level of renewable resources generation, which increases the demand energy from the power grid.

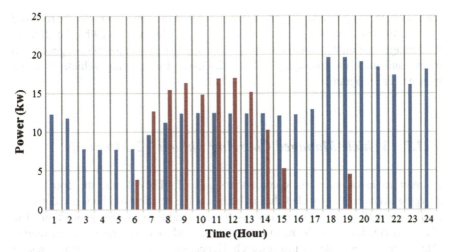

Fig. 5.9 The curve of the comparison of the generation of renewable resources and unmanaged loads

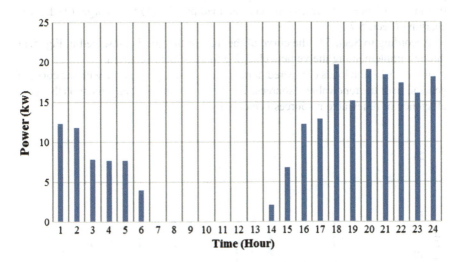

Fig. 5.10 The amount of received energy of the power grid in the unmanaged charge scenario

The Fig. 5.10 presented the house demand energy of power grid. There is indicated that the vehicles enter to charging process at late in the day with low level of renewable resources generation which increases peak load in the distribution grid and the costs. It's assumed that when two vehicle go back to home will be connected to the grid, but in order to complete the capacity of the battery, some of the vehicles enter to charging process at the end of day. For this reason, the rate of the renewable energy permeability drastically reduces at this time moreover the received energy from conventional generation increase to 80 KWh during 24 h. The effect of the management on the amount of received energy from the conventional generation before and after the managing will be discussed by applying vehicle charging management program.

5.11.2 Vehicle Managed Charging without V2G Capacity

In order to investigate and control the status of PHEV loads, the coordinated charging scenario is provided that is based on the relationship between domestic load consumption and the amount of the renewable resources generation. The PHEVs is considered as the main factor of overloads and the penetration of renewable resources controller in this research. Hence, the coordinated charging scenario and time management of vehicle charging are presented to change in charging time to other times to the balance between load consumption and renewable resources generation. In this scenario, the priority of charging vehicles is possible in two ways that in accordance to their account and the load profile: (1) fast charge, (2) delayed charging (Sect. 5.7).

According to Sect. 5.9, the curve of the load of the grid is presented in Fig. 5.11 through implementing the vehicle charging management program.

In Fig. 5.12, the loads compered in random and managed charge with the curve of generation of the renewable resources. The effect of charge management on the more efficiency of renewable resources is recognized.

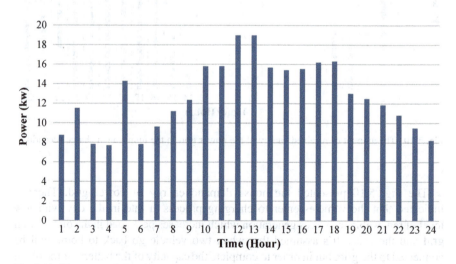

Fig. 5.11 The curve of the grid load with vehicle charging management

5 Optimal Utilization of Solar Energy for Electric Vehicles Charging in... 157

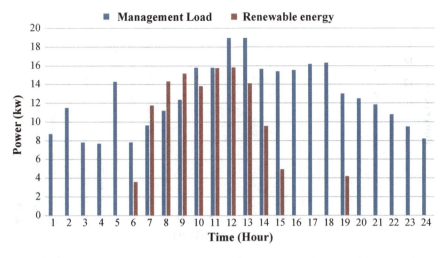

Fig. 5.12 The curve of the manage load without V2G capacity and the curve of the generation of the renewable resources

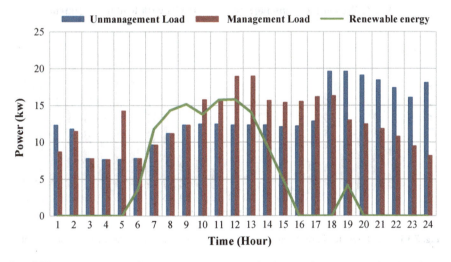

Fig. 5.13 The comparison of generation with domestic load curves in two managed and randomized modes

According to Fig. 5.13, the charging process of some of vehicles that are not in a priority of charge is transferred to 9 and 18 PM. through implementing charging management. As well as due to the high level of the renewable resources generation at these hours, the demanded energy from conventional generation is reduced to 58 kWh which the majority of this energy was cheaper-rate electricity at 3–5 A.M. Certainly, charging management program diminishes the consumers' charges. Figure 5.14 shown received energy from the power grid.

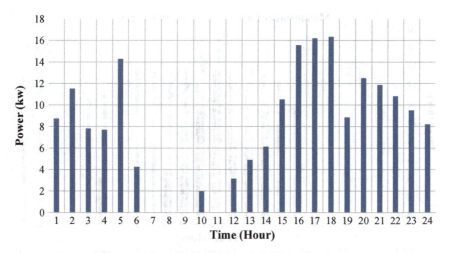

Fig. 5.14 The amount of received energy of power grid in managed and without V2G capacity modes

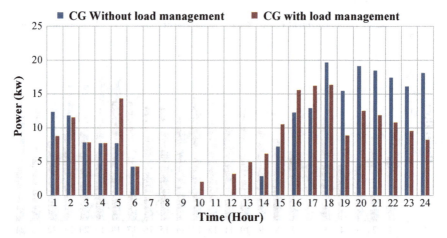

Fig. 5.15 The difference in received energy from the grid before and after charge management

Figure 5.15 is shown the difference in received energy from the power grid with vehicle no/charge management.

5.11.3 Managed Charge with V2G Capacity

In V2G mode, the PHEV may be considered as a grid energy storage system. Every PHEV has three modes: charge, deactivated and discharge that the optimal PHEV dis/charging modes can be reduced the peak load as well as increases the renewable

5 Optimal Utilization of Solar Energy for Electric Vehicles Charging in... 159

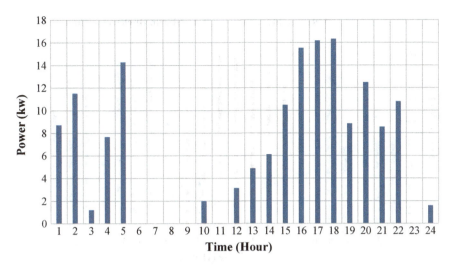

Fig. 5.16 The grid load curve with using V2G

resources penetration in the grid. The main purpose of current project is to minimize the received energy from conventional generation respond in accordance with the consumer needs. The curve of the load grid by use of this feature is presented in Fig. 5.16.

There is compared the new load curve with the amount of renewable generation. After adding V2G to the grid, some of the vehicle's owner are desire to participate in the discharging process at the times with the lower renewable resources generation which increases the penetration of the renewable resources in load response. Figure 5.17 is shown the functionality of the program through comparing generation, load, and V2G capability.

There can be discharged the vehicles depending on the time constraints and the consumers' welfare in use of vehicles which not only reduces received energy from conventional generation, but also increases financial saving for costumers. Considering the vehicle power transmission to grid, the conventional energy generation reduce to 29 kWh. Figure 5.18 shown the conventional generation ratio in two managed and unmanaged modes with V2G capability.

In order to represent the effect of charging management and V2G capability in grid on randomized charge, the graph of the received energy of grid is divided into three modes: (1) unmanaged vehicle charging, (2) managed vehicles charging without V2G capacity, (3) managed cars charging by consideration of V2G capacity with other and the generation of renewable resources curves (Fig. 5.19).

The correlation is used to indicate accuracy of the applied management. Correlation is the statistical indicator which describes the relationship between two dependent variables in other word, it is an statistical indicator that determines various degree of relationship between two dependent variables in a fixed and limited scale.

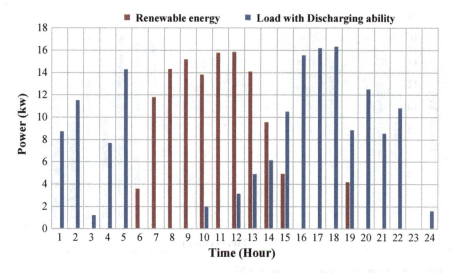

Fig. 5.17 The managed vehicle load with V2G capability and the generation of the renewable resources curve

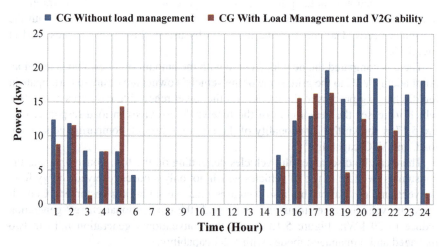

Fig. 5.18 The difference of received energy of grid before and after load management

The Eq. (5.31) indicates correlation between two set of statistic data x and y. The correlation between renewable generation and all household loads in both randomized and managed modes increased from 0.29 to 0.74.

5 Optimal Utilization of Solar Energy for Electric Vehicles Charging in...

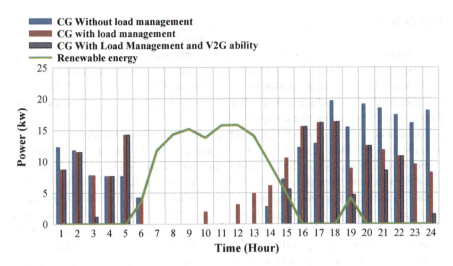

Fig. 5.19 The comparison of the received energy of conventional generation in vehicle charge management modes

$$r_{x,y} = \frac{\operatorname{cov}(x,y)}{\sigma_x \cdot \sigma_y} = \frac{\frac{\sum x_i y_i}{n} - \left(\frac{\sum x_i}{n}\right)\left(\frac{\sum y_i}{n}\right)}{\sqrt{\left(\frac{\sum x_i^2}{n} - \left(\frac{\sum x_i}{n}\right)^2\right)\left(\frac{\sum y_i^2}{n} - \left(\frac{\sum y_i}{n}\right)^2\right)}} \qquad (5.31)$$

The increased correlation confirms the performance of the vehicle charging management program.

$$\text{Corr (Renewable Energy, Load Unmanaged)} = 0.29$$

$$\text{Corr (Renewable Energy, Load managed)} = 0.6821$$

$$\text{Corr (Renewable Energy, Load managed with V2G)} = 0.7474$$

5.12 Conclusion

Due to optimize the use of the renewable energy resources, a model to more efficient of the residential grid usage through the managing PHEV dis/charging during 24 h was provided. The grid consists some smart houses which has photovoltaic panels for generate needed home power. The loads are divided into static loads and flexible loads. As well as, the houses equipped with the smart charge planning system for the

PHEVs that has double-sided energy exchange capabilities with distribution grid. It worth mentioning that a 25 kV transformer powered to the homes.

In the modelling of renewable resources generation, meteorological information was extracted over a period of 1 year based on the time of information used of the static loads of residential consumers' information. On the other hand, the used specifications of the vehicles as the flexible loads in this modelling are extracted from the relevant automotive companies.

The considerable points in this research are the uncertainty on vehicle mileage per day, the use of vehicle at the different time, and the initial capacity of the battery.

For this purpose, used of Monte Carlo simulation algorithm to cover a variety of probable situations then provided a probabilistic variable for each of the decision variables. Additionally, particle swarm optimization is used to optimize grid utilization in this research. In the depend environment, the performance of the issue with the high number of decision-making variables was acceptable and in a relatively short time reaches to the optimal point. The problem with the high number of decision variables and function well in a relatively short time is relative to the optimal point.

- In order to optimize the grid utilization, various scenarios and plans have simulated and is examined that the model presented in this research.
- Carrying out a charge and discharge management plan to cover renewable energy changes that can be reduced consumer costs.
- Decrease the costumers' cost and reduce their dependency to conventional generation through administrating the PHEV discharging and charging management.
- Due to the dependence of this program on people's driving patterns, the use of a definitive planning method is not appropriate and the possible method should be used.
- Drastically decrease the costumers' dependency to generated energy of fossil fuels which is essential for the environmental.
- Increase the grid security through preventing the creation of peak-loads in the grid and transferring the vehicles charging time which prevents of the overload on a distribution transformers and power grid outage.

Acknowledgments Special thanks to department of research and technology of the Islamic Azad University, Damavand branch for their valuable support throughout all the phases of this research.

Appendix

Indexes	
N	Number of solar modules
J	Considered house
I	Considered car
t	Time interval (24 h)
$CR_i(t)$	Charging mode
$DCR_i(t)$	Discharging mode
$BR_i(t)$	Binary of charge sum discharge
Constants	
P_{pv}	Solar generation (kW/m^2)
P_{mpv}	*max* power of solar module (kW/m^2)
G_t	solar radiation intensity (kW/m^2)
SOC	State of charge (%)
λ	maximum distance mileage in electric mode (mile)
d	total distance mileage by the car (mile)
$P_G^j(t)$	total generation of renewable resources (kW)
$P_i^j(t)$	Power consumption of PHEV charging or discharging (kW)
$P_L^j(t)$	static load (kW)
$CG^j(t)$	conventional generation (kW)

References

1. M.G. Kanabar, I. Voloh, A review of smart grid standards for protection, control and monitoring applications, in *Annual Conference for Protective Relay Engineerings* (2012), pp. 281–289
2. S. Kato, T. Naito, H. Kohno, H. Kanawa, T. Shoji, Computer-based distribution automation. IEEE Trans. Power Del. **1**, 265–271 (1986)
3. R.M. Oviedo, Z. Fan, S. Gormus, P. Kulkarni, A residential PHEV load coordination mechanism with renewable sources in smart grids. Electr. Power Energy Syst. **55**, 511–521 (2014)
4. M.K. Meyer, K. Schneider, R. Pratt, Impacts assessment of plug-in hybrid vehicles on electric utilities and regional U.S. power grids part 1: Technical analysis, PNNL Rep (Nov 2007)
5. S. Shafiees, M.F. Firuzabad, Investigating the impacts of plug-in hybrid electric vehicles on power distribution systems. IEEE Trans. Smart Grid **4**(3), 1351–1360 (2013)
6. Z. Darabi, M. Ferdowsi, Aggregated impact of plug-in hybrid electric vehicles on electricity demand profile. IEEE Trans. Sustain. Energy **2**(4), 501–508 (2014)
7. K. Celement, E. Haesaen, J. Driesen, The impact of charging plug-in hybrid electric vehicles on a residential distribution grid. IEEE Trans. Power Syst. **25**(1), 371–380 (2010)
8. J. Taylor, A. Maitra, M. Alexander, D. Brooks, M. Duvall, Evaluation of the impact of plug-in electric vehicle loading on distribution system operations, in *Power & Energy Society General Meeting*, Calgary (July 2009)
9. W. Su, M.Y. Chow, Performance evaluation of an EDA-based large-scale plug-in hybrid electric vehicle charging algorithm. IEEE Trans. Smart Grid **3**(1), 308–315 (2012)
10. Chevy Volt, http://www.chevrolet.com/volt

11. F.A.S. Gil, M. Shafie-Khah, A.W. Bizuayehu, J.P.S. Catalão, Impacts of different renewable energy resources on optimal behavior of plug-in electric vehicle parking lots in energy and ancillary services markets, in *IEEE Power Tech Conference*, Eindhoven (July 2015)
12. M. Shafie-khah, P. Siano, A stochastic home energy management system considering satisfaction cost and response fatigue. IEEE Trans. Industr. Inform. **14**(2), 629–638 (2018)
13. O. Sundstm, C. Binding, Flexible charging optimization for electric vehicles considering distribution grid constraints. IEEE Trans. Smart Grid **3**, 26–37 (2012)
14. H. Morais, T. Sousa, Z. Vale, P. Faria, Evaluation of the electric vehicle impact in the power demand curve in a smart grid environment. Energy Convers Manage **82**, 268–282 (2014)
15. M. Rostami, A. Kavousi-Fard, T. Niknam, Expected cost minimization of smart grids withPlug-in hybrid electric vehicles using optimal distribution feeder reconfiguration. IEEE Trans. Industr. Inform. **11**, 388–397 (2015)
16. Y. Mou, H. Xing, Z. Lin, M. Fu, Decentralized optimal demand-side management for PHEV charging in a smart grid. IEEE Trans. Smart Grid **6**, 726–736 (2015)
17. M.H. Moradi, M. Abedini, S.M. Hosseinian, Improving operation constraints of microgrid using PHEVs and renewable energy sources. Renew. Energy **83**, 543–552 (2015)
18. H. Rahim, A. Khalid, N. Javaid, M. Alhussein, K. Auranzeb, Z. Alikhan, Efficient smart buildings using coordination among appliances generating large data. IEEE Access **6**, 34670–34690 (2018)
19. S.L. Arun, M.P. Selvan, Intelligent residential energy management system for dynamic demand response in smart buildings. IEEE Syst. J. **12**, 1329–1340 (2018)
20. X. Wu, X. Hu, Y. Teng, S. Qian, R. Cheng, Optimal integration of a hybrid solar-battery power source into smart home Nano grid with plug-in electric vehicle. J. Power Sources **363**, 277–283 (2017)
21. L. Zhou, Y. Zhang, X. Lin, C. Li, Z. Cai, P. Yang, Optimal sizing of PV and BESS for a smart household considering different price mechanisms. IEEE Access **6**, 41050–41059 (2018)
22. F.K. Arabul, A.Y. Arabul, C.F. Kumru, A.R. Boynuegri, Providing energy management of a fuel cell, battery, wind turbine, solar panel hybrid off grid smart home system. Int. J. Hydrogen Energy **42**, 26906–26913 (2017)
23. National household travel survey [Online]. http://nhts.ornl.gov
24. C. Camus, C.M. SilvaS, T.L. Farias, J. Esteves, Impact of plug-in hybrid electric vehicles in the Portuguese electric utility system, in *International Conference on Power Engineering, Energy and Electrical Drives*, (2009)
25. Reliability Test System Task force of the IEEE Subcommittee on the Application of Probability Methods, IEEE reliability test system. IEEE Trans. Power Syst. **14**(3), 2047–2054 (1999)

Chapter 6
Integration of Electric Vehicles and Wind Energy in Power Systems

Morteza Shafiekhani and Ali Zangeneh

6.1 Introduction

The most important task of EVs is to meet transportation requirements of their owners. Besides, these vehicles can be used as energy storage units. By optimal charge and discharge of the EVs, energy can be stored in their batteries in some hours (usually off-peak hours) not only to be used as the prime mover of the vehicles but also to discharge energy to the grid in the necessary conditions (usually peak hours).

Many changes will occur in the load profile of distribution systems with the increasing number of the EVs. Various charge strategies and their impacts at load profile of Australian electricity grid are assessed in [1]. The obtained results denote that the appropriate management of EVs charge increases the distribution network capacity. There is no need to install new power plants to supply the additional power for EVs charging if the EVs are coordinately charged during off-peak hours. Similar results are presented in [2] that shows the load curve changes with the variation of the charge pattern.

Tarroja et al. define various scenarios to investigate the effects of EVs charge strategy on the load profile [3]. They indicate that how non-intelligent charge of EVs can increase the power demand in peak hours and intensify the distribution network stress. On the contrary, the coordinated charge of EVs smooth the load curve by filling the valleys. Authors of [4] address the management of EVs in the presence of renewable-solar and wind-powered units, and have shown that by applying the V2G feature of the vehicles, the operation performance of the distribution network has

---

M. Shafiekhani
Department of Electrical Engineering, Faculty of Engineering, Pardis Branch, Islamic Azad University, Pardis, Tehran, Iran

A. Zangeneh (✉)
Electrical Engineering Department, Shahid Rajaee Teacher Training University, Tehran, Iran
e-mail: a.zangeneh@sru.ac.ir

© Springer Nature Switzerland AG 2020
A. Ahmadian et al. (eds.), *Electric Vehicles in Energy Systems*,
https://doi.org/10.1007/978-3-030-34448-1_6

been improved in the presence of renewable generation uncertainties. In [5], the use of renewable energy sources in the presence of electric vehicles has been explored. The results indicate that although the use of electric vehicles as a storage unit requires many costs, there are a lot of benefits to supporting renewable energy in the power system.

By increasing penetration level of EVs in power systems, their effects cannot be ignored. As an example, replacing a quarter of US vehicles with the electric ones would increase the power demand beyond the capacity of existing US power plants [6].

The impact of the transportation system in the presence of electric vehicles has been investigated using the data available in the transportation system. It has been shown that V2G technology has led to an increase in the flexibility of the power system and more optimal utilization of the wind unit [7]. In [8], four different models of charging EVs have been investigated. The obtained results indicate that the smart charging and discharging of EVs lead to increased capacity and reduced costs.

Kiviluoma and Meibom [9] assess the impact of smart and non-smart charging of EVs in the future power system using a model that optimizes the scheduling of EVs participation. The results indicate that the smart charging of EVs is economical in comparison with the non-smart charging.

Furthermore, the coordinated charge and discharge of EVs have some technical benefits such as loss reduction, peak shaving, frequency regulation and load following. The impact of electric vehicles on investment and distribution network losses are investigated in [10]. The results of the study on two different urban and rural areas indicate that the charging of EVs in peak-hours will increase the expansion investment costs of the distribution network by 19%, while charging EVs in other times reduce costs by 70%. In the worst-case scenario, charging EVs increased the loss of distribution network by 40%. It has been shown in [2, 11, 12] that the use of EVs batteries to save energy and deliver it back to the network at some hours along with renewable energy has resulted in more flexible operation of these units. As shown in Fig. 6.1, EVs can be used as consumer (in charge mode) alongside other consumers. On the other hand, they can generate electric power (in discharge mode) and be considered in the power producer side with renewables and non-renewables producers.

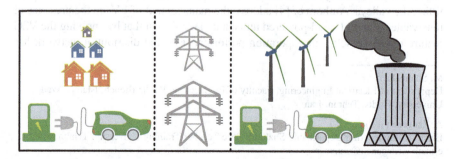

Fig. 6.1 Integration of EVs and wind units

6.2 Problem Formulation

Stochastic programming of dispatchable units with the goal of maximizing profits in the presence of wind and electric vehicles is in the form of the following equations. Equation (6.1) represents the final profit of the problem, which is obtained from the revenue minus the cost. Equation (6.2) represents the revenue obtained by dispatchable units, wind and electric vehicles. Equation (6.3) represents the operation cost, which includes the fuel cost of units, start-up and shut-down cost. The operating cost function of units is a nonlinear expression, so the linear equivalent of that is described as follows:

$$\text{Max, } Profit = \sum_{t}\sum_{w} \pi_{t,w}(Revenue(t,w) - Cost(t,w)) \tag{6.1}$$

$$Revenue(t,w) = \sum_{g} P_{g,t,w}\lambda_t + P_{t,w}^{Wind}\lambda + Revenue_{t,w}^{EV} \tag{6.2}$$

$$Cost(t,w) = \sum_{g} FC_{g,t,w} + SUC_{g,t,w} + SDC_{g,t,w} \tag{6.3}$$

6.2.1 Linear Equivalent of the Cost Function

For this purpose, the cost function, which is a second-order expression, is approximated in a number of lines as shown in Fig. 6.2. The equations related to the linearization of the cost function are Eqs. (6.4, 6.5, 6.6, 6.7, 6.8, 6.9, 6.10, 6.11 and 6.12). The higher the number of lines, the better approximation will be. z represents

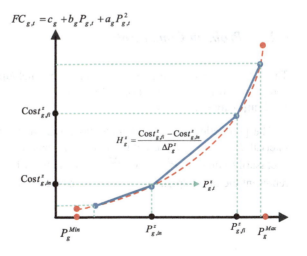

Fig. 6.2 Cost function linearization

the number of cost function intervals that indicate the number of these lines, too. Each interval has an initial power and a final power, which is represented by $P^z_{g,in}$ and $P^z_{g,fi}$ respectively. $P^z_{g,t}$ represents the generated power in each of these intervals. The expression ΔP^z_g represents the ratio of the power variations of each unit to the number of desired intervals. H^z_g represents the ratio of cost variations to power variations in each interval [13].

$$0 \leq P^z_{g,t,w} \leq \Delta P^z_g \alpha_{g,t,w}, \forall z = 1 : n \tag{6.4}$$

$$\Delta P^z_g = \frac{P^{Max}_g - P^{Min}_g}{n} \tag{6.5}$$

$$P^z_{g,in} = (z - 1)\Delta P^z_g + P^{Min}_g \tag{6.6}$$

$$P^z_{g,fi} = \Delta P^z_g + P^k_{g,in} \tag{6.7}$$

$$P_{g,t,w} = P^{Min}_g \alpha_{g,t,w} + \sum_z P^z_{g,t,w} \tag{6.8}$$

$$Cost^z_{g,in} = a_g \left(P^z_{g,in} \right)^2 + b_g P^z_{g,in} + c_g \tag{6.9}$$

$$Cost^z_{g,fi} = a_g \left(P^z_{g,fi} \right)^2 + b_g P^z_{g,fi} + c_g \tag{6.10}$$

$$H^z_g = \frac{Cost^z_{g,fi} - Cost^z_{g,in}}{\Delta P^z_g} \tag{6.11}$$

$$FC_{g,t,w} = a_g \left(P^{Min}_g \right)^2 + b_g P^{Min}_g + c_g \alpha_{g,t,w} + \sum_z H^z_g P^z_{g,t,w} \tag{6.12}$$

6.2.2 Problem Constraints

There are various constraints in this problem, including technical and economic constraints, which are described as follows:

- **Ramp rate constraints:**

The production capacity of units has a minimum and a maximum at any time interval. $\underline{P}_{g,t}$ and $\overline{P}_{g,t}$ represent minimum and maximum, respectively, which are not necessarily equal to P^{Min}_g and P^{Max}_g. Equations (6.13, 6.14, 6.15, 6.16 and 6.17) represent the ramp-up and ramp-down constraints of dispatchable units.

$$\underline{P}_{g,t} \leq P_{g,t,w} \leq \overline{P}_{g,t} \tag{6.13}$$

$$\overline{P}_{g,t} \leq P_g^{Max}\left[\alpha_{g,t,w} - \gamma_{g,t+1,w}\right] + SD_g\gamma_{g,t+1,w} \tag{6.14}$$

$$\overline{P}_{g,t} \leq P_{g,t-1,w} + RU_g\alpha_{g,t-1,w} + SU_g\beta_{g,t,w} \tag{6.15}$$

$$\underline{P}_{g,t} \geq P_g^{Min}\alpha_{g,t,w} \tag{6.16}$$

$$\underline{P}_{g,t} \geq P_{g,t-1,w} - RD_g\alpha_{g,t,w} - SD_g\gamma_{g,t,w} \tag{6.17}$$

In the above equations, $\alpha_{g,\,t,\,w}$ shows the on/off status of the unit g at period t and scenario w. $\beta_{g,\,t,\,w}$ and $\gamma_{g,\,t,\,w}$ also indicate start-up/shut-down status of the unit g at period t and scenario w.

- **Minimum up and down time constraints:**

$$\beta_{g,t,w} - \gamma_{g,t,w} = \alpha_{g,t,w} - \alpha_{g,t-1,w} \tag{6.18}$$

$$\beta_{g,t,w} + \gamma_{g,t,w} \leq 1 \tag{6.19}$$

$$\beta_{g,t,w}, \gamma_{g,t,w}, \alpha_{g,t,w} \in \{0,1\} \tag{6.20}$$

The minimum up time (MUT_g) and the minimum down time (MDN_g) of unit g are described as Eqs. (6.21) and (6.22) [14].

$$\sum_{t=1}^{MUP} \alpha_{g,t+1,w} - 1 \geq MUT_g \forall \beta_{g,t,w} = 1 \tag{6.21}$$

$$\sum_{t=1}^{MDN} 1 - \alpha_{g,t+1,w} \geq MDN_g \forall \gamma_{g,t,w} = 1 \tag{6.22}$$

- **Start-up and shut-down cost:**

The start-up and shut-down cost of the unit g at period t and scenario w are Eqs. (6.23) and (6.24), respectively. The values of CS_g and SD_g are given in Table 6.1.

$$SUC_{g,t,w} = CS_g\beta_{g,t,w} \tag{6.23}$$

$$SDC_{g,t,w} = SD_g\gamma_{g,t,w} \tag{6.24}$$

- **Power Balance constraint:**

The total power produced by the dispatchable units, the wind unit and the discharge of EVs should be equal or greater than sum of the charge of EVs and loads per hour. This Equation is represented as Eq. (6.25).

Table 6.1 Economic and technical information of dispatchable generation units

Unit	a_g ($\$/MW^2$)	b_g ($\$/MW$)	c_g ($\$$)	P_g^{Min} (MW)	P_g^{Max} (MW)	RU_g (MWh^{-1})	RD_g (MWh^{-1})	MUT_g (h)	MDT_g (h)	SD_g ($\$$)	u_0	CS_g ($\$$)
g1	0.0148	12.1	82	8	20	4	4	3	2	9	0	11
g2	0.0289	12.6	49	12	32	6.4	6.4	4	2	13	1	14
g3	0.0135	13.2	100	5	15	3	3	3	2	7	1	8
g4	0.0127	13.9	105	25	52	10.4	10.4	5	3	24	1	25
g5	0.0261	13.5	72	8	28	5.6	5.6	4	2	11	1	13
g6	0.0212	15.4	29	5	15	3	3	3	2	6	0	8
g7	0.0382	14	32	3	12	2.4	2.4	3	2	5	0	6
g8	0.0393	13.5	40	3	11	2.2	2.2	3	2	4.5	0	5.5
g9	0.0396	15	25	2	8	1.6	1.6	0	0	3.5	0	4.5
g10	0.0510	14.3	15	2	6	1.2	1.2	0	0	3	0	4

$$\sum_g P_{g,t,w} + P_{t,w}^{Wind} + \sum_{ev} P_{ev,t,w}^{dch} \eta_{ev} - \sum_{ev} \frac{P_{ev,t,w}^{ch}}{\eta_{ev}} \geq P_t^{Dem} \qquad (6.25)$$

6.2.3 Modeling of Wind Unit

The Weibull probability distribution function [15] is used to model wind speed as shown in Eq. (6.26). In this Equation, k is the shape parameter, C is scale of the shape and v is wind speed. The information about related parameters and wind speed are provided in [16].

$$f_{wind}(v) = \frac{k}{c}\left(\frac{v}{c}\right)^{k-1} \exp\left(\left(-\frac{v}{c}\right)^k\right), 0 < v < \infty \qquad (6.26)$$

Based on Weibull function 10 scenarios are produced for hourly wind speeds and accordingly, hourly wind outputs are determined based on Eq. (6.27).

$$P_{v_{aw}}^{Wind} = \begin{cases} 0 & 0 \leq v_{aw} \leq v_{ci} \\ P_{rated}\dfrac{v_{aw} - v_{ci}}{v_r - v_{ci}} & v_{ci} \leq v_{aw} \leq v_r \\ P_{rated} & v_r \leq v_{aw} \leq v_{co} \\ 0 & v_{co} \leq v_{aw} \end{cases} \qquad (6.27)$$

6.2.4 Modeling of Electric Vehicles

Electric vehicles are as a consumer in charge mode and as a producer in a discharge mode. As long as these vehicles are present in the parking lot, their equations are Eqs. (6.28, 6.29, 6.30, 6.31, 6.32, 6.33 and 6.34). Equation (6.28) shows the revenue/cost of electric vehicles in the parking lot. Equation (6.29) shows the amount of energy stored in batteries of vehicles, which depends on the energy of the previous hour and the amount of charge and discharge of the vehicle. The charging and discharge limits of vehicles are modeled according to Eqs. (6.30) and (6.31). Equation (6.32) shows the minimum and maximum charging limits of vehicles. Equation (6.33) indicates that charging and discharging do not occur simultaneously, and Eq. (6.34) indicates that vehicles have their maximum charge at the departure time from the parking lot [17].

$$Revenue_{t,w}^{EV} = \sum_{ev} P_{ev,t,w}^{ch} \lambda_t^{ch} - P_{ev,t,w}^{dch} \lambda_t^{dch} \tag{6.28}$$

$$E_{ev,t,w}^{EV} = E_{ev,t-1,w}^{EV} + P_{ev,t,w}^{ch} - P_{ev,t,w}^{dch} \tag{6.29}$$

$$P_{ev,t,w}^{ch} \le P_{ev,t}^{ch,Max} \alpha_{t,w}^{ch} \tag{6.30}$$

$$P_{ev,t,w}^{dch} \le P_{ev,t}^{dch,Max} \alpha_{t,w}^{dch} \tag{6.31}$$

$$E_{ev}^{EV,Min} \le E_{ev,t,w}^{EV} \le E_{ev}^{EV,Max} \tag{6.32}$$

$$\alpha_{t,w}^{ch} + \alpha_{t,w}^{dch} \le 1 \tag{6.33}$$

$$\sum_{t=AT}^{t=DT} P_{ev,t,w}^{ch} - P_{ev,t,w}^{dch} + E_{ev,0}^{EV} = E_{ev,cap}^{EV} \tag{6.34}$$

6.3   Simulation Results

Generation scheduling of dispatchable units are studied in the presence of wind turbines and electric vehicles in three different case studies. The typical distribution system includes 10 dispatchable generation units presented in Table 6.1 as well as a wind farm, which its aggregated power is obtained through Weibull distribution function in 10 scenarios. Moreover, 1000 EVs are considered in the distribution network with some simplified assumption as follows:

The arrival and departure time of the EVs to/from the parking is 5:00 pm and 7:00 am respectively. The capacity of each EV is considered 10 kWh and it is assumed that all EVs arrive to the parking with the state of charge equal to 40% of their maximum capacity. In order to encourage EVs owners to participate in V2G mode, the cost rate of charging as shown in Fig. 6.3 is considered 2% higher than their charging price that is equal to electricity market price. The maximum and minimum permissible energy stored in each EV are assumed to be 0.9 and 0.3 of its capacity, respectively. Furthermore, the battery charge and discharge rate of each EV is considered 10% of its stored energy capacity. The amounts of network load and day ahead market price are assumed as a parameter. Figure 6.4 depicts the power demand of distribution network. The values of v_{ci}, v_r and v_{co} in wind unit modeling are 5, 15 and 45 m/s, respectively.

The stochastic behavior of the aggregated wind turbines is modeled in 10 scenarios using Weibull distribution function shown in Fig. 6.5. The scenarios of WTs power generation are presented in Fig. 6.6.

To investigate the effect of the stochastic generation of WTs on EVs, three difference case studies are defined as follows:

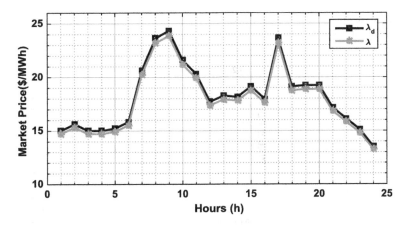

Fig. 6.3 The price of EV charge and discharge

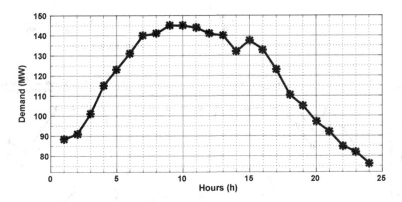

Fig. 6.4 Distribution network power demand

6.3.1 Case Study 1: The Stochastic Generation of WTs Without EVs Participation

In this case, there are only dispatchable generation units and wind turbines, while no EV is considered in the generation scheduling problem. The problem is solved for 10 generation scenarios of wind turbines. To compare different case studies with each other, the output of all case studies are investigated in the fifth scenario.

Figure 6.7 shows the output of the dispatchable generation in the fifth scenario. As shown in this figure as well as Table 6.1, only units 2–5 are generating power at the beginning of the planning period. According to the cost coefficients presented in Table 6.1, the third unit is the most expensive unit among the four generating units. Thus, it will be turned off after its minimum continues working time (3 hours). The first unit has been off at the beginning of the scheduling period and is started 5:00 am. Since the marginal cost of wind turbines are approximately zero, the output

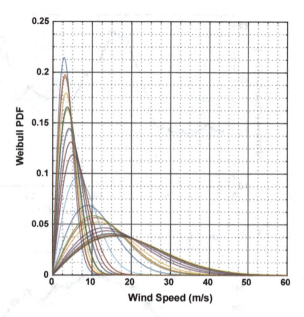

Fig. 6.5 Weibull probability distribution function

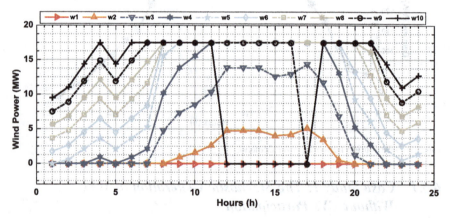

Fig. 6.6 Scenarios of WTs power generation

generation of dispatchable units are reduced by increasing the generation of wind turbines. For example, by significantly increasing the output of wind turbines in the fifth scenario on 8:00 am, the power generation of the fourth dispatchable units reduces. The dispatchable generation units have the largest participation from 7:00 to 13:00 due the peak load conditions of the demand curve. Since the second generating unit is the cheapest one, it generates power at its maximum capacity all day.

6 Integration of Electric Vehicles and Wind Energy in Power Systems

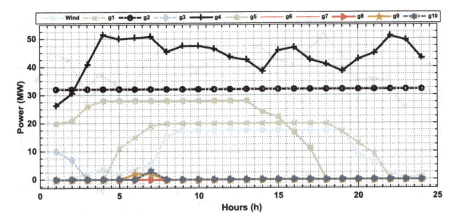

Fig. 6.7 The output power of the dispatchable generating units and the aggregated generation of wind turbines in scenario 5 of the case study 1

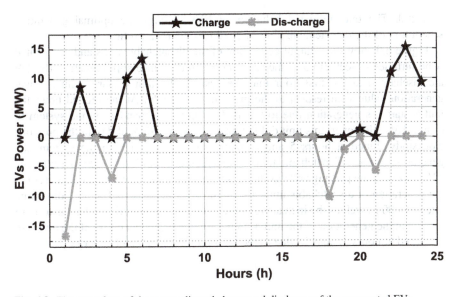

Fig. 6.8 The net values of the uncoordinated charge and discharge of the aggregated EVs

6.3.2 Case Study 2: The Stochastic Generation of WTs with the Participation of the Uncoordinated EVs

In this case, the total number of EVs are considered 1000 with the uncoordinated charge and discharge characteristics. Figure 6.8 shows the net values of the EVs charge and discharge that are considered as known parameters in the generation scheduling problem. Moreover, it is assumed that EVs are in the parking from 7:00

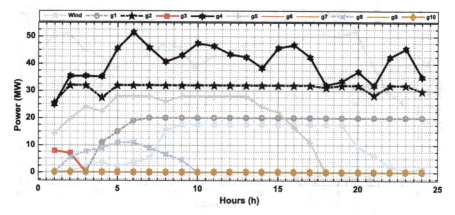

Fig. 6.9 The output power of the dispatchable generating units and the aggregated generation of wind turbines in scenario 5 of the case study 2

to 17:00. The uncoordinated behavior of EVs disrupts the optimal generation scheduling of the dispatchable units and reduces the total profit.

Figure 6.9 shows the output generation of the dispatchable units and aggregated wind turbines in scenario 5 of the second case study. Since approximately 18 MW are discharged at the first hour into the system, the unit 4 starts with a lower generation value with respect to the previous case study. At the second hour, the unit 8 starts its generation due to the 10 MW power demanded by the system to charge EVs. However, this unit did not have any generation all day in the previous case study. Since at hours 5 and 6 much power is required in the system, the generation of most dispatchable units has increased. From hours 7:00 to 17:00, the EVs are out of the parking and thus there is no need to charge or discharge, therefore, the generation of dispatchable units are significantly decreased. The high discharge rate at 18:00 as well as the increasing generation of the wind turbines lead to a sharp decline in the generation of the unit 4. In the contrary, the generation of the units are increased at hours 22 and 23 due to the high power demanded to charge EVs.

6.3.3 Case Study 3: The Stochastic Generation of WTs with the Participation of the Coordinated EVs

In this case study, EVs are charged and discharged based on the coordinated scheduling program. The aggregated power in the charge and discharge mode is significant so that it may affect the generation of the dispatchable units. Since the power demand is low at the hours 1:00 and 2:00, the amount of power generation by dispatchable units is also low and the EVs are charged as shown in Fig. 6.10. On the other hand, when EVs arrive at the parking lot at 18:00, the power demand is low (shown in Fig. 6.4) and EVs starts charging. At the end of the day (hours 23:00 and

6   Integration of Electric Vehicles and Wind Energy in Power Systems   177

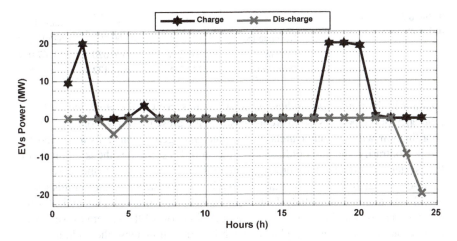

Fig. 6.10 The net values of the coordinated charge and discharge of the aggregated EVs

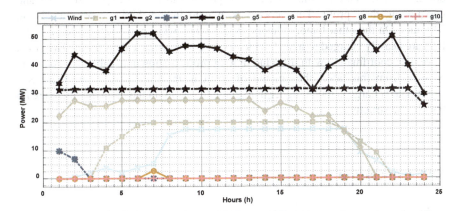

Fig. 6.11 The output power of the dispatchable generating units and the aggregated generation of wind turbines in scenario 5 of the case study 3

24:00), EVs start discharging due to no generation of the wind turbines and low amount of the power available at the power system.

The hourly power generation of the dispatchable units as well as aggregated wind turbines are shown in Fig. 6.11. In the first 2 hours, the units 4 and 5, which are cheaper than the others, increase their generation to provide required energy for EVs charging. By slightly increasing the output power of wind turbines and EVs discharging after that time, the unit 4 reduces its generation. In the mid-day, most units reduce their generation because all EVs are out of the parking lots as well as the output generation of wind turbines are high. At the hours 18:00 and 20:00, the unit 4 increases its generation to participate in charging EVs. Eventually, in the last 2 hours, all units either reduce or stop their generation due to the significant amount of EVs discharge. Table 6.2 summarizes the cost, income and profit of the system

Table 6.2 Comparing different cases

	Case 1	Case 2	Case 3
Revenue ($)	51174.098	51460.191	53608.679
Cost ($)	43901.241	44576.020	44767.097
Profit ($)	7272.858	6884.172	8841.582

operator in the three case studies. As it can be observed in the table, the profit and efficiency of the power system are increased in the presence of EVs, operating in the coordination with the other units, i.e. dispatchable generation units and wind turbines. Contrarily, the un-coordinated operation of EVs imposes the addition cost to the system operator. The dramatic increase of the cost in the case study 2 (un-coordinated EVs) is such that the profit of the system operator in this case is even lower than the case study 1 (without the presence of EVs). This indicates that un-coordinated charging of EVs can add operation cost of the system as well as the need to build new generation plants, which in turn will increase the pollution.

Wind units generate different powers in different scenarios. When generated power is low, vehicles with discharging power cause adjusting load curve and maximizing their profit. This process is evident in the third scenario at 23 and 24 hours. On the other hand, at hours 18–20, when wind power generation is high, EVs are heavily charged to adjust the load curve and increase their profit. The effect of vehicle coordination with the wind unit on profit is shown in Table 6.2. As it can be seen from this table, the amount of profit in the third case where there is a coordinated charge is much higher than in the first case. In the first case, there are no EVs in the structure.

6.4 Conclusion

In this chapter, a model is presented for optimal operation of electric vehicles in the presence of wind turbines and dispatchable generation units. EVs can be charged/discharged in two ways: coordinated and un-coordinated. The stochastic generation of the wind turbines are modeled using Weibull probability distribution function in 10 different scenarios. The generation scheduling of the dispatchable units in the presence of the wind turbines and EVs are conducted in three case studies. In the first case study, it is assumed that no EV is connected to the power system and the dispatchable units are scheduled under the stochastic generation of the wind turbines. The obtained results show that increasing generation of wind turbines decrease the generation of the units. In the case study 2, it is assumed that there are some EVs connected to the grid but they are not coordinated in the operation with other units. The lack of coordination led to an increase in the generation of dispatchable units, an increase in the costs and eventually a reduction in the profits of the system operator. In the last study (case study 3), EVs are charged and discharged in coordination with the other units of the system. The coordinated scheduling of EVs led to the charging at times that the generation of wind turbines

6 Integration of Electric Vehicles and Wind Energy in Power Systems   179

are large or the power demand of the system is low. Contrary, in the peak hours when the power demand is high or power generation is low, the stored energy is discharged into the grid. The coordinated charge and discharge increase the system efficiency and reduce the total cost. An important point is that the un-coordinated operation of EVs not only does not increase the profit but also can increase the power system stress.

Appendix A

Nomenclature	
Indices	
t	Index for time period
w	Index for wind scenario
g	Index for dispatchable units
ev	Index for Electric Vehicles in parking lot
Parameters	
$P_{t,w}^{Wind}$	Wind power generation in scenario w of period t
$\pi_{t,w}$	Probability of scenario w in period t
v_{aw}	Average wind speed in scenario w
P_{rated}	Rated power of wind unit
v_{ci}	Cut-in wind speed
v_r	Rated wind speed
v_{co}	Cut-out wind speed
λ_t	Day-ahead Market price
λ_t^{ch}	Charging price in period t
λ_t^{dch}	Dis-charging price in period t
η_{ev}	Efficiency of EVs
MUT_g	Minimum up time of unit g
$E_{ev}^{EV,Min}$	Minimum allowed energy stored in EV
$P_{ev,t}^{dch,Max}$	Maximum allowed energy stored in EV
$E_{ev,0}^{EV}$	Remained stored energy in EV at arrival hour
MDN_g	Minimum down time of unit g
$SUC_{g,t,w}$	Start-up cost of unit g in period t and scenario w
$SDC_{g,t,w}$	Shut-down cost of unit g in period t and scenario w
P_g^{Min}	Minimum limit of power generation of unit g
P_g^{Max}	Maximum limit of power generation of unit g
$\underline{P}_{g,t}$	Minimum time-dependent operating limit of unit g in period t
$\overline{P}_{g,t}$	Maximum time-dependent operating limit of unit g in period t
P_t^{Dem}	Demand power in period t
Variables	
$P_{g,t,w}$	Power produced by unit g in day-ahead market in period t and scenario w

(continued)

Nomenclature	
$P^{ch}_{ev,t,w}$	Power Charge of EV in period t and scenario w
$P^{dch}_{ev,t,w}$	Power Dis-charge of EV in period t and scenario w
$FC_{g,t,w}$	Cost of unit g in period t and scenario w
$Cost(t,w)$	Total cost of problem in period t and scenario w
$Revenue(t,w)$	Total revenue of problem in period t and scenario w
$Profit$	Total profit of problem
$E^{EV}_{ev,cap}$	Stored energy in EV in period t and scenario w
$\alpha^{ch}_{t,w}$	Binary variable for EV related to charge status in period t and scenario w
$\alpha^{dch}_{t,w}$	Binary variable for EV related to dis-charge status in period t and scenario w
$\alpha_{g,t,w}$	Binary variable, 1 if unit g is on, 0 otherwise
$\beta_{g,t,w}$	Binary variable, 1 if unit g starts up
$\gamma_{g,t,w}$	Binary variable, 1 if unit g shuts down

References

1. J. Mullan, D. Harries, T. Bräunl, S. Whitely, Modelling the impacts of electric vehicle recharging on the Western Australian electricity supply system. Energy Policy **39**, 4349–4359 (2011)
2. A. Foley, B. Tyther, P. Calnan, B.Ó. Gallachóir, Impacts of electric vehicle charging under electricity market operations. Appl. Energy **101**, 93–102 (2013)
3. B. Tarroja, L. Zhang, V. Wifvat, B. Shaffer, S. Samuelsen, Assessing the stationary energy storage equivalency of vehicle-to-grid charging battery electric vehicles. Energy **106**, 673–690 (2016)
4. H. Lund, W. Kempton, Integration of renewable energy into the transport and electricity sectors through V2G. Energy Policy **36**, 3578–3587 (2008)
5. B.V. Mathiesen, H. Lund, K. Karlsson, 100% renewable energy systems, climate mitigation and economic growth. Appl. Energy **88**, 488–501 (2011)
6. W. Kempton, J. Tomić, Vehicle-to-grid power implementation: From stabilizing the grid to supporting large-scale renewable energy. J. Power Sources **144**, 280–294 (2005)
7. R. Atia, N. Yamada, Sizing and analysis of renewable energy and battery systems in residential microgrids. IEEE Trans. Smart Grid **7**, 1204–1213 (2016)
8. N. Rotering, M. Ilic, Optimal charge control of plug-in hybrid electric vehicles in deregulated electricity markets. IEEE Trans. Power Syst. **26**, 1021–1029 (2010)
9. J. Kiviluoma, P. Meibom, Methodology for modelling plug-in electric vehicles in the power system and cost estimates for a system with either smart or dumb electric vehicles. Energy **36**, 1758–1767 (2011)
10. L.P. Fernandez, T.G. San Román, R. Cossent, C.M. Domingo, P. Frias, Assessment of the impact of plug-in electric vehicles on distribution networks. IEEE Trans. Power Syst. **26**, 206–213 (2010)
11. L. Wang, S. Sharkh, A. Chipperfield, Optimal coordination of vehicle-to-grid batteries and renewable generators in a distribution system. Energy **113**, 1250–1264 (2016)
12. J.D. Bishop, C.J. Axon, D. Bonilla, D. Banister, Estimating the grid payments necessary to compensate additional costs to prospective electric vehicle owners who provide vehicle-to-grid ancillary services. Energy **94**, 715–727 (2016)

13. M. Carrión, J.M. Arroyo, A computationally efficient mixed-integer linear formulation for the thermal unit commitment problem. IEEE Trans. Power Syst. **21**, 1371–1378 (2006)
14. M. Shafiekhani, A. Badri, M. Shafie-khah, J.P.S. Catalão, Strategic bidding of virtual power plant in energy markets: A bi-level multi-objective approach. Int. J. Electr. Power Energy Syst. **113**, 208–219 (2019)
15. M. Shafiekhani, A. Badri, A risk-based gaming framework for VPP bidding strategy in a joint energy and regulation market, Iran. J. Sci. Tech. Trans. Electr. Eng. 2019/02/04 (2019)
16. A.Y. Saber, G.K. Venayagamoorthy, Resource scheduling under uncertainty in a smart grid with renewables and plug-in vehicles. IEEE Syst. J. **6**, 103–109 (2011)
17. O. Arslan, O.E. Karasan, Cost and emission impacts of virtual power plant formation in plug-in hybrid electric vehicle penetrated networks. Energy **60**, 116–124 (2013)

Chapter 7
Distributed Charging Management of Electric Vehicles in Smart Microgrids

Reza Jalilzadeh Hamidi

7.1 Introduction

The unmanaged connection of Electric Vehicles (EVs) into power grids possibly results in several problems such as overcurrents, undervoltages, growth in power losses, reduction in power quality, and so forth [1–4]. The mentioned problems occur mainly due to the fact that Distribution Networks (DNs) were designed to supply a specific amount of load. However, the connection of EVs provokes a substantial increase in the aggregate demand, which cannot be readily supplied. Hence, control and management systems are indispensable for governing EV inter-action with DNs to reduce their adverse effects. For the formation of EV-charging control and management systems, various control methods have been utilized that can be broadly divided into two groups as,

(i). prediction-based methods in which the time of EV connection and disconnec-tion, electricity demand, etc. are predicted according to different data such as the driving patterns of the EVs. Then, proper times and perhaps charging rates for the charging or discharging of the EVs are determined [5–20]. Scholars have utilized various parameters to predict different aspects of EV charging. The most common input parameters, objectives, and methodologies for pre-dictions are summarized in Table 7.1 based on [3–20]. Different devices and technologies, including smartphones, The Internet, power line carriers (PLCs), and GPS trackers are often used for collecting the data used for predictions.

(ii). non-prediction-based control methods are essential in Smart Microgrids (SMGs) since, despite the advances in the prediction-based methods, the pre-dictions are still far from high precision, considering that the number of EVs with different features is increasing [17]. At the same time, the technologies for

---

R. Jalilzadeh Hamidi (✉)
Faculty of Electrical Engineering, Arkansas Tech University, Russellville, AR, USA

© Springer Nature Switzerland AG 2020
A. Ahmadian et al. (eds.), *Electric Vehicles in Energy Systems*,
https://doi.org/10.1007/978-3-030-34448-1_7

Table 7.1 The usual inputs, objectives, and prediction methods for prediction-based EV-charging management

Input Parameters	Prediction Objectives	Prediction Method
Daily driving distance Driving cycle (vehicle speed profile versus time) Road grade cycle (road grade profile versus time) Traffic flow data Electricity price profile (temporal variations of electricity cost) Electricity demand profile (temporal variations of electricity consumption) Local generation profile (temporal variations of local electricity production)	Minimization of load variance Peak shading Valley filling Charging cost minimization Household electricity-cost minimization Optimal dispatch schedule Lowering electrical asset decay Temporal EV availability Aggregate Power Capacity (APC) Probability distribution functions for EV's state of charge Prediction of space-time distribution of EVs' charging load	Time series analysis (e.g., ARIMA[a]) Monte Carlo simulations ANN[b] forecasting algorithms Hierarchical or non-hierarchical model predictive controls k-nearest neighbor (kNN) and weighted kNN Nonparametric diffusion-based kernel density estimator (DKDE) Stochastic dynamic programming method

[a]*ARIMA* stands for Auto-Regressive Integrated Moving Average
[b]Artificial Neural Network

Photovoltaic- (PV) and wind-based generation developed into a mature stage, and they are commercially available to domestic customers.

As a consequence, the renewable generation capacity in residential areas is increasing, and the homeowners often install the small-scale distributed generations (DGs) without alerting the utility operator. The main drawback of renewable-based DGs is that their output powers are in nature variable and unpredictable to some extent. This raises the uncertainty level in SMGs and exacerbates the predictions. Furthermore, contingencies (e.g., faults, generation outage, transmission line outage, etc.) are rarely predictable. Thus, although the prediction-based methods are beneficial in the long run [3, 21, 22], they fail to protect the DNs against contingencies and prediction inaccuracy. Accordingly, a control framework must be capable of managing unpredictable situations [23]. Thus, real-time (also referred to as non-prediction-based or myopic) controllers have been designed and introduced in the prior literate. The real-time or semi-real-time control and management systems only consider the present-time state of the system. The real-time controllers do not rely on predictions (i.e., preceding state in the system) although predictions might be valuable for finding optimal or more preferable solutions.

Referring to Fig. 7.1, control frameworks can be broadly divided into the following three categories with respect to their architecture as follows:

(i). Centralized (Central): A single control center governs all the EVs in the DN to charge or discharge according to the data coming from the power network as

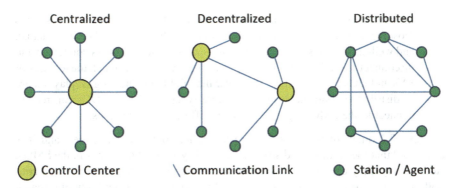

Fig. 7.1 Control architectures

well as EVs, as shown in Fig. 7.1. This control architecture demands reliable and advanced communications. The control center requires managing all the received information in a reasonable time. Therefore, it must be computationally able. If communication links disconnect, the corresponding agents will no longer follow the control objectives, and the entire system may collapse.

(ii). Decentralized (Decentral) Controllers: In decentralized control methods, several smaller control centers collaborate on achieving the system goals. Each group of the EVs is under the jurisdiction of one small control center, and the control centers communicate with one another, as depicted in Fig. 7.1. The charging or discharging rates are determined at each control center based on the received data from the neighboring centers and EVs. In this type of controllers, intelligence is dispersed throughout the network to some extent, and if the link between two centers is cut, the whole system may be able to continue functioning toward fulfilling the control objectives as two separate smaller systems [1]. Less computation-ability and less complicated communications are needed in comparison to the centralized control architectures. It is noteworthy that hierarchal decentralized control methods are widely in use and belong to a subcategory of decentralized controllers. In hierarchal control methods, the centers are ranked and the centers of a higher rank can send control signals to the lower rank centers. However, the inferior centers can only send information to their superiors, but do not commands [24, 25].

(iii). Distributed Controllers: In distributed control systems, the intelligence is fully shared among all the agents (also referred to as stations). A small portion of the system is observable to each agent (i.e., an EV charger). Then, an agent makes its own decision solely based on the observable portion and the received data from the neighboring agents as Fig. 7.1 shows. Thus, the control system requires only sparse communications to enable each agent receiving its neighbors' information [1, 22]. If communication links are cut, there is a good chance for the distributed controllers to keep following the control goals. Each station should be equipped with a local controller; however, the local controllers are far less able in terms of computation and communications compared to the control centers. The connection of the new stations to the whole system is most facilitated in distributed control systems. Thus, they are largely-scalable, and well-suited for SMGs as a random number of DGs and EVs connect to and disconnect from the SMG at any given time. Moreover, the characteristics of EVs and DGs can be widely dissimilar that highlights the plug-&-play feature of distributed controllers, in that diverse types of EVs and DGs can readily connect to the SMG regardless of their different specifications.

Table 7.2 qualitatively compares the foregoing control architectures, and it is concluded that the features of distributed control systems fit the needs in the SMGs. Additionally, they can be established at minimal costs. Therefore, distributed controllers seem to be a promising choice for SMGs and a significant body of research looks into decentralized and distributed EV-charging management methods.

In decentralized charging and discharging control methods, an aggregator receives the information from the DN and EVs. It decides whether EVs charge or discharge, considering Distribution System Owner's/Operator's (DSO) suggestion, Transmission System Operator's (TDO) suggestion, and the electricity market, as shown in Fig. 7.2. The aggregator broadcasts its decision throughout the DN, and each EV finds out its operating mode taking into account the aggregator's decision. Figure 7.2 depicts a general schematic for decentralized control methods. It is noteworthy that each EV must be equipped with a local decision-making unit. The communications between the EVs and the aggregator can be either unidirectional or bidirectional. Some examples of decentralized controllers in the previous literature are as follows: The authors of [26] proposed a controller which relies on the Lagrangian decomposition for scheduling EV charging. This method also compensates the total reactive power demand in the SMG. This method is decentralized since a coordinator or aggregator finds the optimal solution and the EVs using the solution and local data determine their operating modes. Some other decentralized methods are proposed in [27–30] in that EVs send their demands to the aggregator. The aggregator may also receive some information about the voltages and currents in the DN. Then, the aggregator solves an optimization problem and finds an optimal or approximately optimal solution to the EV-charging problem. The outcomes then will be sent to all the EVs, and it is up to the EVs' local controllers to select their operating modes among charging, discharging, and idle modes. This method is also decentralized since the aggregator together with the local controllers find the appropriate answer to any given situation in the SMG. In [31], it is assumed that the charging rates of all the EVs are identical. Then, the aggregator determines the number of EVs to charge and announce it to all the EVs in the SMG. At this moment, any of the EVs makes its own decision to charge or become idle. In some of the decentralized methods, such as [23, 32], the aggregator along with EVs solve an optimization problem in an iterative way. In that, EVs' data transfer to the aggregator, and the optimization problem is solved for one iteration utilizing the most recent data. Then, the results are sent back to the EVs to update their operating modes accordingly. These methods are able to find the optimal solution gradually while the EVs contribute to solve the problem and adjust their modes in step with the

Table 7.2 Comparison of the control architectures

Architecture	Fault Tolerance	Maintenance	Scalability	Ease of Formation	Formation Costs	Optimality of Results	Communication Complexity	User State Information	Reliability
Centralized	Low	Easy	Low	Easy	High	High	High	Public	Low
Decentralized	Rel.[a] High	Moderate	Rel. High	Difficult	Rel. Low	Rel. High	Rel. High	Public	Rel. Low
Distributed	High	Difficult	High	Difficult	Low	Low	Low	Private	High

[a]Relatively

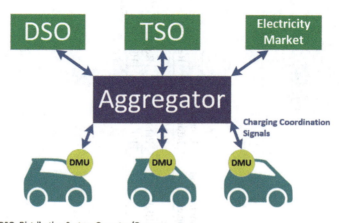

Fig. 7.2 General setup of aggregators for EV-charging coordination

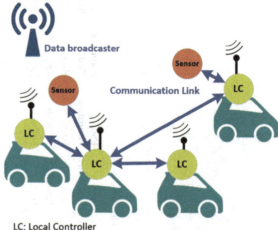

Fig. 7.3 General structure for distributed EV-charging management

convergence of the optimization problem. However, sophisticated and reliable communications must support the management system as a tremendous amount of data transfers between the aggregator and EVs in every single iteration.

With reference to Fig. 7.3, there is no aggregator(s) or control center(s) in distributed charging control systems. All the intelligence, calculation, and decision-making-ability are shared among the EVs in the DN. Each and every EV must be equipped with a local controller, which transmit the data between the EV and its neighbors. The local controllers may also receive sensory data from the DN. It is possible that critical data such as the occurrence of an emergency or electricity price is broadcasted throughout the DN for local controllers. The local controllers are able to process the received data and determine whether the related EV charges, discharges, or becomes idle.

Less number of publications addressed the distributed architecture for EV-charging management compared to the decentralized methods. The authors of [33, 34] developed distributed game theory-based control systems that minimize the charging costs to EV owners. In some works such as [35], the contribution of renewable energies in EV charging is maximized. In [36], a cooperative control-based method is proposed to minimize voltage deviations and balance the generation and demand of distributed generations and EVs in DC grids. The authors of [37] proposed a distributed control method that enables EVs to serve as emergency supplies in islanding mode. The charging or discharging rates of EVs are coordinated by a distributed controller to minimize the power flow at the Point of Common Coupling (PCC) in [38, 39]. Therefore, the power flow fluctuations caused by VERs, loads, and EVs at the PCC will be mitigated [40]. Besides, the SMG becomes independent of the bulk power system, to some extent. The main grid's power quality enhances, as well. Moreover, SMG's ride-through-ability rises since during the islanding mode, the contribution of EVs to supplying the demand becomes maximum. Hence, EVs together with the energy storage system (ESS) will be able to respond to the electricity demand in a DN for a longer time.

In the rest of this chapter, firstly, the need for electric vehicles, as well as their battery and chargers are described. Secondly, the challenges toward the pure renewable-based supply of SMGs are addressed. Thirdly, the control objectives are elaborated. Fourthly, the control design and requirements are explained, and finally, the conclusion is provided.

7.2 Electric Vehicles and Electrical Networks

An electric motor, as an alternative to an internal combustion engine, powers up electric cars. Mostly the limited driving range and long charging time historically inhibited the proliferation of electric cars (also referred to as EVs or Battery Electric Vehicles (BEVs)). As the battery technology continuously improves, the capacity and charging speed drastically increases, while battery cost substantially decreases. Considering the current trend, together with several Socio-Techno-Economic-Political (STEP) factors, the number of commercially manufactured EVs is significantly growing. The main STEP parameters are as follows:

- **Performance benefits:** Electric motors, instead of gas engines in conventional cars, function quietly and smoothly, while providing stronger acceleration and requiring less maintenance compared to the gas engines [41].
- **Energy efficiency:** EVs are able to convert 59–62% of the electrical energy to power at the wheels, while the conventional cars are able to convert as low as 17–21% of the gas energy [41].

- **Environmentally friendly:** EVs emit zero exhaust pipe pollutants [42]. However, the controversy has arisen that the electricity generated at the power plants for charging EVs is environmentally pollutant. In response to the argument, EVs can be charged based on clean energy resources such as wind, hydro, and solar. Thus, the net pollution caused by EVs is far less than conventional cars if green power plants largely supply power systems [42, 43].
- **Energy dependence:** The U.S. alone used almost nine billion petroleum barrels in 2018. Two-third of such a massive fossil fuel utilized for transportation. Therefore, oil shortages and oil-price spikes make economic growth vulnerable. Accordingly, EVs help reduce the reliance on fossil fuels since they can be supplied by renewable energy resources [44].
- **Technical aspects:** EVs are beneficial to the power system resilience and energy efficiency on condition that an expert control and management system coordinates the charging and discharging of EVs in a DN [44].

In the rest of this section, technical features of EV batteries and chargers are elaborated.

7.2.1 Batteries

It is preferable that EV batteries are light, unexplosive, environmentally safe, and inexpensive. They should be also durable, reliable, high-capacity, and able to provide a high power rate if high-acceleration is needed. Lead-acid, NiMH, and Li-ion are commonly utilized in EVs. Zero Emissions Batteries Research Activity (ZEBRA) batteries (also known as Sodium Nickel Chloride or molten salt) are also limitedly used in EVs since it requires an operating temperature as high as 270–350 degrees Celsius. Although Nickel Cadmium (NiCd) and Nickel Metal Hydride (NiMH) batteries provide a higher power and energy density, their high internal- or self-discharging makes them improper for EVs. The life-cycle of batteries shows how many time a battery can be fully charged and discharged. As a study by Department of Energy's (DOE's) National Renewable Energy Laboratory (NREL) shows EV batteries typically last 12–15 years in moderate climates, but they only last for 8–12 years in adverse climates [41].

The other parameter that affects the charging and auxiliary applications of EVs is the charge and discharge rate of the batteries, which is indicated with current rate or C-rate. The batteries with a higher C-rate are able to participate more flexibly in charging coordination. The battery C-rate is effective on its life. Therefore, several methods have been proposed for canceling out the negative influence of high-rate or inconsistent charging. In this regard, Battery Management Systems (BMS) are introduced, which are electronic systems interfaced between the charger and the batteries in EVs. They are recently equipped with large supercapacitors for accepting high charging currents, and then feeding each battery cell with the optimal current rate according to its features, state of charge, temperature, etc. as shown in Fig. 7.4.

7   Distributed Charging Management of Electric Vehicles in Smart Microgrids   191

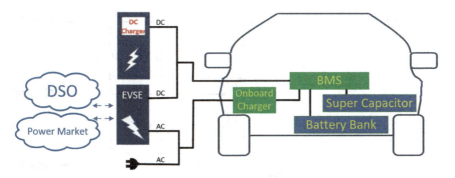

Fig. 7.4 Onboard and off-board chargers and EVSEs

7.2.2 Chargers

Several different types of chargers compatible with different distribution voltage levels and with different output powers have been developed. The breakthrough in EV charging is EV Supply Equipment (EVSE), which is a sophisticated external (off-board) charger station. As shown in Fig. 1.4, it communicates via specific protocols with the EV to determine the maximum current that the EV can receive. On the other hand, the EVSE is able to communicate with DSO and power market to find the optimal charging rate. An EVSE maintains the EV safe during charging since it checks if a charger cable is correctly plugged into the EV. It also ensures that there are no hardware faults. Therefore, EVSEs not only charge the batteries optimally fast, but they also prevent electrical short circuits and damages to the batteries.

The chargers are categorized in several different ways. One way of categorization is their output current in terms of being AC or DC. In this way, the chargers are divided into AC and DC groups.

- AC chargers: The output current of AC chargers is AC and as Fig. 7.4 shows, an onboard charger/rectifier is required for converting the AC current to DC current to be compatible with batteries. As the size and weight of onboard chargers are restricted, their capacities are limited.
- DC chargers: As the output of the charger is DC, an onboard charger/rectifier is not required, and it can be directly used for charging the batteries. As shown in Fig. 7.4, DC chargers are external and can be advantageous of communications and smart technologies. They are capable of adjusting their charging rates based on the EV, DN, and power market parameters.

The installation location of the chargers is another basis for categorizing them into household chargers and charger stations.

- The output power of household chargers is mainly limited by the distribution transformer secondary configuration. In the US, the households are supplied by split-phase transformers that provide two 120-volt lines with a 180-degree phase

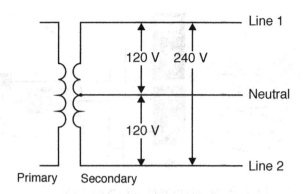

Fig. 7.5 Split-phase topology

Table 7.3 DOE's description of charging levels and their characteristics

Level	Installation Location	Input Voltage [V]	Required DN Topology	Output Power up to [kW]	Output Voltage up to [V]
Level-1	Household	120	US 1-P[b]	1.4	120 AC
Level-2	Household/ Station	240	1-P or SP P-P[c]	9.6	240 AC
Level-3[a]	Station	480	3-P[d]	62.5	600 DC
DC Super-Fast	Station	NF	NF	NF	NF DC

[a]Also known as DC Fast Charger or CHAdeMO charger
[b]1-P means single phase
[c]Split-phase, phase to phase connection
[d]3-P indicates three phase

difference, as depicted in Fig. 7.5. Therefore, it is capable of supplying the loads at 120 or 240 V. The household outlets are connected to one line and neutral that produces 120 V. However, larger appliances such as ovens and air conditioners are connected to lines for being supplied at 240 V. Thus, the maximum power of household chargers is limited to less than 10 kW.

- On the other hand, charger stations can be fed by different transformers connected to DN that enables them to charge EVs at considerably higher rates.

Several different institutions (e.g., DOE, Society of Automotive Engineers (SAE), and International Electrotechnical Commission (IEC)) categorized the chargers into three main groups.

- DOE, as one of the most well-known institutions, categorized the chargers as provided in Table 7.3, [45, 46].
- SAE J1772 Committee developed J1772-plug or J-plug definitions with the specifications given in the below table. It divides the chargers based on their output currents into two main groups, AC and DC charging systems. In AC

7 Distributed Charging Management of Electric Vehicles in Smart Microgrids

Table 7.4 SAE definitions and terminology

AC Charging	Specifications	DC Charging	Specifications
Level-1	Input: 120 V, 1-P AC Output Current: up to 16 A (typically limited to 12 A) Output Power: up to 1.9 kW.	Level-1	Input: 200–450 V DC Output Current: up to 80A Output Power: up to 36 kW
Level-2	Input: 240 V, 1-P AC Output Current: up to 80 A (typically limited to 32 A) Output Power: up to 19.2 kW.	Level-2	Input: 200–450 V DC Output Current: up to 200A Output Power: up to 90 kW
Level-3	Input: 3-P Output Current: More than Level 2 Output Power: up to 43 kW.	Level-3	Input: 200–600 V DC Output Current: up to 400A Output Power: up to 240 kW

Table 7.5 IEC definitions and terminology

Charging Mode	Description
Mode 1	slow charging, household-type socket-outlet
Mode 2	slow charging, household-type socket-outlet with an in-cable protection device
Mode 3	slow or fast charging, specific EV socket-outlet with installed control and protection functions
Mode 4	fast charging, external charger

charging systems, there must be an onboard (internal) charger for rectifying AC to DC. However, in DC chargers, the charger is off-board (external) and supplies the car with DC current, [47] (Table 7.4).

- The International Electrotechnical Commission (IEC) also provided a standard and terminology for the EV chargers as given in Table 7.5, [48, 49].

7.2.2.1 EV Charging Future

It is anticipated that nearly 500 million EVs will be in use until 2030 [50]. Therefore, efficient infrastructure for EV integration is the key factor to make this transition smooth. To this end, the following technologies are coincidently developing: vehicle-to-grid (V2G), renewable charging, and on-road charging. Bidirectional EV chargers are necessary for V2G concept in that EVs are mainly utilized for stabilizing the power grid, improving the power quality, and reducing electricity costs. EV renewable charging paves the way for sustainable and net zero CO_2 emission in e-transportation. On-road EV charging alleviates range anxiety issues [51, 52]. California leads the way in the US with 5% of brand-new vehicle sales, while 58% of new cars are EVs in Norway and the situation is almost the same in the other European countries. Hence, the US adopts EVs relatively slowly. Considering the fast-growing use of EVs in Europe, the future transportation is electric-based in Europe. However, the US needs to pass a long way to this end mostly by investing on on-road charging, which is proper for long travel distances in the US [53].

7.2.3 Aggregators

Aggregators are recently-introduced entities in electrical utilities or networks that perform as mediators among end users, smart components (e.g., smart meters, switches, chargers, etc.), and the utility operator or owner. Aggregators possess the software and hardware technologies necessary to perform their tasks. A wide range of duties have been defined for aggregators in different areas such as demand-response management, electricity market, network voltage and frequency stability, EV charging management, renewable generation, and so on [54, 55].

7.3 Auxiliary Services for EVs

The proliferation of bidirectional chargers along with advanced charging management algorithms provide great potential for addressing the challenges in SMGs. However, before defining the control objectives for solving the issues in SMGs, the undesirable situations in SMGs, and reasons for employing the EVs as an Energy Storage System (ESS) are described.

7.3.1 Stand-Alone Micro-Grids

SMGs must be able to work in both

- Grid-connected mode: when an SMG is connected to the main grid.
- Islanded mode: when an SMG is not connected to the main grid.

In grid-connected mode, the voltage and frequency of an SMG are governed by the main grid, and there is no need for stabilizing them. During islanded mode (also known as stand-alone or autonomous) the voltage and frequency of SMGs must be locally controlled to be in the normal range. If EVs are supposed to contribute to supplying the loads during islanded mode, their output should not unstabilize the DN.

7.3.2 Variable Renewable Energies

The SMGs heavily supplied by VREs, such as wind or solar generation, are unreliable since it is not guaranteed that the loads are supplied at any given time. In addition, large shares of VREs in SMGs result in power system instability and low power quality due to the fluctuations in VREs' outputs [56]. One of the most effective ways for solving such unpreferable issues is to utilize ESSs as they, canceling out the fluctuations, can make SMGs reliable.

7.3.2.1 Electric Vehicles as Distributed ESSs

Although ESSs improve the resilience of SMGs, it seems that DSOs are reluctant to install ESSs due to their high prices. Fortunately, EVs are equipped with batteries and all the battery capacity is not necessarily required all the time. Therefore, the surplus battery capacities can be utilized in line with enhancing DNs as distributed ESSs.

7.4 Charging/Discharging Management of EVs

As one of the promising control methods, the applications of the cooperative control to myopic, technical, and economical EV-charging coordination is addressed in this section. The successfully achieved control objectives are listed and discussed. Then, in the next chapter, the control design and requirements for fulfilling the objectives are detailed.

The cooperative control is selected for setting up the distributed control framework since it is fully distributed, in that each EV receives only local and neighboring data [4, 22], and determines its charging rate accordingly. Figure 7.6 shows a typical SMG with EVs. All EVSEs are equipped with a local controller, which will be elaborated in details later, that receives

- local and neighboring data
- electricity price for reducing charging costs
- voltage, current, and frequency measurements from some places in the DN for stabilizing the DN as well as increasing its power quality and ride-through-ability.

However, particular note should be taken that the DN measurements are sent to at least one of the EVSEs, and it does not make a critical difference which EVSE receives it. The EVSEs utilize the received measurements to determine their charging rates that can be positive, negative, and zero for charging, discharging, and idle operating modes. Also, the measurements are used by EVSEs for adjusting their output frequencies and voltages in discharging mode.

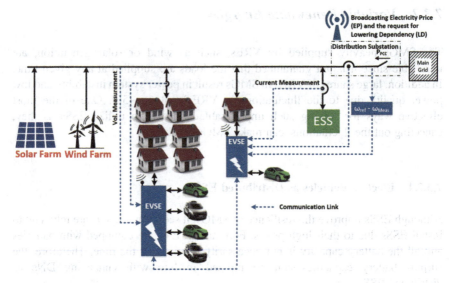

Fig. 7.6 A typical SMG with necessary components for establishment of the cooperative control-based EV-charging management

7.4.1 Fair Charging/Discharging Rate

As mentioned above, there are various charger types with different powers (e.g., Level-1, Level-2, DC Fast Charging, etc.). It is wise that the charging rates of EVs decrease in case of a shortage in the generation or vice versa. It is also reasonable that every single charger changes its output proportional to its capacity. For example, when there are only two 1.4 kW and 6.6 kW chargers (8 kW in total) in the system and the power supply reduces in half from 8 kW to 4 kW, then both chargers are responsible for reducing their charging rates. Therefore, if they cut their charging rates in half, which are respectively 0.7 kW and 3.3 kW, it will be fair. This concept is called "Fair Charging Rates" [4, 37–39], in that the ratios of the "preferable" outputs of the chargers to their maximum capacities will be identical after the system has reached to its equilibrium. This fair charging concept is mathematically defined below, [4, 21, 37, 57].

$$\frac{P_1}{P_1^{Max}} = \frac{P_2}{P_2^{Max}} = \frac{P_3}{P_3^{Max}} = \cdots = \frac{P_i}{P_i^{Max}}, \forall i = 1, 2, 3, \ldots, m \quad (7.1)$$

$$\frac{Q_1}{Q_1^{Max}} = \frac{Q_2}{Q_2^{Max}} = \frac{Q_3}{Q_3^{Max}} = \cdots = \frac{Q_i}{Q_i^{Max}}, \forall i = 1, 2, 3, \ldots, m \quad (7.2)$$

where P_i [pu] and Q_i [pu] are the i-th EVSE's preferable active and reactive output powers, respectively. P_i^{Max} [pu] and Q_i^{Max} [pu] indicate the capacity (maximum output power) of the i-th EVSE, and m is the total number of the EVSEs in the DN.

7.4.2 Reducing SMG Dependency on the Main Grid

One of the objectives that can be realized utilizing the cooperative control is to lower the dependency of an SMG on the main grid and to raise its dependency on the local renewable generation. With reference to Fig. 7.6, the power at the Point of Common Coupling (PCC) should be ideally negative, $P_{PCC} \leq 0$, where P_{PCC} [W] is the power flow from the main grid to the SMG at the PCC. Then a negative P_{PCC} denotes that the SMG sends power to the main grid, and zero P_{PCC} shows that there is no power flow between the main and smart grids. However, this objective is not always possible due to the intermittency in renewable generation. Yet, it is preferable to decrease the unnecessary EV charging in SMGs in case of an electricity shortage in the main grid. To this end, the Boolean variable LD and the following objective are defined.

$$P_i \rightarrow P_j : P_{PCC} \leq 0, \forall (i,j) = 1, 2, 3, \ldots, m \; if \; LD = {'On'} \tag{7.3}$$

where LD becomes 'on' when the DSO detects a shortage in the generation.

7.4.3 Improving Power Flow Fluctuations at PCC

The volatile output of VERs causes deviations in the power flow at the PCC that is shown in Fig. 7.6 with P_{PCC}. A change in P_{PCC} can be detected by the derivative of P_{PCC} with respect to time, $dP_{PCC}(t)/dt$. The charging rates of EVs can be changed to suppress the power flow deviations. This objective is mathematically defined as,

$$P_i \rightarrow P_j : \frac{dP_{PCC}(t)}{dt} \rightarrow 0, \forall i \neq j, \forall (i,j) = 1, 2, 3, \ldots, m \tag{7.4}$$

where the symbol '$\rightarrow$' indicates that the left side approaches the right side, and the symbol ':' means 'such that'.

7.4.4 Frequency Stability

The frequency of an SMG must be locally controlled during islanded mode since the main grid no longer governs it. Therefore, the output frequencies of all EVSEs should be adjusted to match the reference for the frequency ω_{ref} [rad/s], which is set at the distribution substation.

$$\omega_i \rightarrow \omega_{MainGrid}, \forall i = 1, 2, 3, \ldots, m \qquad (7.5)$$

where ω_i [rad/s] denotes the output frequency of the i-th EVSE.

7.4.5 Voltage Stability

The same as the DN frequency during islanding mode, the voltage must also be controlled to stay in the acceptable range. In addition to that, if a large number of EVs are charging at the same time even in grid-connected SMGs, the ends of the feeders are vulnerable to experience an undervoltage. Therefore, the EVSEs reactive power consumption or generation must be adjusted to retain the system voltage as,

$$Q_i \rightarrow Q_j : V^{Min} \leq V_i \leq V^{Max}, \forall i \neq j, (i,j) = 1, 2, 3, \ldots, m \qquad (7.6)$$

V_i [pu] is the output voltage of the i-th EVSE. V^{Min} [pu] and V^{Max} [pu] are the minimum and maximum acceptable voltages in the DN, which are usually 0.95 [pu] and 1.05 [pu] in most of the standards, respectively.

7.4.6 Prevention of Overcurrents

When a significant number of EVs are simultaneously charging, the sending ends of the feeders may experience an overcurrent. The commonsense solution to that is to reduce the charging rates of the EVSEs as expressed below,

$$P_i \rightarrow P_j : I_k \leq I_k^{Max}, \forall i \neq j, (i,j) = 1, 2, 3, \ldots, m, \forall k = 1, 2, 3, \ldots, n \qquad (7.7)$$

where I_k [pu] is the measured current passing through the k-th location, I_k^{Max} [pu] is the maximum allowable current at the k-th location, and n is the number of current sensors in the DN.

7.4.7 Consideration of Minimum SoC

EV owners will not participate in ancillary services if it results in becoming charged less than a preferable level. Thus, no EV should discharge to the SMG when its *SoC* is less than the minimum required. Either the EV owner or an intelligent system can determine the minimum required *SoC* (SoC^{min}). The charging coordinator must secure the *SoC* against falling less than SoC^{min}. This control objective is

$$P_i \geq 0 \; if \; SoC_i < SoC_i^{Min}, \forall i = 1, 2, 3, \ldots, m \tag{7.8}$$

where $P_i \geq 0$ indicates that the EVSE's output-power direction is from the SMG to the EV, SoC_i^{Min} [%] is the minimum acceptable SoC for the i-th EV.

7.4.8 Emergency Charging

In addition to stopping discharging when the SoC of an EV is less than SoC^{Min}, if the EV is in need of charge for functioning, it might be allowed to charge regardless of the DN's situation. This type of charging is defined as emergency charging and it is obvious that the owner should compensate for that. In regard to this objective, the Boolean variable EC is defined and the EV owner is able to set it on or off. The objective of emergency charging is defined as,

$$P_i = P_i^{Max} \; if \; EC_i = 'On' \; and \; SoC_i < SoC_i^{Min}, \forall i = 1, 2, 3, \ldots, m \tag{7.9}$$

where P_i^{Max} is the capacity of i-th charger, EC_i is the emergency charging option for the i-th EV that can be set as on or off since it is a Boolean variable.

7.4.9 Increasing System Ride-through-ability

Referring to Fig. 7.6, the EVs can contribute to supplying the loads together with the SMG's ESS when the SMG is islanded. To this end, the EVSEs start to reduce their charging rates, and if possible, start discharging to the SMG until the output of the ESS becomes zero.

$$P_i \rightarrow P_j : I_{ESS} \rightarrow 0 \; if \; P_{ESS} < 0, \forall i \neq j, (i,j) = 1, 2, 3, \ldots, m \tag{7.10}$$

where I_{ESS} [pu] is the output current of the ESS and $P_{ESS} < 0$ shows that the ESS is supplying the SMG.

7.4.10 Decreasing Charging Costs

It is possible to decrease the charging costs to the EV owners utilizing distributed control methods. In essence, the myopic methods for reducing the charging costs compare the current electricity price with the present charging cost. If the present-time electricity cost is less than the present charging cost, then the EV starts charging and vice versa. As the participation of an EV in the power market may not be attractive to its owner, therefore, another Boolean variable RCC (abbreviation for 'reduce charging costs') is defined that enables or disables the EV to take part in the electricity market. Regarding economical charging, the following objective is defined,

$$ACC_i(t_2) < ACC_i(t_1), \forall(t_2 > t_1, i) \ if \ SoC_i \geq SoC_i^{Min} \tag{7.11}$$

where $ACC_i(t)$ is the i-th EV's average cost of charge at time t, and it is explained in the next chapter. This economic objective not only reduces the charging costs, but it also results in valley filling and peak shifting, and if enough aggregate surplus charge is available, it leads to peak shaving as well.

7.5 Controller Requirements and Design

The special needs and requirements for the establishment of the controller are described in this section. First, the control requirements and then cooperative control design and formulation are detailed.

7.5.1 Control Requirements

The cooperative control is mainly structured on sparse and intermittent communications. Each agent (i.e., EVSE) must be able to exchange data with its direct neighbors. As the system manages the discharge rates of the EVs, the chargers must be bidirectional with a controllable charging and discharging rate. As shown in Fig. 7.7, if the reduction of charging costs is considered as a control objective, the EVSEs should be equipped with a receiver to become aware of the electricity price. This can be realized using a unidirectional price signal, which is announced throughout the DN once a while by the DSO or utility owner.

Furthermore, every EVSE must be equipped with a droop controller to respond to sudden changes in generation and demand in a DN to secure the generation-demand balance as depicted in Fig. 7.7. The droop controller is formulated as,

$$\begin{cases} \omega_i^* = \omega_{ref_i} - \kappa_{p_i}\left(P_{O_i} - P_{ref_i}\right) \\ V_i^* = V_{ref_i} - \kappa_{Q_i}\left(Q_{O_i} - Q_{ref_i}\right) \end{cases} \tag{7.12}$$

where ω_i^* [pu] and V_i^*[pu] are respectively the frequency and voltage commands to the i-th bidirectional EVSE when it is discharging. P_{O_i} and Q_{O_i} are the measured active and reactive power outputs of the i-th EVSE, respectively. ω_{ref_i} and V_{ref_i} are the references for frequency and voltage that are updated via the cooperative control-based controller for stabilization of the DN. κ_{p_i} and κ_{Q_i} are the droop gains.

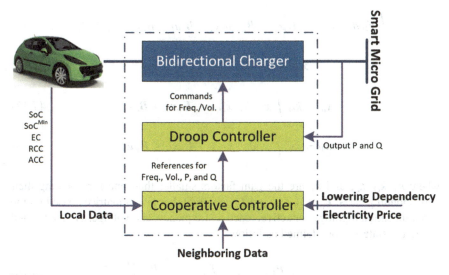

Fig. 7.7 The high-level structure of the local controller

7.5.2 Cooperative Control Formulation

The state equation of the cooperative control is generally presented as [22],

$$\dot{x} = -Lx + Bu \tag{7.13}$$

In that, x indicates the state vector, $(\dot{.})$ is the first derivative with respect to time, B is the control input matrix, u is the control input vector, and $L = [l_{ij}]$ is called the Laplacian matrix, which is built premised on communications between the agents as [37],

$$l_{ij} = \begin{cases} -1, \forall j \in N_i \\ |N_i|, j = i \\ 0, else \end{cases} \tag{7.14}$$

where N_i is a set comprising the i-th EVSE's neighbors (i.e., the EVSEs that directly send their data (i.e., state) to the i-th EVSE). $|N_i|$ denotes the in-degree of the i-th EVSE (i.e., the incoming communication links). The value of each state is updated through an integrator as $x = k \int \dot{x}dt$, and k is the gain factor for the integrator that must be between zero and one, $0 < k < 1$).

Active power, reactive power, frequency, and voltage each must be separately controlled through different communication links, sensory measurements, and control inputs as shown below.

$$x_P = k_p \int \dot{x}_P \, dt, \quad \dot{x}_P = -L_p x_P + B_{ESS} u_{ESS} + B_I u_I + B_{Fluc} u_{Fluc} + B_{LD} u_{LD} \quad (7.15)$$

$$x_Q = k_Q \int \dot{x}_Q \, dt, \dot{x}_Q = -L_Q x_Q \quad (7.16)$$

$$x_\omega = k_\omega \int \dot{x}_\omega \, dt, \quad \dot{x}_\omega = -L_\omega x_\omega + B_\omega u_\omega \quad (7.17)$$

$$x_V = k_V \int \dot{x}_V dt, \quad \dot{x}_V = -L_V x_V + B_V u_V \quad (7.18)$$

where k_p, k_Q, k_ω, and k_V are the gain factors smaller than one for updating their corresponding sates. L_p, L_Q, L_ω, and L_V are graph Laplacian matrices based on the communication topology for active, reactive, frequency, and voltage. x_P, x_Q, x_ω, and x_V are the state vectors defined as follows:

$$x_P = \left[\frac{P_1}{P_1^{Max}}, \frac{P_2}{P_2^{Max}}, \frac{P_3}{P_3^{Max}}, \cdots, \frac{P_i}{P_i^{Max}} \right]^T \quad (7.19)$$

$$x_Q = \left[\frac{Q_1}{Q_1^{Max}}, \frac{Q_2}{Q_2^{Max}}, \frac{Q_3}{Q_3^{Max}}, \cdots, \frac{Q_i}{Q_i^{Max}} \right]^T \quad (7.20)$$

$$x_\omega = \left[\omega_{ref_1}, \omega_{ref_2}, \ldots, \omega_{ref_i} \right]^T \quad (7.21)$$

$$x_V = \left[V_{ref_1}, V_{ref_2}, \ldots, V_{ref_i} \right]^T \quad (7.22)$$

where $(.)^T$ means matrix transposition. The other variables are previously defined. The inputs to the above state equations are detailed as follow,

$$u_{ESS} = \begin{cases} -\dfrac{I_{ESS}}{I_{ESS}^{Max}} & \text{if } P_{ESS} < 0 \\ \\ 0 & \text{if } P_{ESS} \geq 0 \end{cases} \quad (7.23)$$

$$u_I = \begin{cases} \dfrac{I_k^{Max} - I_k}{I_k^{Max}} & \text{if } \left(I_k > I_k^{Max} \right) \\ \\ 0 & \text{if } \left(I_k \leq I_k^{Max} \right) \end{cases} \quad (7.24)$$

$$u_{Fluc} = -\frac{S}{S + \omega_c} P_{PCC} \quad (7.25)$$

$$u_{LD} = \begin{cases} -\dfrac{P_{PCC}}{P_{PCC}^{Max}} & \text{if } \left(I_k > I_k^{Max} \right) \\ \\ 0 & \text{if } (LD = 0) \end{cases} \quad (7.26)$$

7 Distributed Charging Management of Electric Vehicles in Smart Microgrids

$$u_\omega = \omega_{MainGrid} \tag{7.27}$$

$$u_V = \begin{cases} V^{Max} - V \text{ if } \left(V > V^{Max}\right) \\ V - V^{Max} \text{ if } \left(V \leq V^{Max}\right) \end{cases} \tag{7.28}$$

where I_{ESS}^{Max} [pu] is the maximum current of the ESS, ω_c [rad] is the cut frequency of the low-pass filter that supresses the noise in PCC power. The rest of the variables are defined before. The control input vectors related to the state equations are presented as follows,

$$\boldsymbol{B}_{ESS} = [b_{ESS1}, b_{ESS2}, b_{ESS3}, \ldots, b_{ESSi}]^T \tag{7.29}$$

$$\boldsymbol{B}_I = [b_{I1}, b_{I_2}, b_{I_3}, \ldots, b_{I_i}]^T \tag{7.30}$$

$$\boldsymbol{B}_{Fluc} = [b_{Fluc_1}, b_{Fluc_2}, b_{Fluc_3}, \ldots, b_{Fluc_i}]^T \tag{7.31}$$

$$\boldsymbol{B}_{LD} = [b_{LD_1}, b_{LD_2}, b_{LD_3}, \ldots, b_{LD_i}]^T \tag{7.32}$$

$$\boldsymbol{B}_{Fluc} = [b_{Fluc_1}, b_{Fluc_2}, b_{Fluc_3}, \ldots, b_{Fluc_i}]^T \tag{7.33}$$

$$\boldsymbol{B}_\omega = [b_{\omega_1}, b_{\omega_2}, b_{\omega_3} \ldots, b_{\omega_i}]^T \tag{7.34}$$

$$\boldsymbol{B}_V = [b_{V_1}, b_{V_2}, b_{V_3} \ldots, b_{V_i}]^T \tag{7.35}$$

In all of the above control input vectors ($\boldsymbol{B}$'s), if any EVSE receives a control input (u), the corresponding element (b) in the input vector is one, or else the element is zero.

In order to decrease the charging costs to the EV owners, the variable ACC was defined in [4] as a new property of EVs. ACC stands for Average Charging Cost, and it indicates the cost which is already paid for the present amount of charge (i.e., SoC). It is calculated by the moving average of the electricity price and mathematically expressed as below [4],

$$ACC_i(t) = \begin{cases} \dfrac{ACC_i(t_1) \times SOC_i(t_1) \times BC_i + \int_{t_1}^t P_i(\tau) \times EP(\tau)d\tau}{SOC_i(t_1) \times BC_i + \eta \int_{t_1}^t P_i(\tau) \times d\tau}, \text{for charging} \\ \\ \dfrac{ACC_i(t_1) \times SOC_i(t_1) \times BC_i - \int_{t_1}^t P_i(\tau) \times EP(\tau)d\tau}{SOC_i(t_1) \times BC_i - (1/\eta)\int_{t_1}^t P_i(\tau)d\tau}, \text{for discharging} \end{cases} \tag{7.36}$$

where $ACC_i(t)$ [\$/kWh] is the i-th EV's ACC at a given time t, t_1 shows the beginning of a charging/discharging, BC_i [kWh] is the capacity of the i-th EV's battery, $P_i(\tau)$ [kW] is the charging rate, $EP(t)$ [\$/kWh] is the electricity price that is announced by the DSO or electricity market as shown in Figs. 1.6. and 1.8. Finally, η indicates EVSE efficiency. The below numerical example is provided for the elaboration of

Table 7.6 The situation and parameters for a typical EV

Charging Process		Discharging Process	
$ACC(0)$ [\$/kWh]	0.120	$ACC(2)$ [\$/kWh]	0.105
$ACC(2)$ [\$/kWh]	0.105	$ACC(3.62)$ [\$/kWh]	0.102
$SoC(2)$ [%]	19	$SoC(3.62)$ [%]	10
$SoC(0)$ [%]	10	$SoC(2)$ [%]	19
P [kW]	1	P [kW]	1
EP [\$/kWh]	0.080	EP [\$/kWh]	0.120
η_c [%]	90	η_d [%]	90

the ACC concept. Please assume that an EV is connected to a DN. Its features are given in Table 7.6. The EV's battery capacity is 20 kWh (i.e., $BC = 20$ kWh). The charging process starts at $t_1 = 0$ and takes long for 2 hours. Utilizing the parameters given in Table 7.6, $ACC(2) = 0.12 \times 0.1 \times 20 + \int_0^2 1 \times 0.08 \, dt / 0.1 \times 20 + 0.9 \int_0^2 1 \, dt = 0.105$ \$/kWh, [4]. Therefore, the EV's ACC lessens by $0.12 - 0.105 = 0.015$ \$/kWh. Now, EP rises by \$0.004 and becomes 0.12 \$/kWh. Therefore, as the SoC of the EV is greater than its SoC^{Min}, it is able to discharge and sell electricity for making financial benefits. After 1.62 hours of discharging, the EV's SoC decreses to 10% which is identical to its SoC^{Min}. At this moment, the EV stops discharging since it is not desired to deplete the batteries to a level less than the minimum acceptable amount (SoC^{Min}). During this charging and discharging cycle, the EV started charging from an SoC equals 10% and again discharged to the same SoC. However, the new ACC is $ACC(3.62) = 0.105 \times 0.19 \times 20 - \int_2^{3.62} 1 \times 0.12 \, dt / 0.19$- $\times 20 - (1/0.9) \int_2^{3.62} 1 \, dt = 0.102$ \$/kWh. As the new ACC is less than its initial value (at $t_1 = 0$) for the same amount of charge, the EV contribution in the electricity market was economically beneficial to its owner.

The active-power output equation of the controller is

$$P_{ref} = h_p \left(x, SoC, SoC^{Min}, EC, P^{Max}, EP, LD \right) \tag{7.37}$$

where $P_{ref} = \left[P_{ref\,1}, P_{ref\,2}, \ldots, P_{ref\,i} \right]^T$ is the control output vector for active power, which consists of the power references for the local droop controllers. The active power references are determined by the function $h_p(.)$ that is described in Table 7.7. $SoC = [SoC_1, SoC_2, \ldots, SoC_i]^T$ is a vector that consists of all the EVs' SoCs. $SoC^{Min} = \left[SoC_1^{Min}, SoC_2^{Min}, \ldots, SoC_i^{Min} \right]^T$ is a vector comprising the minimum SoCs of the EVs in the DN. $P^{Max} = \left[P_1^{Max}, P_2^{Max}, P_3^{Max} \ldots, P_i^{Max} \right]^T$ is the vector of all the EVSEs' capacities. The control output function (h_p) is formulized in Table 7.7.

The reactive-power output equation of the controller is

7 Distributed Charging Management of Electric Vehicles in Smart Microgrids

Table 7.7 The control output function for active power

if $(EC_i = = 'On' \ \& \ SoC_i < SoC_i^{Min})$
$P_{ref_i} = P_i^{Max}$; % **Charging Mode** %
elseif $(EC_i = = 'Off'$ or $(EC_i = = 'On' \ \&$ $SoC_i > SoC_i^{Min}))$
if $(LD = 'On' \ \& \ SoC_i > SoC_i^{Min})$
$P_{ref_i} = -P_i^{Max}$; % **Discharging Mode** %
elseif $(LD = 'Off' \ \& \ SoC_i < SoC_i^{Min})$
$P_{ref_i} = x_i \times P_i^{Max}$; % **Charging Mode** %
elseif $(LD = 'Off' \ \& \ SoC_i > SoC_i^{Min})$
if $(ACC_i > EP)$
$P_{ref_i} = x_i \times P_i^{Max}$; % **Charging Mode** %
elseif $(ACC_i < EP)$
$P_{ref_i} = -P_i^{Max}$; % **Discharging Mode** %
endif
else
else
$P_{ref_i} = 0$; % **Idle Mode** %
end

$$Q_{ref} = h_Q(x, Q^{Max}) \tag{7.38}$$

where $Q_{ref} = \left[Q_{ref_1}, Q_{ref_2}, \ldots, Q_{ref_i}\right]^T$ is the control output vector for setting the reactive-power references in the local droop controllers, $h_Q(x) = x_Q \odot Q^{Max}$ is the control reactive-power function in which $\odot$ means elementwise matrix multiplication.

The output equation of the controller for the voltage and frequency are respectively $V_{ref} = x_V,$ and $\omega_{ref} = x_\omega$, where $V_{ref} = \left[v_{ref_1}, v_{ref_2}, \ldots, v_{ref_i}\right]^T$ is the control output vector for the voltage references that are applied to the local droop controllers. $\omega_{ref} = \left[\omega_{ref_1}, \omega_{ref_2}, \ldots, \omega_{ref_I}\right]^T$ is the control output vector for the frequency references which are applied to the local droop controllers.

7.6 Conclusion

As an enormous number of Electric Vehicles (EVs) will be in service in future, the current electric networks should be developed in a way that they can host EVs. This chapter, having compared prediction-based and myopic (also referred to as real-time or non-prediction-based) EV charging coordination methods, focused on the myopic methods for EV charging as the predictions are to some extent inaccurate and also the fact that contingencies are hard to predict. The control architectures are divided into the following three categories in terms of communication topology: centralized, decentralized, and distributed. These architectures were compared in this chapter and the pros and cons of each were discussed in line with the Smart Microgrids' (SMGs) requirements. Accordingly, it is concluded that the distributed control architectures are well-suited for SMG setups since they are:

- Highly scalable
- Plug-&-playable
- Less costly
- More robust against communication failures and cyber attacks

Among the distributed architectures, greedy (e.g., Game Theory) and consensus algorithms (e.g., Cooperative Control) are compared. As the parameters such as the Distribution Network (DN) frequency are identical throughout SMGs, all the agents must follow the same objective. In addition, the objectives such as increasing the system ride-through-ability are realized only when all the EVs recognize the control objectives prior to their individual benefits. Thus, consensus algorithms are proper for EVs' ancillary services. Hence, the cooperative control was selected for the establishment of the controller. The control objectives, including technical and economic ones, were defined, considering the abilities of the cooperative control, EVs, and EV Supply Equipment (EVSE). Then, the requirements for setting up the control framework were discussed. Finally, the control design was elaborated.

Appendix A

The nomenclature is shown below.

P_i	i-th EVSE's preferable active output power
P_j	j-th EVSE's preferable active output power
P_i^{Max}	Maximum active output power of the i-th EVSE
Q_i	i-th EVSE's preferable reactive output power
Q_j	j-th EVSE's preferable reactive output power
Q_i^{Max}	Maximum reactive output power of the i-th EVSE
m	Number of the EVSEs existing in a DN
ω_i	i-th EVSE's output frequency
$\omega_{MainGrid}$	Main grid frequency
V^{Min}	Minimum acceptable voltage in a DN
V^{Max}	Maximum acceptable voltage in a DN
V_i	Output voltage of the i-th EVSE
I_k	Measured current passing through the k-th location
I_k^{Max}	Maximum allowable current at the k-th location
n	Number of current sensors in a DN
SoC_i	i-th EV's state of charge

(continued)

SoC_i^{Min}	Minimum state of charge for the i-th EV
EC	Emergency charging
I_{ESS}	Output current of the energy storage system
P_{ESS}	Output power of the energy storage system
ACC_i	Average cost of charge
ω_i^*	Frequency command from the i-th local controller to the i-th charger
ω_{ref_i}	Frequency reference for the i-th droop controller
κ_{p_i}	Droop gain for active power-frequency
P_{ref_i}	Active power reference for the i-th droop controller
P_{O_i}	i-th EVSE measured active output power
V_i^*	Voltage command from the i-th local controller to the i-th charger
V_{ref_i}	Voltage reference for the i-th droop controller
κ_{Q_i}	Droop gain for reactive power-voltage
Q_{ref_i}	Reactive power reference for the i-th droop controller
Q_{O_i}	i-th EVSE measured reactive output power
L	Laplacian matrix
x	State vector
B	Control input matrix
u	Control input vector
N_i	i-th EVSE's neighbors
$|N_i|$	Indegree of the i-th EVSE
P_{PCC}	Active power flow at PCC
k	Integrator gain for updating controller's states
P_{PCC}^{Max}	Maximum active power flow at PCC
LD	Boolean variable defined for reducing the SMG dependency on the main grid
x_P	State vector for active power
x_Q	State vector for reactive power
x_ω	State vector for frequency
x_V	State vector for voltage
L_p	Laplacian matrix for active power
L_Q	Laplacian matrix for reactive power
L_ω	Laplacian matrix for frequency
L_V	Laplacian matrix for voltage
k_p	Integrator gain for active power
k_Q	Integrator gain for reactive power
k_ω	Integrator gain for frequency
k_V	Integrator gain for voltage
ω_c	Cut frequency of the derivative of the power flow at PCC
$\omega_{MainGrid}$	Frequency reference for the grid in islanded mode
EP	Electricity price
P_{ref}	Control output vector composed of active power references for the local droop controller
SoC	Vector consisting of all the EVs' SoCs
SoC^{Min}	Vector consisting of all the minimum acceptable EVs' SoCs

(continued)

EC	Vector comprising the EVs' emergency charging choice
P^{Max}	Vector of the maximum active power output of EVSEs
Q_{ref}	Control output vector of reactive power references for the local droop controller
Q^{Max}	Vector of the maximum reactive power output of EVSEs
V_{ref}	Control output vector consisting of voltage references for the local droop controller
ω_{ref}	Control output vector containing frequency references for the local droop controller

References

1. S. Han, S. Han, K. Sezaki, Development of an optimal vehicle-to-grid aggregator for frequency regulation. IEEE Trans. Smart Grid **1**(1), 65–72 (2010)
2. N. Leemput, J. Van Roy, F. Geth, P. Tant, B. Claessens, J. Driesen, Comparative analysis of coordination strategies for electric vehicles, in *Proc. ISGT Europe*, (Manchester, 2011), pp. 1–8
3. A. Mohsenian-Rad, V. W, S. Wong, J. Jatskev, R. Schrober, A. Leon-Garcia, Autonomous demand side management bases on game-theoretic energy consumption scheduling for the future smart grid. IEEE Trans. Smart Grid, 320–331 (2010)
4. R. Jalilzadeh Hamidi, H. Livani, Myopic real-time decentralized charging management of plug-in hybrid electric vehicles. Electric Power Syst. Res. **143**, 522–543 (2017)
5. M. Majidpour, C. Qiu, P. Chu, R. Gadh, H.R. Pota, Fast prediction for sparse time series: demand forecast of EV charging stations for cell phone applications. IEEE Trans. Ind. Inf. **11**(1), 242–250 (2015)
6. B. Khaki, Y. Chung, C. Chu, R. Gadh, Nonparametric user behavior prediction for distributed EV charging scheduling, in *Proc. 2018 IEEE Power & Energy Society General Meeting (PESGM)*, (Portland, 2018), pp. 1–5
7. J.J. Valera, B. Heriz, G. Lux, J. Caus, B. Bader, Driving cycle and road grade on-board predictions for the optimal energy management in EV-PHEVs, in *Proc. 2013 World Electric Vehicle Symposium and Exhibition (EVS27)*, (Barcelona, 2013), pp. 1–10
8. L. Agarwal, W. Peng, L. Goel, Probabilistic estimation of aggregated power capacity of EVs for vehicle-to-grid application, in *Proc. 2014 International Conference on Probabilistic Methods Applied to Power Systems (PMAPS)*, (Durham, 2014), pp. 1–6
9. G. Hilton, M. Kiaee, T. Bryden, B. Dimitrov, A. Cruden, A. Mortimer, A stochastic method for prediction of the power demand at high rate EV chargers. IEEE Trans. Transport. Electrif. **4**(3), 744–756 (2018)
10. Y. Liao, C. Lu, Dispatch of EV charging station energy resources for sustainable mobility. IEEE Trans. Transport. Electrif. **1**(1), 86–93 (2015)
11. X. Long, J. Yang, Y. Wang, X. Dai, X. Zhan, Y. Rao, A prediction method of electric vehicle charging load considering traffic network and travel rules, in *Proc. 2018 International Conference on Power System Technology (POWERCON)*, (Guangzhou, 2018), pp. 930–937
12. S. Zhao, X. Lin, M. Chen, Robust online algorithms for peak-minimizing EV charging under multistage uncertainty. IEEE Trans. Autom. Control **62**(11), 5739–5754 (2017)
13. Y. Xiong, C. Chu, R. Gadh, B. Wang, Distributed optimal vehicle grid integration strategy with user behavior prediction, in *Proc. 2017 IEEE PES MG*, (Chicago, 2017), pp. 1–5
14. S. Ai, A. Chakravorty, C. Rong, Household EV charging demand prediction using machine and ensemble learning, in *Proc. 2018 IEEE International Conference on Energy Internet (ICEI)*, (Beijing, 2018), pp. 163–168
15. M.H.K. Tushar, A.W. Zeineddine, C. Assi, Demand-side management by regulating charging and discharging of the EV, ESS, and utilizing renewable energy. IEEE Trans. Indust. Infor. **14**(1), 117–126 (2018)

16. J.D. Hamilton, *Time Series Analysis* (Princeton Univ. Press, Princeton, 1994)
17. G.E.P. Box, G.M. Jenkins, G.C. Reinsel, *Time Series Analysis: Forecasting and Control, 4th ed* (Wiley, Hoboken, 2013)
18. Y.Q. Li, Z.H. Jia, F.L. Wang, Y. Zhao, Demand forecast of electric vehicle charging stations based on user classification. Appl. Mech. Mater. **291–294**, 855–860 (2013)
19. K.N. Kumar, P.H. Cheah, B. Sivaneasan, P.L. So, D.Z.W. Wang, Electric vehicle charging profile prediction for efficient energy management in buildings, in *Proc. IEEE Conference Power Energy*, (2012), pp. 480–485
20. F. Kennel, D. Gorges, S. Liu, Energy management for smart grids with electric vehicles based on hierarchical MPC. IEEE Trans. Indust. Infor. **9**(3), 1528–1537 (2013)
21. R.J. Hamidi, H. Livani, S.H. Hosseinian, G.B. Gharehpetian, Distributed cooperative control system for smart microgrids. Electric Power Syst. Res. **130**, 241–250 (2016)
22. A. Bidram, A. Davoudi, F.L. Lewis, J.M. Guerrero, Distributed cooperative secondary control of microgrids using feedback linearization. IEEE Trans. Power Syst. **28**(3), 3462–3470 (2013)
23. S. Deilami, A.S. Masoum, P.S. Moses, M.A.S. Masoum, Real-time coordination of plug-in electric vehicle charging in smart grids to minimize power losses and improve voltage profile. IEEE Trans. Smart Grid **2**(3), 456–467 (2011)
24. H. Myoken, Hierarchical decentralized control and its application to macro econometric systems. IFAC Proc. Vol. **10**(6), 73–80 (1977)
25. J. Lian, J. Hansen, L.D. Marinovici, K. Kalsi, Hierarchical decentralized control strategy for demand-side primary frequency response, in *Proc. IEEE PES GM*, (Boston, 2016)
26. B. Jiang, Y. Fei, Decentralized scheduling of PHEV on-street parking and charging for smart grid reactive power compensation, in *Proc. ISGT*, (Shanghai, 2013), pp. 1–6
27. C.K. Wen, J.C. Chen, J.H. Teng, P. Ting, Decentralized energy management system for charging and discharging of plug-in electric vehicles, in *Proc. WCSP*, (Huangshan, 2012)
28. C.K. Wen, J.C. Chen, J.H. Teng, P. Ting, Decentralized plug-in electric vehicle charging selection algorithm in power systems. IEEE Trans. Smart Grid **3**(4), 1779–1789 (2012)
29. M. Gillie, G. Nowell, The future for EVs: reducing network costs and disruption, in *Proc. HEVC*, (London, 2013), pp. 1–5
30. E. Saunders, T. Butler, J. Quiros-Tortos, L.F. Ochoa, R. Hartshorn, Direct control of EV charging on feeders with EV clusters, in *Proc. CIRED*, (Lyon, 2015)
31. K. Turitsyn, N. Sinitsyn, S. Backhaus, M. Chertkov, Robust broadcast-communication control of electric vehicle charging, in *Proc. Smart Grid Comm*, (2010), pp. 203–207
32. M.G. Vaya, G. Andersson, S. Boyd, Decentralized control of plug-in electric vehicles under driving uncertainty, in *Proc. ISGT-Europe*, (Istanbul, 2014), pp. 1–6
33. M. Zhongjing, D. Callaway, I. Hiskens, Decentralized charging control for large populations of plug-in electric vehicles: application of the nash certainty equivalence principle, in *Proc. CCA*, (2010), pp. 191–195
34. Z. Ma, D.S. Callaway, I.A. Hiskens, Decentralized charging control of large populations of plug-in electric vehicles. IEEE Trans. Control Syst. Technol. **21**(1), 67–78 (2013)
35. I. Harrabi, M. Maier, Performance analysis of a real-time decentralized algorithm for coordinated PHEV charging at home and workplace with PV solar panel integration, in *Proc. IEEE PES GM*, (2014), pp. 1–5
36. R. Mahmud, A. Nejadpak, R. Ahmadi, Cooperative load sharing in V2G application, in *Proc. EIT*, (2015), pp. 451–456
37. R. Jalilzadeh Hamidi, R. Heidarykiany, T. Ashuri, Decentralized control system for enhancing smart-grid resiliency using electric vehicles, in *Proc. PECI*, (Champaign, 2019)
38. R. Jalilzadeh Hamidi, R.H. Kiany, Decentralized control framework for mitigation of the power-flow fluctuations at the integration point of smart grids, in *Proc. IEEE PES-GM*, (Atlanta, 2019)
39. R. Jalilzadeh Hamidi, T. Ashuri, R.H. Kiany, Reducing smart microgrid dependency on the main grid using electric vehicles and decentralized control systems, in *Proc. eNergetics*, (Nis, 2018)

40. A. Zakeri, O. Asgari Gashteroodkhani, I. Niazazari, H. Askarian-Abyaneh, The effect of different non-linear demand response models considering incentive and penalty on transmission expansion planning. Eur. J. Electr. Comput. Eng. **3**(1) (2019)
41. US Department of Energy, Office of Energy Efficiency & Renewable Energy, the official US government source for fuel economy information, Available: https://www.fueleconomy.gov/feg/evtech.shtml
42. US Department of Energy, Office of Energy Efficiency & Renewable Energy, Reducing Pollution with Electric Vehicles, Available: https://www.energy.gov/eere/electricvehicles/reducing-pollution-electric-vehicles
43. A. C. Z. de Souza, D. Q. Oliveira, P. F. Ribeiro, Overview of plug-in electric vehicles technologies, in *Plug-In Electric Vehicles in Smart Grids, Energy Management*, 1st edn. (Springer Singapore, Singapore, 2015), ch 1, sec. 1, pp. 1–24
44. US Department of Energy, Office of Energy Efficiency & Renewable Energy, Electric vehicle benefits, Available: https://www.energy.gov/eere/electricvehicles/electric-vehicle-benefits
45. US Department of Energy, Office of Energy Efficiency & Renewable Energy, Vehicle charging, Available: https://www.energy.gov/eere/electricvehicles/vehicle-charging
46. ABB, Electric vehicle infrastructure Terra 54 and Terra 54HV UL DC fast charging station, available: https://new.abb.com/ev-charging/products/car-charging/multi-standard
47. SAE International, Std. J1773_201406, SAE electric vehicle inductively coupled charging, 2014
48. IEC, Std. IEC 62196-1, Plugs, socket-outlets, vehicle connectors and vehicle inlets - conductive charging of electric vehicles-Part 1, 2014
49. IEC, Std. IEC 62196-2, Plugs, socket-outlets, vehicle connectors and vehicle inlets - conductive charging of electric vehicles-Part 2, 2014
50. International Energy Agency (IEA), Global EV outlook 2016 beyond one million electric cars, available: https://www.iea.org/publications/freepublications/publication/Global_EV_Outlook_2016.pdf
51. G.R.C. Mouli, P. Venugopal, P. Bauer, Future of electric vehicle charging, in *Proc. 19th International Symposium Power Electronics Ee2017*, (Novi Sad, 2017)
52. O.H. Hannisdahl, H.V. Malvik, G.B. Wensaas, The future is electric! The EV revolution in Norway – explanations and lessons learned, in *Proc. 2013 World Electric Vehicle Symposium and Exhibition (EVS27)*, (Barcelona, 2013)
53. J. Wolfe, What is holding America's EV future back while the world charges ahead? available: https://cleantechnica.com/2019/04/21/what-is-holding-americas-ev-future-back-while-the-world-charges-ahead/
54. L. Gkatzikis, I. Koutsopoulos, T. Salonidis, The role of aggregators in smart grid demand response markets. IEEE J. Sel. Areas Commun. **31**(7), 1247–1257 (2013)
55. S. Rahnama, S.E. Shafiei, J. Stoustrup, H. Rasmussen, J. Bendtsen, Evaluation of aggregators for integration of large-scale consumers in smart grid, in *Proc. 19th World Congress the International Federation of Automatic Control*, (Cape Town, 2014)
56. M. Jafari, A. Ghasemkhani, V. Sarfi, H. Livani, L. Yang, H. Xu, Biologically inspired adaptive intelligent secondary control for MGs under cyber imperfections. IET Cyber-Phys. Syst. Theory Appl., 1–12 (2019)
57. S.D.J. McArthur et al., Multi-agent systems for power engineering applications—Part I: concepts, approaches, and technical challenges. IEEE Trans. Power Syst. **22**(4), 1743–1752 (2007)

Chapter 8
Optimal Energy and Reserve Management of the Electric Vehicles Aggregator in Electrical Energy Networks Considering Distributed Energy Sources and Demand Side Management

Mehrdad Ghahramani, Morteza Nazari-Heris, Kazem Zare, and Behnam Mohammadi-ivatloo

Nomenclature

Indices	Index for:
t	Time
j	DG
i	PHEV
k	WT
Parameters	
P_R^k	The supplied power of kth WT at the rated speed of the wind
$P_W^{k,t}$	The supplied power of kth WT at time t
V_c^k	The lower bound of speed for kth WT
V_R^k	The nominal speed of kth WT
V_F^k	The upper bound of speed for kth WT
V^t	The predicted speed of wind at tth time
$P_{PV}^{p,t}$	The manufactured power of PV p at tth time
η^p	Arrays efficiency for PV
s^p	Surface size of PV

(continued)

M. Ghahramani · K. Zare
Faculty of Electrical and Computer Engineering, University of Tabriz, Tabriz, Iran

M. Nazari-Heris (✉)
Faculty of Electrical and Computer Engineering, University of Tabriz, Tabriz, Iran

Department of Architectural Engineering, Pennsylvania State University, State College, PA, USA

B. Mohammadi-ivatloo
Faculty of Electrical and Computer Engineering, University of Tabriz, Tabriz, Iran

Department of Energy Technology, Aalborg University, Aalborg, Denmark

© Springer Nature Switzerland AG 2020
A. Ahmadian et al. (eds.), *Electric Vehicles in Energy Systems*,
https://doi.org/10.1007/978-3-030-34448-1_8

Indices	Index for:
T_a	Temperature of ambient of the PV
G^t	Radiation on the surface of PV
a^j, b^j	The coefficients of supplied power cost of DG j
$P^j_{LDG,max}$	The upper bound of power supply of DG j
$P^j_{LDG,min}$	The lower bound of power supply of DG j
MUT_j/MDT_j	Minimum up/down-time for DG j
$t^{j,t}_{ON}, t^{j,t}_{OFF}$	Time interval of continuous on/off status of the jth DG at time t
UDC^j	Startup Cost of the jth DG
RD^j, RU^j	The rate of increase/decrease of the jth DG
$\psi^{j,t}_{LDG}$	The spinning reserve cost of DG j at time t
$\psi^{i,t}_{EV}$	The spinning reserve cost of EV i at time t
$TU^{j,n}/TD^{j,n}$	The minimum on/down-time of jth DG
π^t_{UG}	The energy cost of UG at time t
N_{Ev}	The amount of EVs parked in the EVs aggregator
P^{max}_{UG}	The exchangeable energy between the DN and the UG
Δt	Time for computing available EV in the EVs aggregator
$P^i_{Ch,max}/$ $P^i_{Dch,max}$	The maximum amount of charge/ discharge of charger i
$SOC^i_{max}/$ SOC^i_{min}	The max/min amounts of SOC for the EV i
ΔSOC^i_{max}	The upper bound of allowable charging/discharging rate of EV i
T^i_p	Estimated time of presence EV i in the EVs aggregator
$\pi^i_{Ch,Ev}$	The optimal EV charging price in the EVs aggregator
$\pi^i_{Dch,Ev}$	The optimal cost of discharging for ith EV in the EVs aggregator
η_{V2G}	The efficiency of discharging EV battery
η_{G2V}	The efficiency of charging EV battery
$SOC^{i,t}_{Arrival}$	The initial SOC for ith EV at leaving from EVs aggregator at time t
N_{max}	The upper restriction for switching among the states of charge/discharge
ω_W	The forecasted error of wind speed
ω_{Pv}	The forecasted error of solar radiation
$load^t_0$	The base load at time t
$M^{i,t}$	A binary variable that is equal to 1 if the EV i is in the EVs aggregator at time t otherwise it is 0
α	The ratio of the EV discharged power participating in spinning reserve market
t^i_a	Approximate entrance time of ith EV to EVs aggregator
t^i_d	Approximate leaving time of ith EV from EVs aggregator
Variables	
P^t_{UG}	Power exchange between DN and UG
$C^{j,t}_{LDG}$	Power supply by DG j
$SC^{j,t}_{LDG}$	Startup cost of DG j
$SR^{j,t}_{LDG}$	Scheduling of spinning reserve of DG j at time t

(continued)

Indices	Index for:
$P^{i,t}_{Ch,Ev}$	The charge power of EV i at time t
$P^{i,t}_{Dch,Ev}$	The discharge power of EV i at time t
$SR^{i,t}_{Ev}$	The scheduling of EV i spinning reserve at time t
$P^{j,t}_{LDG}$	The scheduled power of DG j at time t
$SOC^{i,\ t}$	SOC of ith EV at tth time
$\Delta SOC^{i,\ t}$	Energy change between two consecutive time intervals
$SOC^{i,t}_{Departure}$	Final SOC of the EV i when leaving the EVs aggregator
$Up_{j,\ t}/Dn_{j,\ t}$	Auxiliary variables for linearization modelling of DG j minimum up/down-time
$load^t$	The load considering effect of DRP
DR^t	The participation value of DRP at time t
idr^t	Shifted demand from a time to another one
$load^t_{inc}$	The value of increased demand at time t
inc^t	The size of increased demand at time t
$U^{j,\ t}$	Binary variable that is equal to 1 if DG j is on at time t otherwise it is 0
$W^{i,t}_{ch}$	Binary variable that is equal to 1 if the EV i in the EVs aggregator is in charging mode
$W^{i,t}_{Dch}$	Binary variable that is equal to 1 if the EV i in the EVs aggregator is in discharging mode
$SRS^{i,\ t}$	Binary variable that is equal to 1 if the EV i participates in spinning reserve at time t otherwise it is 0

8.1 Introduction

The power generation and transmission companies were legislators of the electricity industry in traditional power grids. In fact, electricity consumers were only interested in getting electricity at fixed prices without considering the volatility of the electricity market. In addition to the reduction of efficiency and system performance, environmental pollution and an increase in the average global temperature forced policymakers to restructure the power systems [1–3]. Some strategies have been selected in the field of restructuring electricity industry to performance modification and the environmental pollution reduction, which can be noted as practical ways in making smart DNs and PHEV parking lots [4, 5].

DRP is known as an effective instrument to create the possibility of load-side coordination in optimal scheduling of the electricity grid. It has facilitated the involvement of loads in critical condition in order to decrease network load in a short time [6]. Two types of DRP used in this chapter are incentive-based programs [7]. Incentive DRPs offer money to consumers for decreasing their consumption or shift it to none-peak periods.

A series of challenges such as the increasing requirement for power demand and reduction of fossil-fueled power plants increase utilization of RESs [8, 9]. By modifying the structure of the power network and moving to make smart electric networks, smart DNs have a significant characteristic in providing energy consumption [10]. The presence of distributed power plants in the DN will create important benefits such as higher efficiency, fewer environmental challenges, and lower economic problems. The presence of distributed units like WT [11], PV [12, 13], MT [14, 15], FC [16, 17] are studied through various studies.

There are valuable studies done by researchers in the field of PHEVs aggregator energy scheduling with the aim of sufficient charge/discharge of EVs. For example, in [18] the parking lots make a connection between grid and EVs. The authors have proposed a game theory-based model for charging EV in the parking lot in [19]. In [20], the optimal charging schedule of EVs is investigated in two scenarios, where one of them is related to the parking lot service on the day and besides the commercial consumers, and the other service at night and besides the residential consumers. In [21], a scheduling system for EVs in the parking lot is presented by using the real movements and park patterns of the EVs with a focus on the personal parking lot. In [22], an optimal energy and reserve scheduling model is presented for EVs, where DRP and the satisfaction of the EV owner are considered. A method based on probabilities is presented in [23], which uses the estimated points to specify the optimal placement of EV parking lot in the DN and obtain the optimal capacity of the EVs parking. In [24], a fuzzy method is presented for coordination of EVs in DN. In [25], a parking lot with a PV roof is studied by a mathematical method, where the aim is estimating discharge capacity of parking. In [26], EVs battery in the aggregator is taken into account as a source of the energy storage in the multifunctional networks. Optimal placement of the EVs aggregator in DNs is studied in [27], where the objective function is decreasing costs, decreasing losses and increasing reliability of the DN. In [28], a common parking lot is converted to the EVs aggregator and therefore, charging and discharging of EVs will take place in huge amounts. For enhancing the sales of the stored energy in batteries of the PHEVs, some proceedings have been studied such as aggregator settings, disconnecting laws, free charging, and other conditional laws, which are obtained from scientific publications and experimental observations [29]. In [30], a model of energy resources of the DN management is proposed considering energy supply constraints in the DN and constraint of EVs and EV owners. In [31], a model of scheduling and intelligent management for an EVs aggregator considering constraints of the EV battery and capacity of EV batteries is presented. A stochastic planning model is proposed for charging and discharging of PHEVs inside the aggregator. Two optimizations are done in [32], where the first optimization is related to the optimum size and location of installed distributed generators in the DN and the second optimization is related to the optimal size of the combined RESs and the appropriate number of the decision variables. A multi-objective approach is presented in [33] to ascertain the best location and the optimal size of the EVs aggregator while contributing to supplying the demand of the DN. In [34], a multi-objective algorithm is proposed to specify the optimum number, location and size of the EVs aggregator in DN and determine the produced power by each energy source in the DN.

8 Optimal Energy and Reserve Management of the Electric Vehicles... 215

In this chapter, an energy management model for a DN is presented, in which the EVs are coordinated in energy and reserve providing of DN, and an aggregator is a connection between DN and EVs. In addition, DRPs is used for decreasing the performance cost of DN. The proper amount for charge and discharge scheduling of EVs, as well as contribution amount of DGs, FC, WT, and DRP are solved in various cases for showing the advantages of the EVs aggregator contribution in energy and reserve providing and decreasing costs.

To sum up, the innovations of this chapter are as:

1. Integrated modeling for DGs, WT, FC, DRP and EVs aggregator in a DN.
2. Utilizing the EVs aggregator to comfort connection among the EVs and DN.
3. Using the DRP in order to reduce operation costs.
4. A full mathematical scheme for EVs aggregator and energy networks.
5. A novel scheme for simultaneous energy and reserve planning of a DN

The structure of this chapter is organized in four sections. A mathematical scheme has provided in Sect. 8.2. The proposed method is applied to a 33-bus DN and their results were compared with each other to analyze the role of the EVs aggregator in Sect. 8.3. Section 8.4 as the last section makes the main conclusions.

8.2 Formulation

The proposed DN in this chapter includes various distributed sources such as WT, DG, and EVs aggregator. To increase the reliability of the network, DN is connected to UG to supply energy according to necessities of the DN. The EVs aggregator has an important role in decreasing costs, and it has the characteristic of a power plant with discharging EVs. When EVs arrive in the aggregator, some of the information is received from their owners. Then, EVs aggregator sends the received information to the operator in order to reach optimal scheduling and reduce the costs.

8.2.1 Objective Function

The main goal of this chapter is reducing the operation costs of the DN, and the objective function is mathematically modeled as follows:

$$OBJ = \sum_{t=1}^{T} \left[\begin{pmatrix} P_{UG}^t \times \pi_{UG}^t + \\ \sum_{j=1}^{G} (C_{LDG}^{j,t} + SC_{LDG}^{j,t} + (SR_{LDG}^{j,t} \times \psi_{LDG}^{j,t})) + \\ + \sum_{n=1}^{N_{LIL}} \{ C_{LIL}^E(n,t) + C_{LIL}^R(n,t) \} + \sum_{l=1}^{N_{DRA}} \{ C_{DRA}^E(l,t) + C_{DRA}^R(l,t) \} \\ \sum_{i=1}^{N} \left(-W_{Ch}^{i,t} \times P_{Ch,EV}^{i,t} \times \pi_{Ch,EV}^i + W_{Dch}^{i,t} \times P_{Dch,EV}^{i,t} \times \pi_{Dch,EV}^i + SRS^{i,t} \times SR_{EV}^{i,t} \times \psi_{EV}^{i,t} \right) \end{pmatrix} \times \Delta t \right]$$

$$(8.1)$$

Equation (8.1) consists of four parts. The first and second parts include the exchanged power cost between DN and UG, and the start-up and operation costs of DG in the DN. The third part is the cost of DRP including industrial large loads (LIL) and demand response aggregators (DRA) in supplying energy and reserve. Also, the fourth part is the charging/discharging cost and exchange power between DN and EV. The costs and constraints of the costs will be discussed in the following.

8.2.2 WT

The mathematical formulation of the WT power is based on air speed as [35]:

$$P_W^{k,t} = \begin{cases} 0 & V^t < V_c^k \, or \, V^t \geq V_F^k \\ \dfrac{V^t - V_c^k}{V_R^k - V_c^k} \times P_R^k & V_c^k \leq V^t < V_R^k \\ P_R^k & V_R^k \leq V^t < V_F^k \end{cases} \qquad (8.2)$$

8.2.3 PV

The relation between PV and solar radiation and temperature is as [36]:

$$P_{PV}^{p,t} = G^t \times s^p \times \eta^p \times (1 - 0.005 \times (T_a - 25)) \qquad (8.3)$$

8.2.4 DG

The operation and start-up costs of DGs are as follows:

$$C_{LDG}^{j,t} = a^j \times U^{j,t} + b^j \times P_{LDG}^{j,t} \qquad (8.4)$$

$$SC_{LDG}^{j,t} \geq \left(U^{j,t} - U^{j,t-1} \right) \times UDC^j \qquad (8.5)$$

$$SC_{LDG}^{j,t} \geq 0 \qquad (8.6)$$

Constraints (8.7, 8.8, 8.9, 8.10, 8.11 and 8.12) covers the DG restrictions, which are formulated as follows:

$$P_{LDG}^{j,t} + SR_{LDG}^{j,t} \leq P_{LDG,\,max}^{j} \times U^{j,t} \qquad (8.7)$$

$$P_{LDG}^{j,t} \geq P_{LDG,\,min}^{j} \times U^{j,t} \qquad (8.8)$$

$$P_{LDG}^{j,t} - P_{LDG}^{j,t-1} \leq RU^j \times U^{j,t} \qquad (8.9)$$

$$P_{LDG}^{j,t-1} - P_{LDG}^{j,t} \leq RD^j \times U^{j,t-1} \qquad (8.10)$$

$$U^{j,t} - U^{j,t-1} \leq U^{j,t+Up_{j,f}} \qquad (8.11)$$

$$U^{j,t-1} - U^{j,t} \leq 1 - U^{j,t+Dn_{j,f}} \qquad (8.12)$$

$$Up_{j,f} = \begin{cases} f & f \leq MUT_j \\ 0 & f > MUT_j \end{cases} \qquad (8.13)$$

$$Dn_{j,f} = \begin{cases} f & f \leq MDT_j \\ 0 & f > MDT_j \end{cases} \qquad (8.14)$$

Power supply by each DG should be limited to its restrictions as (8.7) and (8.8). Equations (8.9) and (8.10) indicate the constraints of increase/decrease of power production of each DG in consecutive time intervals. Also, the minimum up/down-times of each DG are as (8.11) and (8.12), respectively. Linear models of minimum up/down-times of each DG are as (8.13) and (8.14), respectively.

8.2.5 The Constraint of the UG

Equation (8.15) restricts the power exchange between DN and UG.

$$\left| P_{UG}^t \right| \leq P_{UG}^{max} \qquad (8.15)$$

8.2.6 EVs Aggregator of the EVs

EVs in the aggregator should consider the following constraints for exchanging power with the EVs aggregator. Constraints (8.16) and (8.17) are used to limit the highest rate of charge/discharge of each charger.

$$P_{Ch,EV}^{i,t} \leq P_{Ch,\max}^{i} \times W_{ch}^{i,t} \times M^{i,t} \tag{8.16}$$

$$P_{Dch,EV}^{i,t} + SR_{EV}^{i,t} \leq P_{Dch,\max}^{i} \times W_{Dch}^{i,t} \times M^{i,t} \tag{8.17}$$

Constraint (8.18) aims to avoid charge and discharge of EV batteries simultaneously.

$$W_{ch}^{i,t} + W_{Dch}^{i,t} \leq 1 \times M^{i,t} \tag{8.18}$$

Equation (8.19) allows operators to consider restrictions for changing from charge state to discharge state and vice versa.

$$\sum_{t=t_a^i}^{t_d^i} W_{ch}^{i,t} + W_{Dch}^{i,t} \leq N_{\max} \tag{8.19}$$

Constraints of the spinning reserve of the EVs, which only participate in discharge mode, are presented in (8.20) and (8.21).

$$SR_{EV}^{i,t} \leq \alpha \times P_{Dch,\max}^{i} \times SRS^{i,t} \times M^{i,t} \tag{8.20}$$

$$SR_{EV}^{i,t} \leq \alpha \times P_{Dch,\max}^{i} \times W_{Dch}^{i,t} \times M^{i,t} \tag{8.21}$$

At each period of the time, the stored energy of EV obtains from the EV charging and discharging plus the efficiency of the EVs charge/discharge as (8.22).

$$SOC^{i,t} = SOC^{i,t-1} + \eta_{G2V} \times P_{Ch,EV}^{i,t} - 1/\eta_{V2G} \times P_{Dch,EV}^{i,t} \tag{8.22}$$

The amount of stored energy in the EV should be limited to its lower and upper bounds as (8.23).

$$SOC_{\min}^{i} \leq SOC^{i,t} \leq SOC_{\max}^{i} \tag{8.23}$$

Constraint (8.24) allows the operators to consider the charge/discharge limits of the EV through the operation. Accordingly, the difference between the speed of the charge and discharge for various EVs are considered by using the following equation.

$$-\Delta SOC^i_{\mathrm{max}} \leq SOC^{i,t} - SOC^{i,t-1} \leq \Delta SOC^i_{\mathrm{max}} \tag{8.24}$$

Constraint (8.25) specifies the SOC of the EVs at leaving time from the aggregator. Also, constraint (8.26) specifies the amount of EV energy when EVs enter to the aggregator.

$$SOC^{i,t}_{\mathrm{Departure}} = SOC^i_{\mathrm{max}} \tag{8.25}$$

$$SOC^{i,t} \geq SOC^{i,t}_{Arrival} \tag{8.26}$$

8.2.7 DR Program

The proposed methodology considers two types of programs for consumers to involve in DR programs. DRAs in the first type creates an opportunity for the participation of small consumers. So, the small consumers will be connected to the DSO for cooperating in DR programs and receiving rewards by such cooperation, which can be modeled as (8.27, 8.28, 8.29 and 8.30). DRAs are responsible for aggregating the responses of consumers and delivering it to the DSO.

$$\hbar^d_{\mathrm{min}} \leq H^d_1 \leq \hbar^d_1 \tag{8.27}$$

$$0 \leq H^d_k \leq \left(\hbar^d_{k+1} - \hbar^d_k \right) \quad \forall\, k = 2, 3, \ldots, K \tag{8.28}$$

$$P^{DAS}_{DRA}(d,t) = \sum_k H^d_k \tag{8.29}$$

$$cost^{DAS,E}_{DRA}(d,t) = \sum_k \omega^{k,d}_{DRA} \times H^d_k \tag{8.30}$$

Equation (8.27) provides the limitation of the accepted amount of decrement by lth aggregator (h^l_1) between lower bound of the decreaseable amount (h^l_{min}) and the proposed reduction by aggregator in the first step (H^l_1). The presented acceptance of the lth aggregator can be between zero and the amount of load decrement proposed in each of the other steps, which is satisfied by (8.28). The sum of decreased power by lth aggregator at time t (P_{DRA}) is the same with the sum of whole reducing offers accepted at that time, which is formulated in (8.29). In addition, the load reduction cost through the aggregator is computed by (8.30) based on the cost of reducing energy consumption ($\omega^{k,l}_{DRA}$) in the accepted reduction of the consumer d. The total value of energy scheduled (P_{DRA}) and reserve scheduled (R_{DRA}) will be restricted to the upper bound of accepted demand decrement (P^{max}_{DRA}), which is satisfied by (8.31). Equation (8.32) computes the cost of providing reserve by DRAs based on the cost of each reserve unit KR_{DRA}.

$$P^{DAS}_{DRA}(d,t) + R^{DAS}_{DRA}(d,t) \le P^{\max}_{DRA}(d,t) \tag{8.31}$$

$$cost^{DAS,R}_{DRA}(d,t) = R^{DAS}_{DRA}(d,t) \times \Omega_{DRA}(d,t) \tag{8.32}$$

Decrement of load by LILs is the second classification of DR utilized in this chapter, which can reduce their load when required by the DSO. Equation (8.33) derives the limitation of reserve and energy schedule, and (8.34) and (8.35) calculates the cost of energy and reserve supply.

$$P^{DAS}_{LL}(i,t) + R^{DAS}_{LL}(i,t) \le P^{\max}_{LL}(i,t) \tag{8.33}$$

$$cost^{DAS,E}_{LL}(i,t) = P^{DAS}_{LL}(i,t) \times \omega_{LL}(i,t) \tag{8.34}$$

$$cost^{DAS,R}_{LL}(d,t) = R^{DAS}_{LL}(i,t) \times \Omega_{LL}(i,t) \tag{8.35}$$

The sum of the energy (P^{DAS}_{LL}) and the reserve (R^{DAS}_{LL}) supplied by LILs should be lower than the maximum decreaseable value ($P^{\max}_{LL}$) as mentioned by (8.33). Equation (8.34) states that the cost of decreasing energy by LILs (CE_{LL}) is the product of the decreased energy consumption of the LILs (P_{LL}) and the power reduction cost per unit (KR_{LL}).

8.2.8 Spinning Reserve of the DN Constraints

If any problem occurs and wind turbines are not capable of power injection to the distribution system or the predicted load change from prediction, DGs, EVs aggregator, LILs and DRA should provide electrical energy to the distribution system and cause balance between produced power and power consumed in the distribution network, for which (8.36) is expressed. This constraint guarantees that a 20% variation in load and wind will be addressed by the distribution network.

$$\sum_{l=1}^{L} R_{DRA}(l,t) + \sum_{n=1}^{N} R_{LIL}(n,t) + \sum_{j=1}^{G} SR^{j,t}_{LDG} + \sum_{i=1}^{I} SR^{i,t}_{EV}$$
$$\ge \left(0.2 \times P^{k,t}_{W} + 0.2 \times load^t\right) \tag{8.36}$$

8.2.9 Power Balance

The balance between generated energy and load of the DN, which is mathematically modeled by (8.37).

$$P^t_{UG} + \sum_{k=1}^{K} P^{k,t}_W + \sum_{p=1}^{P} P^{p,t}_{PV} + \sum_{j=1}^{G} P^{j,t}_{LDG} + \sum_{i=1}^{N} P^{i,t}_{Dch,EV} - \frac{1}{\eta^{dis}} \times P_{BESS,dis}(b,t)$$

$$= load^t + \sum_{i=1}^{N} P^{i,t}_{Ch,EV} + \eta^{ch} \times P_{BESS,ch}(b,t)$$

$$(8.37)$$

8.3 Case Study

In this chapter, an objective function is proposed for investigating the effect of the EVs aggregator on a DN and the obtained results are compared through the two cases. The main aim of this chapter is reducing operation costs of the DN considering constraints of DGs, LILs, DRAs, EVS aggregator and UG.

8.3.1 Input Data

Table 8.1 provided the required information for WTs and PV [37]. The required data of the DGs are provided in Table 8.2, which contains MT and FC. The prediction of

Table 8.1 WT and PV data

Photovoltaic system			Wind turbine		
Parameter	Value	Unit	Parameter	Value	Unit
η	15.7	%	P_R	500	MW
s	1500	m^2	V_C	3	m/s
T_a	25	C°	V_R	12	m/s
ω_{PV}	20	%	V_F	30	m/s
			ω_W	20	%

Table 8.2 DGs data

Unit	a_i ($)	b_i ($/MWh)	c_i ($/MWh2)	Startup cost($)	Minimum up/down time (h)	Maximum ramp up/down rate (MW/h)	Pmax (MW)	Pmin (MW)
MT 1	27	87	0.0025	15	2	1.8	3.5	1
FC 1	25	87	0.0035	25	1	1.5	3	0.75
FC 2	28	92	0.0035	28	1	1.5	3	0.75
MT 2	26	81	0.184	26	2	1.8	4.1	1

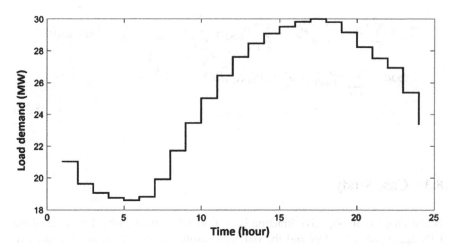

Fig. 8.1 Predicted demand

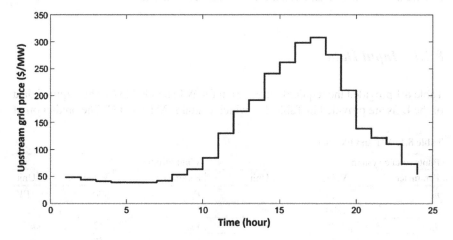

Fig. 8.2 Predicted UG price

the demand curve and upstream day-ahead power price are presented in Figs. 8.1, 8.2, 8.3 and 8.4 are related to WTs and PV [38]. The spinning reserve cost of the ith EV is considered as 10% of the ith EV desirable discharge price. The spinning reserve cost of the jth DG is considered 10% of the open market electricity prices at each period of the time. The capacity of the EVs aggregator is 220 EV and the SOC of EVs at leaving time from the EVs aggregator are considered a random value between 0.15 and 0.75. The charged and discharged price of EVs in the aggregator are surmised accidentally from 0.1 to 0.4. Some other required parameters of EVs are mentioned in Table 8.3. Exchanged power between the DN and UG is also limited to 1000 KW. DRPs have the capacity of shifting 20% of the demand for 24 hours. Exchanged energy among the DN and UG is also restricted by 100 MW.

8 Optimal Energy and Reserve Management of the Electric Vehicles... 223

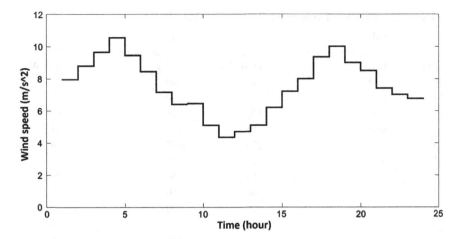

Fig. 8.3 Predicted speed of wind

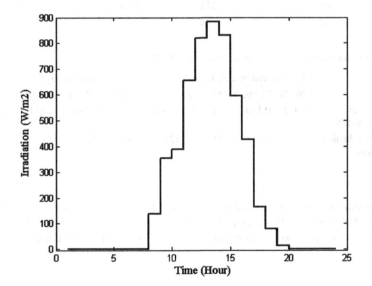

Fig. 8.4 Predicted radiation for solar

Table 8.3 Parameters of the EVs

α	T_P^i	$P_{Ch,\max}^i$	$P_{Dch,\max}^i$	$SOC_{\max}^i$	$SOC_{\min}^i$	$\Delta SOC_{\max}^i$	η_{G2V}	η_{V2G}	$N_{\max}$
0.2	2~8	5~10	5~10	10~20	0	5~10	0.9	0.8	10

Table 8.4 Bid-quantity offers of power decrement by LILs

Hour	LIL1		LIL 2	
	Maximum Decrease (MW)	Cost ($/MWh)	Maximum Decrease (MW)	Cost ($/MWh)
10	0.85	43	0.40	18
11	0.90	77	0.40	30
12	0.90	122	0.45	53
13	0.95	108	0.45	43
14	1	273	0.45	79
15	1	122	0.45	43
16	1	404	0.50	79
17	1	304	0.50	73
18	1	126	0.50	67
19	1	118	0.45	47
20	0.95	84	0.45	40
21	0.90	104	0.45	33
22	0.90	318	0.40	32
23	0.85	72	0.40	16

Table 8.5 Bid-quantity offers of power decrement by DRAs

DRA	The covered area by DRA 1				The covered area by DRA 2			
Buses	26, 27, 28, 29, 30, 31, 32,33				9, 10, 11, 12, 13, 14, 15, 16, 17,18			
Quantity of reduction (MW)	0–0.1	0.1–0.7	0.7–1.2	1.2–1.5	0–0.3	0.3–0.8	0.8–1.3	1.3–1.8
Cost of reduction ($/MWh)	9	68	102	136	20	51	88	108

The day-ahead offers of reduction for the power of the LILs and DRAs are proposed in step-by-step status that is provided in Tables 8.4 and 8.5. The reserve cost accepted by responder demands is 10% of accepted maximum decrement of power.

8.3.2 Simulation Results

To evaluate the influence of EVs aggregator in power and reserve supply of DN, two case studies are selected for testing the presented framework. Through the first case, the model has been investigated regarding the constraints of the UG, DGs, LILs, and DRPs. In case two, the EVs aggregator participates in the energy and reserve management of DN while constraints of the UG, DGs, LILs, and DRPs are considered, and operating costs of the DN have been investigated. The output power of the PV and the WTs are shown in Figs. 8.5 and 8.6. It should be mentioned that because

8 Optimal Energy and Reserve Management of the Electric Vehicles... 225

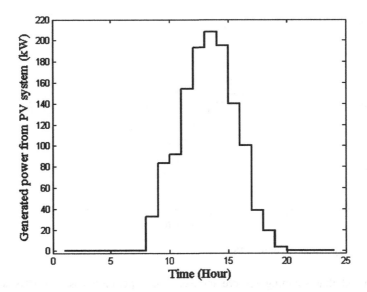

Fig. 8.5 PV output power

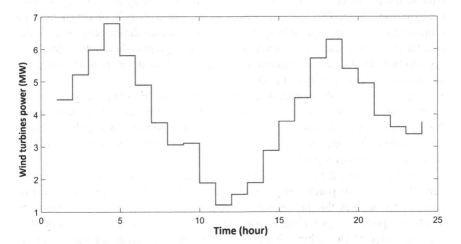

Fig. 8.6 WTs output power

the short-term operation of PV and WTs has not cost, PV and WTs are working in their maximum capacity in both cases.

Also, purchasing power from UG is getting lower in comparison with the first case. In other words, less power purchasing at high price hours cause a decrement in operation costs. Decreasing the provided energy by the first FC and increment of the power supplied by the second FC lead to a reduction in operation costs. In the second case, the presented objective function is investigated considering the aggregator for observing the participation influence of the EVs on the operating costs of the DN. It

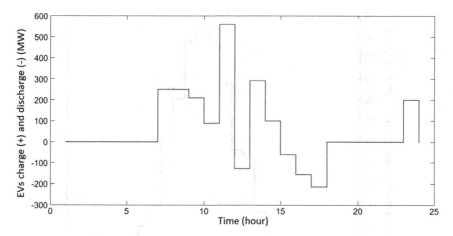

Fig. 8.7 Charge/discharge of the EVs

is admitted that spinning reserve increases the operation costs of DNs lead to reliability costs. These increases in the operation cost is due to considering 20% error in the predicted weather conditions and the probability of reducing the produced power by the WT. By comparing cases 1 and 2, it is observed that the contribution of the aggregator in spinning reserve releases the capacity of DGs and help them to contribute to energy supply at market high price hours. In general, the aggregator had reduced the operation costs of the DN. In addition to its advantages, it has increased the flexibility of the DN.

Figures 8.7 shows the discharging and charging of the EVs in the aggregator. As it can be observed, the price of the upstream grid is high at some time intervals and this means more discharging occurred at these time periods. In addition, there is more charging in aggregator at low price hours, which helps the operator to reduce the operation costs of the DN.

Figure 8.8 shows the energy exchange between the DN and UG. In the second case, the aggregator purchases power from the UG at off-peak periods with low price, and it utilizes this power for charging the EVs. In addition, it utilizes the purchasing power for providing the required energy of DN during on-peak periods when the price of UG is high. Accordingly, this process causes a reduction in the operation costs of the DN.

The power output of the DGs is demonstrated in Figs. 8.9 and 8.10, respectively. By comparison of Figs. 8.9 and 8.10, it can be known that considering the presence of EVs in case 2, the capacity of DGs is released, and such plants are capable to generate more electrical energy at on-peak hours with respect to case 1, and hence less power has been supplied from the main grid. In addition, the reserve participation of DGs in cases 1 and 2 are shown in Fig. 8.11. It is clarified that the responsibility of reserve supply is given to EVs in case 2, where the reserve cooperation of DGs has been decreased.

8 Optimal Energy and Reserve Management of the Electric Vehicles... 227

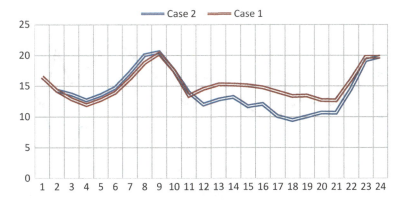

Fig. 8.8 Energy exchange with the UG

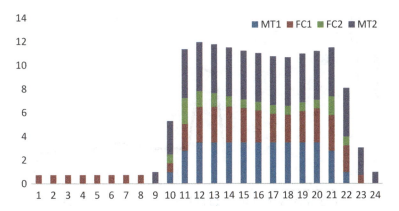

Fig. 8.9 Output power of the MTs and FCs in case 1

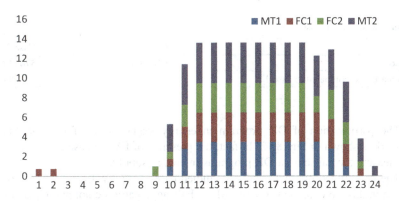

Fig. 8.10 Output power of the MTs and FCs in case 2

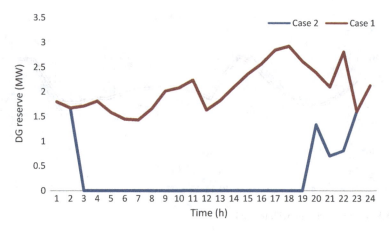

Fig. 8.11 Reserve participation of DGs in cases 1 and 2

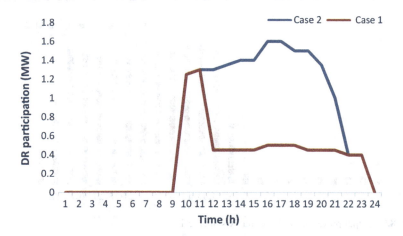

Fig. 8.12 The participation of DR program in energy management of the distribution network

Table 8.6 operation costs

	Case 1	Case 2
Costs	62819.6	57414.637
Decrease cost compared to case 1	–	8.6%

The participation of DR program in energy management of the distribution network is shown in Fig. 8.12. In fact, the presence of EVs in case 2 was effective in releasing the capacity of DR programs in energy supply.

The determined optimal results of studied cases are provided in Table. 8.6. As shown in this table, in the first case, where EVs aggregator did not participate in the energy market and the reserve market, the operation cost was $62819.6. In case

2, the model has been investigated to examine the impact of the aggregator on reducing the operation costs of the DN. By comparing the first and second cases, it can be seen that aggregator Causes 8.6% reduction in costs.

8.4 Conclusion

In this chapter, optimal scheduling for operation of the EVs aggregator is presented, which has two various performance as a load and power source for the DN. The main aim of the model is reducing operational costs of the DN, which contains EVs aggregator and RESs such as MT, FC, WT, and PV. In this chapter, the effect of EVs aggregator are studied, and the results are compared to show that EVs had a positive influence on decreasing the costs and realizing the capacity of DGs. Additionally, by investigating the simulation results, it can be observed that EVs aggregator plays a collector role, which prevents the system from occurrence the overload and decreases the risk of increasing demand in peak hours by managing charging/discharging of the EVs. Also, the aggregator may produce power balance and cost challenges for DN.

References

1. D. Zhang, S. Evangelisti, P. Lettieri, L.G. Papageorgiou, Economic and environmental scheduling of smart homes with microgrid: DER operation and electrical tasks. Energy Convers. Manag. **110**, 113–124 (2016)
2. M. Nazari-Heris, B. Mohammadi-Ivatloo, G.B. Gharehpetian, A comprehensive review of heuristic optimization algorithms for optimal combined heat and power dispatch from economic and environmental perspectives. Renew. Sust. Energ. Rev. **81**, 2128–2143 (2018)
3. J. Song, W. Yang, Y. Higano, X.E. Wang, Introducing renewable energy and industrial restructuring to reduce GHG emission: Application of a dynamic simulation model. Energy Convers. Manag. **96**, 625–636 (2015)
4. D.S. Chai, J.Z. Wen, J. Nathwani, Simulation of cogeneration within the concept of smart energy networks. Energy Convers. Manag. **75**, 453–465 (2013)
5. O. Sadeghian, M. Nazari-Heris, M. Abapour, S.S. Taheri, K. Zare, Improving reliability of distribution networks using plug-in electric vehicles and demand response. J. Modern Power Syst. Clean Energy. **7**(5), 1189–1199 (2019)
6. E. Heydarian-Forushani, M.P. Moghaddam, M.K. Sheikh-El-Eslami, M. Shafie-Khah, J.P. Catalão, A stochastic framework for the grid integration of wind power using flexible load approach. Energy Convers. Manag. **88**, 985–998 (2014)
7. M. Ghahramani, M. Nazari-Heris, K. Zare, B. Mohammadi-Ivatloo, Energy and reserve management of a smart distribution system by incorporating responsive-loads/battery/wind turbines considering uncertain parameters. Energy. **183**, 205–219 (2019)
8. S. Nikolova, A. Causevski, A. Al-Salaymeh, Optimal operation of conventional power plants in power system with integrated renewable energy sources. Energy Convers. Manag. **65**, 697–703 (2013)

9. M. Ghahramani, M. Nazari-Heris, K. Zare, B. Mohammadi-Ivatloo, Energy management of electric vehicles parking in a power distribution network using robust optimization method. J Energy Manage. Technol. **2**(3), 22–30 (2018)
10. G.R. Aghajani, H.A. Shayanfar, H. Shayeghi, Presenting a multi-objective generation scheduling model for pricing demand response rate in micro-grid energy management. Energy Convers. Manag. **106**, 308–321 (2015)
11. M. Ghahramani, M. Nazari-Heris, K. Zare, B. Mohammadi-ivatloo, Robust short-term scheduling of smart distribution systems considering renewable sources and demand response programs, in *Robust Optimal Planning and Operation of Electrical Energy Systems*, (Springer, Cham, 2019), pp. 253–270
12. M. Rahimi, M. Banybayat, Y. Tagheie, P. Valeh-e-Sheyda, An insight on advantage of hybrid sun–wind-tracking over sun-tracking PV system. Energy Convers. Manag. **105**, 294–302 (2015)
13. M.M. Rahman, M. Hasanuzzaman, N.A. Rahim, Effects of various parameters on PV-module power and efficiency. Energy Convers. Manag. **103**, 348–358 (2015)
14. S. Nojavan, M. Majidi, A. Najafi-Ghalelou, M. Ghahramani, K. Zare, A cost-emission model for fuel cell/PV/battery hybrid energy system in the presence of demand response program: ε-constraint method and fuzzy satisfying approach. Energy Convers. Manag. **138**, 383–392 (2017)
15. S.F. Rodrigues, R.T. Pinto, M. Soleimanzadeh, P.A. Bosman, P. Bauer, Wake losses optimization of offshore wind farms with moveable floating wind turbines. Energy Convers. Manag. **89**, 933–941 (2015)
16. B. Vural, O. Erdinc, M. Uzunoglu, Parallel combination of FC and UC for vehicular power systems using a multi-input converter-based power interface. Energy Convers. Manag. **51**(12), 2613–2622 (2010)
17. I.S. Han, S.K. Park, C.B. Chung, Modeling and operation optimization of a proton exchange membrane fuel cell system for maximum efficiency. Energy Convers. Manag. **113**, 52–65 (2016)
18. S. Rezaee, E. Farjah, B. Khorramdel, Probabilistic analysis of plug-in electric vehicles impact on electrical grid through homes and parking lots. IEEE Trans. Sust Energy **4**(4), 1024–1033 (2013)
19. Z. Lei, Y. Li, A game theoretic approach to optimal scheduling of parking-lot electric vehicle charging. IEEE Trans. Veh. Technol. **99**, 1–10 (2016)
20. L. Zhang, Y. Li, Optimal management for parking-lot electric vehicle charging by two-stage approximate dynamic programming. IEEE Trans. Smart Grid **99**, 1–9 (2015)
21. M.S. Kuran, A.C. Viana, L. Iannone, D. Kofman, G. Mermoud, J.P. Vasseur, A smart parking lot management system for scheduling the recharging of electric vehicles. IEEE Trans. Smart Grid **6**(6), 2942–2953 (2015)
22. M. Shafie-khah, E. Heydarian-Forushani, G.J. Osorio, F.A. Gil, J. Aghaei, M. Barani, J.P. Catalao, Optimal behavior of electric vehicle parking lots as demand response aggregation agents. IEEE Trans. Smart Grid (2015)
23. M.J. Mirzaei, A. Kazemi, O.A. Homaee, Probabilistic approach to determine optimal capacity and location of electric vehicles parking lots in distribution networks. IEEE Trans. Ind. Inf. **12**(5), 1963–1972 (2015)
24. E. Akhavan-Rezai, M.F. Shaaban, E.F. El-Saadany, F. Karray, Online intelligent demand management of plug-in electric vehicles in future smart parking lots. Syst. J. IEEE. **10**(2), 483–494 (2015)
25. U.C. Chukwu, S.M. Mahajan, V2G parking lot with PV rooftop for capacity enhancement of a distribution system. IEEE Trans. Sustain. Energy **5**(1), 119–127 (2014)
26. M. Yazdani-Damavandi, M.P. Moghaddam, M.R. Haghifam, M. Shafie-khah, J.P. Catalão, Modeling operational behavior of plug-in electric vehicles' parking lot in multienergy systems. IEEE Trans. Smart Grid. **7**(1), 124–135 (2015)

8 Optimal Energy and Reserve Management of the Electric Vehicles... 231

27. N. Neyestani, M. Yazdani Damavandi, M. Shafie-Khah, J. Contreras, J.P. Catalão, Allocation of plug-in vehicles' parking lots in distribution systems considering network-constrained objectives. IEEE Trans. Power Syst. **30**(5), 2643–2656 (2015)
28. M. Jannati, S.H. Hosseinian, B. Vahidi, A significant reduction in the costs of battery energy storage systems by use of smart parking lots in the power fluctuation smoothing process of the wind farms. Renew. Energy **87**, 1–4 (2016)
29. H.A. Bonges, A.C. Lusk, Addressing electric vehicle (EV) sales and range anxiety through parking layout, policy and regulation. Transp. Res. A Policy Pract. **83**, 63–73 (2016)
30. M. Honarmand, A. Zakariazadeh, S. Jadid, Integrated scheduling of renewable generation and electric vehicles parking lot in a smart microgrid. Energy Convers. Manag. **86**, 745–755 (2014)
31. M. Honarmand, A. Zakariazadeh, S. Jadid, Optimal scheduling of electric vehicles in an EVs aggregator considering vehicle-to-grid concept and battery condition. Energy **65**, 572–579 (2014)
32. F. Fazelpour, M. Vafaeipour, O. Rahbari, M.A. Rosen, Intelligent optimization to integrate a plug-in hybrid electric vehicle smart parking lot with renewable energy resources and enhance grid characteristics. Energy Convers. Manag. **77**, 250–261 (2014)
33. M. Moradijoz, M.P. Moghaddam, M.R. Haghifam, E. Alishahi, A multi-objective optimization problem for allocating parking lots in a distribution network. Int. J. Electr. Power Energy Syst. **46**, 115–122 (2013)
34. A. El-Zonkoly, L. dos Santos Coelho, Optimal allocation, sizing of PHEV parking lots in distribution system. Int. J. Electr. Power Energy Syst. **67**, 472–477 (2015)
35. M.A. Mirzaei, A. Sadeghi-Yazdankhah, M. Nazari-Heris, B. Mohammadi-ivatloo, IGDT-based robust operation of integrated electricity and natural gas networks for managing the variability of wind power, in *Robust Optimal Planning and Operation of Electrical Energy Systems*, (Springer, Cham, 2019), pp. 131–143
36. A. Yona, T. Senjyu, A. Y. Saber, T. Funabashi, H. Sekine, C. H. Kim, Application of neural network to one-day-ahead 24 hours generating power forecasting for photovoltaic system. InIntelligent Systems Applications to Power Systems, 2007. ISAP 2007. International Conference on 2007 Nov 5 (pp. 1-6). IEEE
37. S.X. Chen, H.B. Gooi, M. Wang, Sizing of energy storage for microgrids. IEEE Trans. Smart Grid **3**(1), 142–151 (2012)
38. Iran's Meteorological Organization. Historical wind speed and solar radiation data. http://www.weather.ir. Accessed 10.08.13

Chapter 9
An Interactive Model for the Participation of Electric Vehicles in the Competitive Electricity Market

Mohammad Reza Fallahzadeh and Ali Zangeneh

9.1 Introduction

Over the last century, consumedly emission of green gases produced by the factories and internal combustion engines lead to significant changes in earth climate. Global warming may be the most crucial change, which endangered herbal and animal life on the planet. Also, fossil fuel energy is a limited energy source whose price fluctuated during the time by several factors. From the last decade, governments and environmental advocates have been working to reduce air pollution and its extensive harmful effects on climate. Some part of these efforts is to use less internal combustion engine vehicle and replace them with electric vehicle [1, 2].

Undoubtedly, having a trustworthy source of energy, which obtains reliably is an integral part of the current and future needs of societies. Endangered human life is another motivation that leads scientists to do research and develop another source of energy instead of using fossil fuel, especially in transportation. Transportation is a significant and central part of consuming oil and its derivatives. Instead, electric energy is a good, accessible, and clean alternative for fossil fuel, which can also be used in transportation effectively.

Nowadays, the electric vehicles have less acceptance in public views due to the high purchase price of electric vehicles, lack of charge and discharge infrastructure, and high cost of batteries [3]. Besides, low distance traveled by a fully charged battery in comparison with conventional vehicle is another essential factor, which leads to reducing their public satisfaction. On the other hand, it is usually difficult for people to change their regular customs, for example, using EVs instead of fossil-fueled vehicles. However, there are plenty of ways to persuade them to use EV and reduce the harmful effects of the current vehicles. To this end, economic incentive

---

M. R. Fallahzadeh · A. Zangeneh (✉)
Department of Electrical Engineering, Shahid Rajaee Teacher Training University, Tehran, Iran
e-mail: a.zangeneh@sru.ac.ir

© Springer Nature Switzerland AG 2020
A. Ahmadian et al. (eds.), *Electric Vehicles in Energy Systems*,
https://doi.org/10.1007/978-3-030-34448-1_9

will be an excellent characteristic to encourage people covering the cost of EVs charge and earn money.

One way to persuade people to use electric vehicles is to show their advantages. For instance, the participation of EVs in the power market with the ability to store energy and release it in the required times, help them to earn profit [4].To this aim, EV battery can be considered as a load with an elasticity that has positive energy consuming (charging mode) and negative energy consuming (discharging mode). Therefore, it is possible to use the stored energy in an appropriate time. This feature is known as vehicle to grid (V2G), which capable EV to take part in the energy, regulation, and reserve market, and some other ancillary services. On the other hand, it is helpful for the distribution system to permanently change its generating and consuming energy as well as provide some financial and technical benefits for the other market participants [5, 6].

If each EV wants to connect to grid uncoordinatedly in its favorite time and start charging/discharging, the distribution system may be faced with challenges and limitations. With the increasing number of the EVs, this individual behavior could have destructive impacts on distribution system such as load profile, line overload, increasing grid losses, and exceed bus voltages. Thus, the lack of an effective scheduling program to manage charge and discharge of the high number of the EVs may have irreparable damages to the grid [7–9].

To solve these problems, a coordination method for the charging and discharging of electric vehicles has been investigated to shift the load charging to non-peak hours by setting multi-tariff prices [10]. In this method, the EV owner is free to charge her/his vehicle at any time and this behavior could face grid with its limitations. In [11], an online scheduling strategy is presented using categorization of EVs in priority groups and coordinate their charging at non-peak hours to reduce costs. However, the energy derived from V2G has not been studied in [11]. Askari et al. [12] have examined the role of an EV parking lot to participate in an ancillary service market and propose regulation up/down in the day-ahead and spot market. However, they do not consider the behavior and interactions with the aggregator.

Although an individual EV does not have a considerable effect on the distribution system, the aggregated impacts of numerous EVs are significant. Therefore, existing an agent to coordinate many EVs are quite substantial to avoid disruptions in the distribution system. EV parking is a new agent to provide offers for aggregator to inject/absorb energy and ancillary services. It is assumed that, EV parking act as a distributed generation (DG) which have both positive and negative energy consuming. This fact enables parking owner to purchase energy during off-peak hours and sell it in the peak hours at a higher price and consequently earns more profit.

Due to the advent of the retail markets and the existence of various agents, it is necessary to consider a decentralized decision-making model that includes the behavior of various market players, such as EVs parking with the ability to buy and sell energy under the supervision of the aggregator [13, 14]. In this chapter, the market interaction between aggregator and EVs parking is formulated as a bi-level programming model. It consists of a parking with many standby EVs and an aggregator. In this model, the price offered by EVs parking and the amount of energy that aggregator purchases from EVs parking could be determined.

9.2 The General Framework of the Bi-Level Programming Approach for EV Parking Price Bidding

The bi-level scheduling model is presented in Fig. 9.1, which shows the competitive interaction between two market players in distribution networks: EVs parking and aggregator. At the upper level, the EV parking aims to maximize its profit by optimal scheduling the charge and discharge of vehicles. It purchases energy during off-peak hours and stores it in EV batteries to sell it back during peak hours in the offered price, which is usually higher than off-peak hours. The schedule is influenced by the market price, the initial SOC of vehicles and the arrival and departure time. At this level, the EV parking has an objective function, which determines two factors: V2G

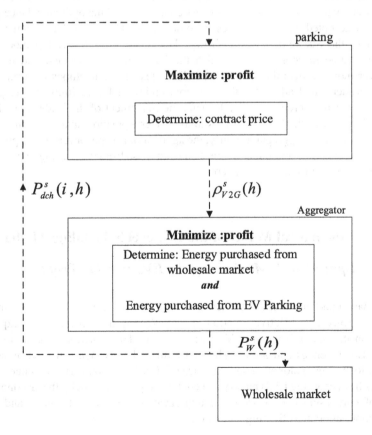

Fig. 9.1 The bilevel framework to model the interaction between aggregator and EV parking

bidding price and required energy for EVs charging per hour. These variables are considered as parameters in the lower level. On the other hand, at the lower level, the aggregator is the sole of the distribution system, which is responsible for providing energy for subscribers and minimizing its cost. Aggregator can purchase all its required energy from either the wholesale market or some part of it from the wholesale market and the remaining through the EV parking. The lower level determines two variables according to the wholesale energy price and the parking -bidding price: the amount of energy purchased from the wholesale market and the EVs parking. As shown in Fig. 9.1, purchasing energy from EVs parking variable (V2G) is sent as a parameter to the parking at the upper level. The advantage of the scheduling model is to consider the behavior of both players based on their viewpoints. This model helps each player to make a more accurate decision.

In this compromise, the parking should offer the optimal V2G price to aggregator to maximize its profits. If the proposed price is high, the aggregator does not accept it and provides all its required energy through the wholesale market. Contrariwise, if the proposed price is low, the parking cannot earn maximum profit. The equilibrium point represents the accepted bidding price of EV parking and quantity of energy purchased by the aggregator. This process continues until the optimal balance point, which is accepted by both entities become stable. The bi-level program is very similar to the Stackelberg game theory. The Stackelberg leadership model is a strategic game in economics in which the leader firm moves first and then the follower moves sequentially. The Stackelberg leader is sometimes referred as the market leader. The leader takes the first move and waits for the follower's response. The follower then proceeds to understand the movement of the leader [15]. In the scheduling model, the EVs parking is in the leader's position, and offers the V2G's optimal price to the aggregator hourly, the aggregator accepts or rejects the proposed price, in terms of the amount and time of the purchase from the parking owner or the wholesale market to minimize its cost.

9.3 Mathematical Model of the Bi-Level Scheduling Model

9.3.1 Upper-Level: Maximizing the EVs Parking Profit

Each market participant wants to increase its profit, by cutting costs and maximizing revenue. Thus, the objective function of the EVs parking is presented in Eq. (9.1) based on the maximizing the EVs parking profit. This function consists of three terms: two revenues and a cost. The first term is the revenue achieved by selling energy to the aggregator at an optimal price (V2G). The second is the revenue, which obtains from charging EV's battery at a contracted price and finally the last one is the cost of purchasing energy from the aggregator to feed EVs batteries and other consumptions such as lighting and so on.

9 An Interactive Model for the Participation of Electric Vehicles in...

$$f_{upper} : max_{P_{ch}^s \rho_{V2G}^s} \sum_{h=1}^{H} \sum_{i=1}^{N}$$

$$\times \left[P_{dch}^s(i,h).\rho_{V2G}^s(h) + P_{ch}^s(i,h).\rho_{co}^s(h) - P_{ch}^s(i,h).\rho_w^s(h) \right] \qquad (9.1)$$

where $P_{dch}^s(i,h)$ is the discharging power from ith EV at hour h and scenario s, $\rho_{V2G}^s(h)$ is the bidding price for V2G at hour h, $P_{ch}^s(i,h)$ is the required power for charging ith EV at hour h, scenario s, $\rho_{co}^s(h)$ is the contracted price for charging EVs which is assumed in three values (35–45-50 \$/kw) for non-peak, mid and peak hours respectively. N is the total number of the EVs under the control of the parking and H is the number of the hours in a planning horizon (24 hours). The parking is responsible to keep the state of charge (SOC) of the batteries on a satisfying quantity (for instance 85% of the EV battery) when EVs want to departure the parking (9.2).

$$\sum_{h=arrival}^{departure} [ch^s(i,h) - dch^s(i,h)] + soc_0^s(i) = 0.85 \times C_{EV}(i) \qquad \forall h \in H, \forall i \in N \quad (9.2)$$

where $ch^s(i,h)$ and $dch^s(i,h)$ are charge and discharge of the ith EV at hour h respectively, $soc_0^s(i)$: is the initial SOC of the EVs when they arrival to the parking and $C_{EV}(i)$: is the capacity of the ith EV.

It is supposed that charge, discharge and SOC of EVs at each hour should not exceed from a certain amount. The maximum value of SOC must be equal to the battery capacity and its minimum value depends on the type of the battery that is usually considered to be around 5% of battery capacity. In the simulations, the battery capacity is considered 32 kWh and the charging and discharging rates are same and equal to 3.2 kW/h. Due to the operational constraints of the battery, there are three constraints for the battery charging and discharging behaviors (9.3, 9.4, and 9.5):

$$0 \leq ch^s(i,h) \leq \overline{ch}(i,h) \forall h \in H, \forall i \in N \qquad (9.3)$$

$$0 \leq dch^s(i,h) \leq \overline{dch}(i,h) \forall h \in H, \forall i \in N \qquad (9.4)$$

$$soc_{min} \leq soc^s(i,h) \leq C_{EV}(i) \forall h \in H, \forall i \in N \qquad (9.5)$$

where $\overline{ch}(i,h)$ and $\overline{dch}(i,h)$ are the upper level of charge and discharge respectively and soc_{min} is the minimum amount of SOC.

It is essential to obtain soc. in each hour and has to be checked which not overstep its limit. Therefore, Eq. (9.6) represents the vehicle's SOC at h, which is computed using the SOC of the previous hour, plus (minus) charging (discharging) of ith EV at hour h. Converter which transfers energy from EV to grid or vice versa has a specific efficiency. Equations (9.7) and (9.8) state charge and discharge power regarding the efficiency (η).

$$soc^s(i,h) = soc^s(i,h-1) + ch^s(i,h) - dch^s(i,h) \forall h \in H, \forall i \in N \qquad (9.6)$$

$$P_{ch}^s(i,h) = ch^s(i,h)/\eta \qquad (9.7)$$

$$P_{dch}^s(i,h) = dch^s(i,h).\eta \qquad (9.8)$$

9.3.2 Lower Level: Minimize the Cost of the Aggregator

The objective function of the aggregator is represented in Eq. (9.9), which is optimized in the lower level of the problem. It consists of two cost terms: cost of purchasing energy from the wholesale market for 24 hours and EVs parking as V2G program for all presented EV according to their present time. In Eq. (9.9), there are two decision variables: the amount of energy purchased from the wholesale market ($P_w^s(h)$) and amount of energy, which is bought from the parking ($P_{dch}^s(i,h)$) in an acceptable price.

$$f_{lower} : \ Min_{P_{dch}^s, P_w^s} \sum_{h=1}^H \sum_{i=1}^N P_{dch}^s(i,h).\rho_{V2G}^s(h) + \sum_{h=1}^H P_w^s(h).\rho_w^s(h) \qquad (9.9)$$

where $\rho_w^s(h)$ and $P_w^s(h)$ are the price of the wholesale market and amount of the power purchase from it respectively.

The balance between power generation and consumption in the distribution system is shown in Eq. (9.10). This means that all generated energy must be equal with consumed energy. In this equation, generation have to parts including $P_{dch}^s(i,h)$ and $P_W^s(h)$, on the other hand, the consumption parts are $P_D^s(h)$ and $P_{ch}^s(i,h)$. Since the distribution system is modeled as a single bus, and the network is not modeled, power losses are not included in Eq. (9.10).

$$P_D^s(h) + \sum_{i=1}^N \left[P_{ch}^s(i,h) - P_{dch}^s(i,h) \right] = P_w^s(h) \forall h \in H, \forall i \in N : \sigma^s(h) \qquad (9.10)$$

where $P_D^s(h)$ is the total demand of the distribution system at h, and $\sigma^s(h)$ is the Lagrange coefficient related to this constraint.

Equation (9.11) indicates the constraint of the upstream transformer capacity to feed distribution network $P_T(h)$. This power purchased from the wholesale market ($P_w^s(h)$) is limited, according to Eq. (9.11). $\xi^s(h)$ is the Lagrange coefficient regarding this constraint.

$$P_w^s(h) - P_T(h) \leq 0 \forall h \in H, \forall i \in N \qquad (9.11)$$

9.3.3 Single-Level Equivalent Problem

There are various methods to solve bi-level problems. One of the most common ways is using the Karush–Kuhn–Tucker (KKT) optimality conditions which transform bi-level optimization problem into an equivalent single-level problem [16–18]. In this method, the lower-level of the optimization problem is replaced by its equivalent KKT conditions. So, the model is transformed into a single-level optimization problem [19]. The KKT conditions are applied if the lower level of the optimization problem is convex in the continuous variables. To implement, all unequal constraints have to transfer to 'equal to zero' form with its proportional Lagrange coefficient (9.13) and (9.14). Afterwards they are added to the lower-level objective function; finally differentiate the manufactured equation according to its variables (9.15) and (9.16). In this level of optimization, variables are $P_{dch}^s(i, h)$ and $P_W^s(h)$ which considered as input parameter in the upper-level.

$f_{upper:profit\ of\ parking}$

$$\max_{\substack{P_{ch}^s, \rho_{V2G}^s, \\ dch(i,h), P_w(h)}} \sum_{h=1}^{H} \sum_{i=1}^{N} \left[P_{dch}^s(i,h).\rho_{V2G}^s(h) + P_{ch}^s(i,h).\rho_{co}^s(h) - P_{ch}^s(i,h).\rho_w^s(h) \right] \tag{9.12}$$

Subject to Eqs. (9.3, 9.4, 9.5, and 9.6) and (9.10 and 9.11) and KKT condition as follows:

$$\delta^s(h).\left\{ P_D^s(h) - P_w^s(h) + \sum_{i=1}^{N} \left[P_{dch}^s(i,h) - P_{ch}^s(i,h) \right] \right\} = 0 \tag{9.13}$$

$$\xi^s.\left\{ P_T(h) - P_w^s(h) \right\} = 0 \tag{9.14}$$

$$\rho_{V2G}^s(h) - \delta^s(h) = 0 \tag{9.15}$$

$$\rho_w^s(h) - \delta^s(h) + \xi^s(h) = 0 \tag{9.16}$$

9.4 Numerical Results

It could be assumed that optimal placement of EV parking has been done, and charge and discharge actions have no harmful effects on distribution network busbars. Also, it is proposed that network losses are small and have no considerable impact on the usual functions. Based on these assumptions and to avoid Complexity and nonlinearity of the bi-level problem, it has skipped to Utilizing power flow and considering distribution network losses. With regarding drivers' habits, the daily distance which is traveled by each vehicle and arrival and departure time are vary during days. Thus, the soc. of vehicles and time needed to charge EVs fully are changing by vehicle to vehicle (Scenario 4). Although the proposed model is simulated for fixed arrival and departure time and daily traveled distance (Scenarios 1–3). Three different EV penetration (100, 300, and 500) are used to show the effectiveness of the proposed model.

To analyze the performance of the proposed model, the distribution network is modeled as a single bus connected to the power grid via a transformer. The total equivalent demand of the distribution system and EVs parking are connected at this bus. The parking has a maximum capacity of connecting 500 EVs. Each EV is equipped with a battery, which has 32 kWh capacity. The maximum charge and discharge rates of all EVs are assumed 3.2 kW/h, and the converter efficiency is considered 90% in both charge and discharge mode [12]. Other input data such as arrival and departure time, initial SOC at the moment of arrival, the amount of distribution system demand, the market price and the number of EVs in the parking in each hour are introduced as different scenarios. Data related to demand and energy price are extracted from Nordpool market data [20].

Four different scenarios are defined to analyze the scheduling model. Each scenario is simulated for three separate days (with a different price of the wholesale market and network demand) and a different number of vehicles (100, 300, and 500). These typical days are characterized using off-peak, mid and on-peak hours, and consequently, the market price has three different curves. Demand curve of the distribution grid and the wholesale market price curve are shown in Figs. 9.2 and 9.3, respectively based on three different days from the NordPool market [21].

Scenario 1: The arrival/departure time of EVs to/from the parking and the initial value of SOC at the time of entrance are other vital parameters to determine the

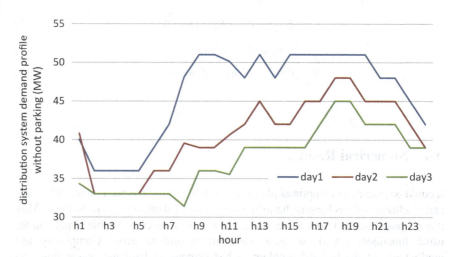

Fig. 9.2 The daily curve of distribution network demand in three different days without the parking

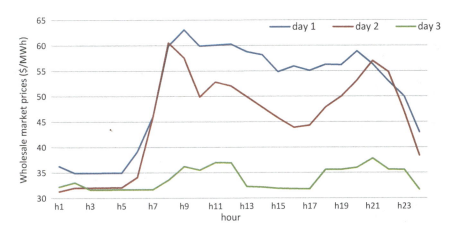

Fig. 9.3 Wholesale market prices in three different days

strategy of charge and discharge. In this scenario, the arrival and departure time of all EVs are considered equal at 4:00 p.m. and 8:00 a.m., respectively. It is also assumed that the traveled distance for all EVs is 25 km on a specified day.

Scenario 2: This scenario is similar to Scenario 1, but the arrival and departure time are 8:00 p.m. and 10:00 a.m., respectively. The initial SOC for all EVs is 16.3 kWh and traveled distance is 25 Km.

Scenario 3: The arrival and departure time of EVs are at 11:00 p.m. and 12:00 p. m., respectively. The initial SOC is 16 kWh, and traveled distance is 35 km.

Scenario 4: Arrival and departure time and initial SOC of EVs are random variables that depend on EVs' behavior. In this scenario, they are chosen randomly using a normal distribution function.

The EV parking decides how to plan for the optimal charging and discharging of EVs to maximize its profit. This decision is made by using the initial SOC of EVs, the arrival and departure time, the battery capacity of each vehicle, the time needed to fully charge of EVs' battery and the price of energy. By introducing the parking as a new entity to gather the capacity of the small distributed generation/storage units, a rebate in peak hours Will be observed, and valley areas can also be filled and smoother. Figure 9.4 illustrates this issue very well in the first scenario. In the second and third scenarios, the load profile of the distribution network improves due to the length of time that EVs are available. In the second scenario, the demand decreases in peak hours; it is approximately near 3 MW; while the valley areas have been filled by 3.6 MW. In the third scenario, the demand in peak areas reduced by 2.9 Mw and increased by 3.56 MW in the valley.

Figure 9.5 shows the demand curve of the distribution network with the presence of a parking with a capacity of 500 EVs for different days in the first scenario. The parking predicts the peak and valley area of distribution demand and accordingly, the maximum and minimum price hours have been obtained. Then plan to purchase and sale energy at peak and valley areas, respectively.

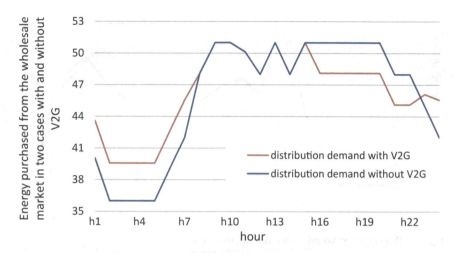

Fig. 9.4 Energy purchased from the wholesale market in two cases with and without parking with a capacity of 300 EVs in the first scenario

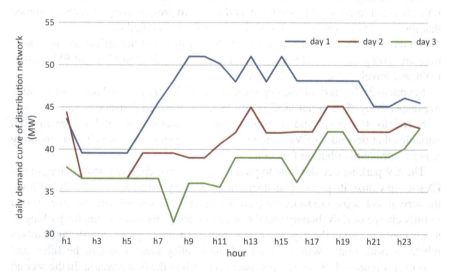

Fig. 9.5 Daily demand curve of the distribution network in three different days with the presence of the parking with 500 number of EVs and in Scenario 1

By appropriately scheduling EVs in the distribution network, not only the peak load of the distribution network did not increase, but also the peak and valley areas of the power demand curve are improved and smoothed. Comparison between Figs. 9.2 and 9.5 show that the load profile of the distribution network has better feature after the presence of EV parking. It can be realized that since by adding the great number of EVs, the power demand curve will be increased, so need an entity that coordinate EVs charge and discharge is quite indispensable. Table 9.1 shows the load factor in

Table 9.1 Load profile in different scenarios and days (with 500 EVs)

Scenario	Day 1	Day 2	Day 3
No EV	1.11	1.18	1.20
Scenario 1	1.10	1.11	1.11
Scenario 2	1.15	1.24	1.23
Scenario 3	1.16	1.23	1.22
Scenario 4	1.17	1.20	1.18

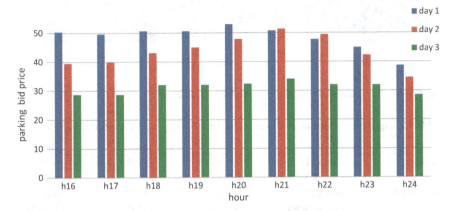

Fig. 9.6 The parking proposed bid price to aggregator at different times and days, and in Scenario 4

different scenarios and typical days. The first row shows the load factor with no EV while, the second row (Scenario 1) shows this factor, which has the desired reduction in comparison with the first row. In other cases (Scenarios 2 to 4), there is a small increase in the load factor values due to the time of arrival and departure, in other words, parking does not have enough time to fill all valley area. Therefore, the load factor will have a slight increase.

Based on the interaction between aggregator and the parking in the bi-level model, the proposed optimal V2G price is accepted, and the parking will sell energy from discharging of EVs to the aggregator. Figure 9.6 shows the parking -bidding price at some hours of three typical days, in scenario 4, with a total number of 500 EVs. Given that on the first day, the cost of the wholesale market is more than other 2 days. Consequently, the parking -bidding price will be more than the other 2 days. Therefore, the parking-bidding price depends on the wholesale market price, and it must be less than the wholesale price due to aggregator goal to minimize its total cost. Parking follows the wholesale electricity price and offers an optimal rate in the peak hours. This bidding is lower than the wholesale price and makes an incentive scheme for consumers to buy energy from the parking.

This biding price is a vital factor that directly affects the parking profit. As it can be shown in Fig. 9.6, parking bids its proposed V2G rate in peak hours; while in other hours, parking tries to buy power for filling up EVs battery at minimum cost.

The parking tends to maximize its profit by buy and sell energy among EVs. This issue can be observed in Fig. 9.7. When an EV enter the parking, it has some initial

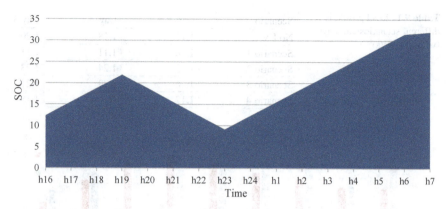

Fig. 9.7 SOC of a sample EV during the first day in Scenario 4

Table 9.2 Aggregator operation cost in different scenarios with a total number of 300 EVs ($)

		Day 1	Day 2	Day 3
Scenario 1	with V2G	157943.5	116176.0	93889.2
	without V2G	163854.2	120521.3	98526.4
Scenario 2	with V2G	166136.2	120266.3	96512.7
	without V2G	173235.5	126352.6	103478.6
Scenario 3	with V2G	167084.9	120090.6	96762.5
	without V2G	173894.3	125698.5	102422.8
Scenario 4	with V2G	174799.3	127078.4	101899.9
	without V2G	180329.2	132211.3	108566.9

SOC; if the entrance time is during off-peak hours, the parking tries to purchase energy and store it in EV batteries to sale it back in the peak hours with a higher price to obtain more profit. Otherwise if the entrance time is in the peak-hours, and also if the EV battery has enough SOC, parking decides to sell its' V2G power.

Also, the repetitive charge and discharge action are harmful to EV batteries, and it may have destructing effect and cost. The lower price for charging the vehicles is a beneficial factor, which encourages EVs to participate in the program.

Table 9.2 shows the cost of aggregator in different scenarios and different day patterns in two cases: with the presence of V2G and without V2G. Because of the lower price in the second and third days, aggregator operation cost in the first day is more than the other 2 days. In Scenario 1, EVs are more available. Thus, the parking will be more likely to use the V2G capabilities of EVs. In this scenario, the aggregator reduces its costs by purchasing more energy from the parking at a price lower than the wholesale market price. By smoothing the demand curve (cutting the peak hours and filling the valley area), aggregator can minimize its total cost and thus maximize its benefit. It can be observed that in all cases, aggregator cost will be reduced with V2G capability. With the presence of the EV parking, all market players such as aggregator and customers benefit financially. This helpful feature

9 An Interactive Model for the Participation of Electric Vehicles in... 245

Table 9.3 Aggregator operation cost with different number of EVs ($)

Number of EV	V2G capability	Day 1	Day 2	Day 3
100 EV	with V2G	58527.1	42401.7	33984.1
	without V2G	59832.2	43510.6	34125.5
300 EV	with V2G	174799.3	127078.4	101899.9
	without V2G	180329.2	132211.3	108566.9
500 EV	with V2G	291394.8	211749.9	169787.0
	without V2G	298455.8	221654.9	175087.2

Table 9.4 Parking profit in different EV penetrations and days ($)

Number of EV	Day 1	Day 2	Day 3
500 EV	796.8	767.3	683.3
300 EV	477.5	459.4	408.9
100 EV	156.4	152.3	137.8

is a professional privilege that leads to better performance and postpones the restructuring of the distribution system.

Due to the reduction of the wholesale market price in the second and third typical days, aggregator costs will decrease accordingly. If the number of EVs increases, aggregator has to spend more to purchase energy for EVs charging. Thus, if EVs do not support the V2G ability, the addition of EVs to the distribution system, acted as an extra load for the network and the aggregator cost increases. Table 9.3 shows the effect of these factors in forth Scenario.

Scenario 4 is more similar to a real status that EVs arrival and departure time and traveled distance are randomly simulated and applied in the simulation. The parking profit of this scenario is presented in Table 9.4 for different typical days. In the first day, despite the higher price of the wholesale market, the parking profit is more than the second and third days. Given that, the offered V2G rate is floating; because it is strongly depends on the wholesale market price. The V2G price, in the first day is higher than other days, and consequently the parking profit in the first day is more than the rest. According to the availability time of EVs, the gain may vary, and if this time is more during off-peak hours, the parking can earn more profit. In this table, the contracted price for charge EVs is close to the average rate, of the wholesale market. This feature is a motivation for the EV to charge its EV under the parking-scheduling program with lower price. Besides, the parking uses EVs battery to store and release energy in the desired time. By increasing the number of EVs, the parking profit rises accordingly, because of the parking sale more power to aggregator (discharging mode) and EVs (charging mode). For maximizing the benefit, the number of EVs is an essential factor that can be obtained by persuading the owners of the EVs to charge their vehicles in a particular place.

9.5 Conclusion

In this chapter, a distributed decision-making model based on a bi-level programming approach, including a parking and aggregator, is presented. Parking decides about the optimal time of the EVs' charging as well as choosing an optimal bidding strategy according to the aggregator reaction to the price and V2G rate. The purpose of this chapter is, modelling the behavior of the parking and aggregator in an interactive and competitive market. By adding lots of coordinated EVs through a parking in a distribution network, not only better features such as improved load factor and reliability are obtained, but also the concern of adding EVs as an extra load is reduced.

Appendix A

The nomenclature is shown below.

f_{upper}	The upper-level objective function
$P^s_{dch}(i, h)$	Discharging power from ith EV at hour h in scenario s
$\rho^s_{V2G}(h)$	Bidding price for V2G at hour h
$P^s_{ch}(i, h)$	The required power for charging ith EV at hour h in scenario s
$\rho^s_{co}(h)$	Contracted price for charging EVs
N	The total number of the EVs under the control of the parking
H	The number of hours in a planning horizon (24 hours)
i	EV number
h	Hour number
SOC	State of charge of EV battery
$ch^s(i, h)$	Charge of the ith EV at hour h
$dch^s(i, h)$	Discharge of the ith EV at hour h
$soc_{0s}(i)$	The initial SOC of the EVs when they arrive to the parking
$C_{EV}(i)$	The capacity of the ith EV
$\overline{ch}(i, h)$	The upper-level of charge-discharge
$\overline{dch}(i, h)$	The upper-level of discharge
soc_{min}	The minimum amount of SOC
η	The efficiency of charge and discharge converter
$\rho^s_W(h)$	Price of the wholesale market
$P^s_W(h)$	Amount of the power purchase from the wholesale market
$P^s_D(h)$	The total demand of the distribution system at hour h
$\delta^s(h)$	Lagrange coefficient related to its constraint.
$P_T(h)$	Upstream transformer capacity
$\xi(h)$	Lagrange coefficient related to its constraint.

References

1. K. Parks, P. Denholm, T. Markel, Cost and emissions associated with plug-in hybrid vehicle charging, National Renewable Energy Laboratory (NREL), Technical Report, 2007
2. G.K. Venayagamoorthy, G. Braband, Carbon reduction potential with intelligent control of power systems. IFAC Proc. Vol. **41**, 13952–13957 (2008)
3. IEA (OECD/IEA), Technology Roadmap Electric and plug-in hybrid electric vehicles, 2009, http://www.iea.org/about/copyright.asp
4. K.J. Yunus, Plug-in electric vehicle charging impacts on power systems, Chalmers University of Technology Göteborg, Sweden, 2010
5. A.d. Guibert, Batteries and supercapacitors cells for the fully electric vehicle, 2009, http://www.smart-systemsintegration.org/public/electricvehicle/batteryworkshopdocuments/presentations/Anne%20de%20Guibert%20Saft.pdf/download
6. W. Kempton, V. Udo, K. Huber, K. Komara, S. Letendre, S. Baker, D. Brunner, N. Pearre, A test of vehicle-to-grid (V2G) for energy storage and frequency regulation in the PJM system, 2008
7. E. Akhavan-Rezai, M.F. Shaaban, E.F. El-Saadany, A. Zidan, Uncoordinated charging impacts of electric vehicles on electric distribution grids: Normal and fast charging comparison, IEEE Conference on Power and Energy Society General Meeting (IEEE, San Diego, 2012)
8. P. Denholm, W. Short, An evaluation of utility system impacts and benefits of optimally dispatched plug-in hybrid electric vehicles, National Renewable Energy Laboratory, 2006
9. K. Clement-Nyns, E. Haesen, J. Driesen, The impact of vehicle-to-grid on the distribution grid. Electr. Power Syst. Res. **81**(1), 185–192 (2011)
10. M. Singh, P. Kumar, I. Kar, A multi Charging Station for electric vehicles and its utilization for load management and the grid support. IEEE Trans. Smart Grid **4**(2), 1026–1037 (2013)
11. S. Deilami, A.S. Masoum, P.S. Moses, M.A.S. Masoum, Real-time coordination of plug-in electric vehicle charging in smart grids to minimize power losses and improve voltage profile. IEEE Trans. Smart Grid **2**(3), 456–467 (2011)
12. F. Askari, M.R. Haghifam, J. Zohrevand, A. Khoshkholgh, An economic model for power exchange of V2Gs in parking lots, Integration of Renewables into the Distribution Grid, CIRED, 2012
13. C. Hutson, G.K. Venayagamoorthy, K.A. Corzine, Intelligent scheduling of hybrid and electric vehicle storage capacity in a parking lot for profit maximization in grid power transactions, IEEE Conference on Energy (IEEE, Atlanta, 2008)
14. G. Zhang, G. Zhang, Y. Gao, J. Lu, Competitive strategic bidding optimization in electricity markets using Bilevel programming and swarm technique. IEEE Trans. Indust. Electron. **58**(6), 2138–2146 (2011)
15. M.A. Amouzegar, A global optimization method for nonlinear bilevel programming problems. IEEE Trans. Syst. Man Cybernet. **29**(6), 771–777 (1999)
16. M. Carrion, J.M. Arroyo, A.J. Conejo, A bilevel stochastic programming approach for retailer futures market trading. IEEE Trans. Power Syst. **24**(3), 1446–1456 (2009)
17. J.M. Arroyo, Bilevel programming applied to power system vulnerability analysis under multiple contingencies. IET Gener. Transm. Distr. **4**(2), 178–190 (2009)
18. M.J. Rider, J.M. Lopez-Lezama, J. Contreras, A. Padilha-Feltrin, Bilevel approach for optimal location and contract pricing of distributed generation in radial distribution systems using mixed-integer linear programming. IET Gener. Transm. Distr. **7**(7), 724–734 (2012)
19. R.E. Rosenthal, *GAMS: A user's guide* (GAMS Development Corporation, Washington, DC, 1998). Available at: http://www.gams.com/dd/docs/bigdocs/GAMSUsersGuide.pdf
20. Nordpool marker website. Available at: http://www.nordpoolspot.com/Marketdata1
21. J.M. Lopez-Lezama, A. Padilha-Feltrin, J. Contreras, J.I. Munoz, Optimal contract pricing of distributed generation in distribution networks. IEEE Trans. Power Syst. **26**(1), 128–136 (2011)

Chapter 10
Optimal Scheduling of Smart Microgrid in Presence of Battery Swapping Station of Electrical Vehicles

Mohammad Hemmati, Mehdi Abapour, and Behnam Mohammadi-ivatloo

10.1 Introduction

According to global reports, a significant share of air pollution is due to the use of fossil fuels in the transport fleet [1, 2]. One of the solutions to this challenge is the utilization of electrical vehicles (EVs) and development of them in transport fleets. Despite the development of EVs in recent years, the challenge of long charging time of EV's battery failed to attract the satisfaction of consumers [3]. Furthermore, the use of EVs in the power network, due to the uncertainties at the arrival time of them, will face the electrical network with several technical issues. One of the new ideas in this field is a fast charging station that reduces battery charging time for EVs [4]. However, the fast charging station is not widely used for emerging technology [5]. Also, these stations cannot overcome the uncertainty and the high power consumption challenges that occur at peak times.

In recent years, the idea of battery swapping station (BSS) has been presented as a convenient way to manage the charge of EVs. The basic of BSS is to replace the depleted battery with full-charge ones. The BSS can be considered as a bank of batteries, which purchases the power at off-peak times and charges batteries to be ready at any moment to serve EV's owner. Therefore, any consumer which faced with the depleted battery, referring to the BSS and in return for a fee, swap a depleted battery with a full one at the shortest time [6].

In [5, 7], the charging of electric buses is scheduled by BSS. Using the BSS for this purpose is much more efficient than the method that each EV's owner charges EV in different parking lots. In [8], stochastic programming based on Monte-Carlo simulation was proposed for modeling the intermittent change of energy consumption of BSS. In [9–11], BSS scheduling problem by considering power market is studied. In [11], an operation model for BSS is presented. In this model, BSS is

---

M. Hemmati (✉) · M. Abapour · B. Mohammadi-ivatloo
Faculty of Electrical and Computer Engineering, University of Tabriz, Tabriz, Iran
e-mail: m.hemmati@tabrizu.ac.ir; abapour@tabrizu.ac.ir; bmohammadi@tabrizu.ac.ir

© Springer Nature Switzerland AG 2020
A. Ahmadian et al. (eds.), *Electric Vehicles in Energy Systems*,
https://doi.org/10.1007/978-3-030-34448-1_10

scheduled in day-ahead clearing process for profit maximization goal while the market price is considered as an uncertain parameter and robust-optimization is implemented to model it. In [12], a dynamic operation strategy for scheduling of BSS to participate in power market in short term is proposed. The proposed model tries to maximize the profit of BSS, considering the uncertainty of power market. In [5], the scheduling of BSS by considering the energy efficiency and emission is studied. This project was investigated in China and implemented on electric taxies.

The Microgrid is a small/medium-scale distribution network consists of multiple loads (controllable and non-controllable), generation units (dispatchable and non-dispatchable) and energy storage systems (ESS) that are working under the master controller supervision [13, 14]. Despite the fact that the attention to the BSS has increased for various purposes, in recent years, the development of this idea in microgrid is rarely studied. In the other worlds, if the BSS can be considered as a new service that can support the microgrid in emergency conditions, besides the serving the EV's owner, it would have different benefits for microgrid.

The BSS can operate as an agent between EV's owner and MG. When the stored energy in EVs battery does not sufficient for the next trips, BSS by providing the swap service, helps the driver to continue their trips by changing the empty battery with the full one at the short time. The empty batteries will be charged and appropriated to participate in the swap service or reserve, for the next periods. Briefly, some of the main advantages that BSS follows can be summarized:

- Enabling drivers to drive long distances due to the high-speed swapping of the battery.
- Reduction of replacement cost and battery life cycle issues.
- Reduction of power congestion and causing peak shaving due to operation as an energy storage system.
- Utility support in an emergency condition.
- Management of microgrid by centralizing the charging process at night that can provide technical and economic benefits.

In [14], a multi-objective model of optimal charging management of EVs, through BSS is proposed. Minimizing the power loss and battery charging cost with voltage profile flattening are the main objectives of this work. In [6], optimal scheduling of islanded microgrid with BSS considering demand response programs is presented. While the operation cost of islanded microgrid and profit maximization of the problem are formulated as a bi-level model, no exchanging between the microgrid and upstream network leads to the unrealistic operation of the proposed approach. In [15], the operation of the microgrid in both connected and islanded modes in presence of BSS is investigated. In the grid-connected mode, the proposed model is intent on reducing the BSS charge cost, in other words, tries to maximize the profit of BSS, in islanded mode, fuzzy control model is used to determine the price of serving to the microgrid. The scheduling and operation of BSS in active distribution networks and reliability analyzing are studied in [16, 17]. According to [16, 17], for reliability improvement purpose in presence of BSS, firstly, the behavior of EV's owner must be extracted.

Given that microgrid separates from the upstream network in an emergency conditions, need support in islanded mode, so it could continue to the operation at a minimum load interruption. In this situation, BSS as a bank of batteries can be considered as a reserve and support the microgrid in an emergency conditions. Then the microgrid operator and BSS owner are interacting and exchange the power to get their goals.

One of the unique features of microgrids is islanding operation. Microgrid can spate from the main grid in faulty and emergency conditions. In this situation, the need for a suitable reserve that guarantees the local load is a critical challenge for microgrid owner [18]. Due to operation of BSS as an energy storage system, microgrid owner have new option that can rely on it for successful islanding operation. Although microgrid scheduling with different types of uncertainty is reported in the literature, there are few research works, which consider the scheduling of microgrids with islanding capability. In [19], a chance-constrained energy management model for an islanding microgrid is developed following the objective of minimizing the generation cost, ESS degradation cost and emission cost. Generated power by RES is considered as an uncertain parameter and novel ambiguity set is proposed to capture the uncertainty. In [20], microgrid scheduling with multi-period islanding constraints is proposed. To identify the microgrid capability in operating in islanded condition, the $T - \tau$ criterion is introduced which also facilitates the scheduling process for multiple hours during the examined mode.

Probabilistic nature of RESs often leads to power fluctuations. Therefore, the accurate scheduling of renewable-based microgrids becomes much more difficult, especially in islanded mode [21]. Therefore, reliability and load shedding problem will be the critical challenges in islanded operation. Authors of [22] stated that load variation and high penetration of renewable resources raise the instability of islanded microgrid. Therefore, in this chapter, for the coordination of responsive loads and distributed energy resources in islanded microgrids, a two-stage stochastic model is proposed. The proposed method optimized by considering the voltage and frequency security constraints. To quantify the probability of microgrid meeting local loads and to maintain an adequate reserve in islanded mode, the probability of successful islanding (PSI) criteria is proposed in [23]. The proposed strategy is developed with chance-constrained islanding capability to ensure successful islanding of a microgrid with a specified probability. In [21], to demonstrate the microgrid capability for operating in islanded mode, a probability-based concept is proposed. The proposed method is analyzed in presence of forecast errors of demand and wind power generation. In [24], by considering the uncertainties of load demand, the price of electricity and renewable energy, a risk-constrained stochastic framework for autonomous microgrids is proposed. By using conditional value at risk (CVaR), the risk of low profits in the worst scenarios is modeled. In [25], a robust optimization–based model is developed for optimal scheduling of microgrid operation with islanding capability constraints. With the proper robust level, the proposed microgrid scheduling model with islanding constraints ensures the successful off-grid operation with minimum load shedding. Therefore, we consider new constraint shows the successful islanding operation of microgrid for multiple hours. Actually, BSS beside other resources can guarantee the successful islanding.

In this chapter, a novel microgrid operation with BSS is proposed. The problem is formulated as bi-level model. In the upper-level, microgrid operation costs including generation cost, the cost of purchasing power from upstream network and purchasing cost, BSS to participate in reserve market is minimized. In the lower-level, BSS profit by maximizing the revenue from day-ahead and reserve markets and minimizes charging cost is formulated. Beside the BSS, microgrid have wind, PV and micro-turbine as local resources. Wind and PV power generation and arrival time of EVs are considered as uncertain parameters and to address them, scenario-based approach is implemented.

The rest of this chapter is organized as follows: Sect. 10.2 describes the components of MG. In this section, modeling of the uncertainty is proposed. Section 10.3 provides the problem formulation, the MG scheduling in presence of the BSS is formulated by bi-level model. The formulation of upper/lower levels with corresponding constraints is presented. In addition, in this section real-time pricing based demand response program is presented. Simulation results of the proposed method on the 10-bus MG test system are presented in Sect. 10.4. Finally, the chapter is concluded in Sect. 10.5.

10.2 The MG's Component

10.2.1 Uncertainty Modeling

As mentioned, the penetration of renewable energy resources (RESs) into MG effects on the operation and planning of the power system. PVs and WTs are useful RESs in active distribution networks and MGs. The generated power by PVs and WTs are caused by solar irradiation and wind speed as a prime energy source, respectively. Because of probabilistic nature of wind speed and sun irradiance, generated power of those resources sustains significant uncertainties. Furthermore, daily load behavior and arrival time of EVs in BSS are considered uncertain parameters. The probabilistic analysis at presence of multiple uncertainties is a powerful tool for scheduling and operation of power network. To address the uncertain parameters, a new probabilistic scenario-based framework is presented in this section. In addition, the operation of BSS is presented in this section.

10.2.1.1 Wind Power Output

The generated power by WT depends on wind speed. To model the uncertainty of wind speed, we assume that the wind speed variation is subject to a Weibull distribution. If V_{mean} and σ are mean and standard deviation of forecasted wind speed, respectively, the parameters of Weibull distribution are calculated as:

10 Optimal Scheduling of Smart Microgrid in Presence of Battery Swapping...

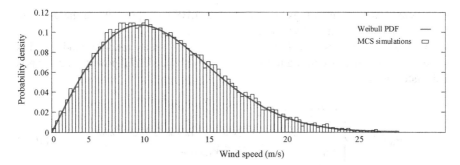

Fig. 10.1 The Weibull PDF fit by MCS

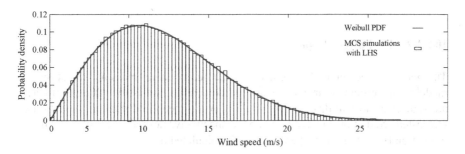

Fig. 10.2 The Weibull PDF fir by MCS with LHS

$$r = \left(\frac{\sigma}{V_{mean}}\right)^{-1.086} \qquad c = \frac{V_{mean}}{Gamma\left(1 + \frac{1}{r}\right)} \qquad (10.1)$$

According to the Weibull parameters (r,c), the Weibull probability distribution function (PDF) is calculated as:

$$f(V) = \frac{r}{c}\left(\frac{V}{c}\right)^{r-1} \exp\left[-\left(\frac{V}{c}\right)^r\right] \qquad (10.2)$$

Monte Carlo simulation (MCS) generates a high number of scenarios subject to the Weibull distribution which each of them is assigned a probability that is equal to one divided by the number of generated scenarios [26]. In each scenario, random wind speed is considered for each hour. Because MCS generates a large number of scenarios, the variance of scenarios is too much. Latin hypercube sampling (LHS) is an adequate technique, which can reduce the number of runs for MCS to achieve a precise random distribution. Furthermore, LHS can reduce the variance of MCS scenarios [26]. To demonstrate the effect of LHS for the variance of MCS sampling reduction, an example is presented for wind speed subject to Weibull distribution with $V_{mean} = 12 \ m/s$ and $\sigma = 4 \ m/s$. Figures 10.1 and 10.2 display the ordinary MCS and the MCS with LHS respectively. In both simulations, 1000 samples are considered. According to Figs. 10.1 and 10.2, LHS can approximate the accurate Weibull PDF much better than the ordinary MCS.

According to the assigned PDF, in each scenario, hourly random wind speed is generated, therefore, according to the random wind speed, WT power generation is calculated as [27]:

$$P_W(V) = \begin{cases} 0 & 0 \leq V \leq V_{cut-in} \\ (k_1 + k_2 V + k_3 V^2) P_{W_{rated}} & V_{cut-in} \leq V \leq V_{rated} \\ P_{W_{rated}} & V_{rated} \leq V \leq V_{cut-out} \\ 0 & V_{cut-out} \leq V \end{cases} \quad (10.3)$$

10.2.1.2 PV Power Output

The generated power of PV depends on air temperature and solar radiation. To model the uncertainty of irradiation and air temperature, we assume those parameters are subject to normal distribution. If μ and σ are mean and standard deviation of forecasted irradiation (air temperature), respectively, normal distribution PDF for irradiation (G_{ING}) or air temperature (T_r) is calculated as:

$$f(G_{ING}, T_r) = \frac{1}{\sqrt{2\pi} \times \sigma} \exp\left(-\frac{((G_{ING}, T_r) - \mu)^2}{2 \times \sigma^2}\right) \quad (10.4)$$

As wind speed simulation, a large number of scenarios are generated by MCS and LHS is employed to decrease the variance of MCS samples. Figures 10.3 and 10.4 show the generated irradiation example for ordinary MCS and MCS with LHS subject to Normal distribution, respectively. In both of simulations, 1000 samples are considered with $\mu = 800$ W/m^2 and $\sigma = 150$ W/m^2.

As can be seen, LHS can obtain accurate Normal PDF much better than ordinary MCS.

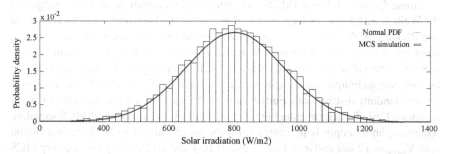

Fig. 10.3 Normal PDF fit by MCS

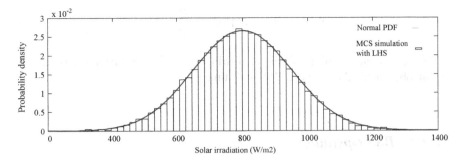

Fig. 10.4 Normal PDF fit by MCS with LHS

According to the assigned PDF, in each scenario, hourly random air temperature and irradiation are generated. Therefore, PV power generation output is calculated as:

$$P_{pv} = P_{STC} \times \frac{G_{ING}}{G_{STC}} \times (1 + K(T_c - T_r)) \tag{10.5}$$

where G_{ING} is hourly irradiation, G_{STC} is standard irradiation (1000W/m^2), T_c and T_r are cell and air temperature, P_{STC} is rated power of PV and K is maximum power temperature coefficient [28].

10.2.1.3 Load Demand

Due to load variation during the day, the probabilistic behavior of load should be considered as the uncertain parameter. The uncertainty of load demand is subject to the normal distribution [29] like Eq. (10.4). Therefore, MCS generates a large number of scenarios and LHS reduces the variance of MCS samples. As previous uncertain parameters, according to the assigned PDF, in each scenario, hourly random load demand is generated.

10.2.1.4 Arrival Time of EV Model

In line with former researches, Poisson distribution is used for modeling the incident of waiting events. In the other word, the number of EVs that arrive at t^{th} time at BSS follow the Poisson distribution. The probability of arriving of N numbers of EVs can be formulated as following [6, 30]:

$$\text{Prb}\{N_{ev}(t) = n_{ev}\} = \frac{e^{-\lambda_{ev} \cdot t}(\lambda_{ev} \cdot t)^{n_{ev}}}{n_{ev}!} \tag{10.6}$$

Where λ_{ev} is a rate of arrival of Poisson distribution at t^{th} time at BSS, $N_{ev}(t)$ denotes the number of EV that arrive at BSS at t^{th} time.

10.2.2 BSS Operation

BSS as a stand-alone unit provides battery swap service for its consumers during a day. When the stored energy in EV's battery is not sufficient for next trip, battery swap option provides for EV's user to replace the empty battery with full one in the least possible time through BSS. The empty battery is delivered to the station. When the delivered battery charged, in addition to being ready for swapping in the next period, provides reserve capacity for the MG. To simplify the analysis in this chapter, some assumptions are considered:

1. The time of battery swapping is not considered. In addition, we assume that the empty batter will be charged at the next period.
2. The capacity of battery of station is constant and each battery can charge/discharge with constant power.
3. To relieve the computational burden, we assumed that EV' user can accomplish battery swapping operation by swapping only one battery at each period.

Figure 10.5 shows the proposed model of MG contains of renewable energy resources, dispatchable units and loads, which exchange the power with BSS. During low price times, BSS purchases the power from MG, named grid to BSS (G2B) mode. During high price times, BSS sells power to MG, named BSS to grid (B2G) mode. In addition, BSS serves the consumers, in return for a fee. Whole of the system is in connection with upstream network. It should be noted that MG central

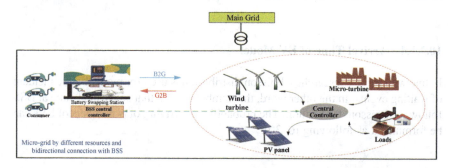

Fig. 10.5 The schematic of proposed model of MG with BSS while in connection with upstream network

controller and BSS controller are interacting with each other and any exchanging between MG and BSS are determined by these two central controllers.

10.3 Problem Formulation

As previously discussed, the MG scheduling with BSS problem is formulated as bi-level model. In the lower-level problem, the operation of BSS to get maximum profit under real-time pricing mechanism as [6, 31]. In the upper-level, MG operation cost is minimized. In this section, based on real-time pricing mechanism, in the lower-level problem, the schedule of charging/discharging process of BSS is determined, then, it will give feedback to upper-level problem. Then, the real-time electricity price is determined by upper-level.

10.3.1 The Upper-Level Formulation

As previously discussed, in the upper-level, MG operation cost is minimized. The MG cost consists of fuel cost, startup and shutdown cost of MTs, purchasing power cost from the main grid, exchange power cost with BSS during charging/discharging and spinning reserve cost. It should be noted that MG can sell power to main grid as well as BSS during charging. Therefore, the MG revenue must be subtracted from total cost. The upper-level objective function can be formulated as follows:

$$F_1 : \min \sum_t \sum_{i=1}^{N_g} (F(P_{i,t})U_{i,t} + SU_{i,t} + SD_{i,t}) + \sum_t^T \rho_t P_{M,t} + \sum_t \xi_t^{gr} P_{i,t}^r$$
$$+ r \sum_t \lambda_t P_{BSS,t} + (1 - r) \sum_t \lambda_t^r P_{BSS,t}^r \tag{10.7}$$

The first term of Eq. (10.7) represents the MT cost beside startup and shutdown cost, where N_g is the number of MTs, $P_{i,t}$ is generated power by i^{th} MT. $F(P_{i,t})$ is the cost function of ith MT, which is calculated as following:

$$F(P_{i,t}) = a_i + b_i \times P_{i,t} + c_i \times P_{i,t}^2 \tag{10.8}$$

Where, a, b and c are cost coefficients of ith MT.

The cost /revenue of purchasing /selling power from/to main grid, is presented in the second term. $P_{M,t} > 0$ is exchanged power between MG and main grid, when MG purchases the power from main grid, $P_{M,t} > 0$, otherwise, $P_{M,t} < 0$ to appear as revenue and minus from MG cost, ρ_t is the power exchanged price between MG and upstream network. The third term shows the spinning reserve cost for i^{th} MT as

[6]. The fourth term shows the exchange cost between MG and BSS. $P_{BSS, t}$ is power exchange which is charged (purchase by BSS owner from MG) or discharge (sell to the MG) at t^{th} time. During charging $P_{BSS, t} < 0$ and MG get revenue, otherwise $P_{BSS, t} > 0$. In the fourth term, λ_t is the power exchanged price between MG and BSS that is obtained as a real-time pricing mechanism and will be calculated later. r is binary variable that describes the BSS operation. When BSS service the EV and MG through charging/discharging $r = 1$, otherwise, when BSS participate in reserve $r = 0$. The last term of Eq. (10.7) represents the cost of reserve provided by BSS and λ_t^r is the price of reserve when the BSS participates in the reserve market.

10.3.2   Constraints of Upper-Level Problem

The upper-level problem has multiple limitation which described as following:

$$P_{i,t}^{\min} \leq P_{i,t} \leq P_{i,t}^{\max} \qquad t \in T, \ i \in N_g \qquad (10.9)$$

$$MUT_i(X_{i,t} - X_{i,t-1}) \leq T_i^{on} \qquad t \in T, \ i \in N_g \qquad (10.10)$$

$$MDT_i(X_{i,t-1} - X_{i,t}) \leq T_i^{off} \qquad t \in T, \ i \in N_g \qquad (10.11)$$

$$P_{i,t} - P_{i,t-1} \leq UR_i \qquad t \in T, \ i \in N_g \qquad (10.12)$$

$$P_{i,t-1} - P_{i,t} \leq UD_i \qquad t \in T, \ i \in N_g \qquad (10.13)$$

$$0 \leq P_{M,t} \leq P_{M,t}^{\max} \qquad t \in T \qquad (10.14)$$

$$\text{Prob}\left\{P_{i,t}^r + P_{BSS,t}^r \geq P_t^L - P_t^{PV} - P_t^{wind}\right\} \qquad (10.15)$$

$$\sum_{i=1}^{N_g} P_{i,t} + P_{M,t} \pm P_{BSS,t} = P_t^{TOTAL} \qquad (10.16)$$

Constraint (10.9) represents the region of ith MT power output which limited by minimum and maximum power output, where $P_{i,t}^{\min}$ and $P_{i,t}^{\max}$ are the minimum and maximum power output of ith MT, respectively. The minimum up and down time constraints are represented in Eqs. (10.10) and (10.11), where T_i^{on} and T_i^{off} are the number of hours when i^{th} MT is on/off, respectively. $X_{i, t-1}$ and $X_{i, t}$ are the ith MT status at $t-1th$ and t^{th} periods, respectively. The limitation of how much the power output of ith MT may be increased or decreased between two consecutive time simples like t and $t-1$ are described in Eqs. (10.12) and (10.13), where UR_i and UD_i show the ramp up and ramp down value of ith MT, respectively. Constraint (10.14) represents the limit of power exchanged between MG and the upstream network, where $P_{M,t}^{\max}$ is the maximum power exchanged between MG and upstream network. Constraint (10.15) shows the probability of successful islanding operation. If the sum of the reserve provided by MTs and BSS is more than forecast error of

10 Optimal Scheduling of Smart Microgrid in Presence of Battery Swapping... 259

uncertain parameters, that means MG can operate in islanded mode. Constraint (10.16) represents the power balance limitation, means that the sum of power output of MTs, power exchange with the upstream network and power exchange with the BSS must be equal to the equivalent load of system. P_t^{TOTAL} in Eq. (10.17) represents the equivalent load that defined as a minus of the load and power output of renewable units (PV and wind) that is calculated as follows:

$$P_t^{\text{TOTAL}} = P_t^L - P_t^{PV} - P_t^{WT} \tag{10.17}$$

10.3.3 Lower-Level Formulation

In the lower-level, BSS profit consist of minus of revenues and costs is maximized as follow:

$$F_2 : \max \ \lambda_{sw} \times C_{bat} \times \sum_{t=1}^{T} N_{EV,t} + (1 - r) \sum_{t=1}^{T} \lambda_t^r P_{BSS,t}^r + r \sum_{t=1}^{T} \lambda_t P_{BSS,t}$$
$$- \delta \sum_{t=1}^{T} N_{EV,t} \times P_{BSS,t}^{dis} \frac{\Delta t}{C_{bat}} \tag{10.18}$$

The first term of Eq. (10.18) represents the revenue from serving the EV's, where C_{bat} shows the rated capacity per battery (kWh), $N_{EV,t}$ shows the number of EVs that arrived at BSS at t^{th} time and λ_{sw} is swap price. The second term of Eq. (10.18) shows the revenue of participation in the reserve service. The third term of (10.18) represents the revenue/cost of BSS caused by exchanging power with MG, when BSS sells the power to MG, this term considered as revenue, otherwise, considered as a cost and subtracted from the total BSS incomes. The last term of Eq. (10.18) shows the depreciation cost, where, δ is depreciation coefficient of discharge-charge schemes of a battery [6].

10.3.4 Constraints of Lower-Level Problem

$$C_{BSS,t+1} = C_{BSS,t} + \eta_{BSS}^{ch} P_{BSS,t}^{ch} \times \Delta t - \frac{P_{BSS,t}^{dis}}{\eta_{BSS}^{dis}} \times \Delta t \tag{10.19}$$

$$0 \leq P_{BSS,t}^{ch} \leq P_{BSS,t}^{ch,\max} \tag{10.20}$$

$$P_{BSS,t}^{dis, min} \leq P_{BSS,t}^{dis} \leq P_{BSS,t}^{dis, max} \tag{10.21}$$

$$P_{BSS,t}^{r} \leq \eta_b^{dis} \frac{(C_{BSS,t} - C_{BSS, min})}{\Delta t} \qquad \forall t \in T_{reserve} \tag{10.22}$$

$$N_{bat,ful,t} + N_{bat,emp,t} + N_{bat,t}^{ch} + N_{bat,t}^{dis} = N_{bat} \tag{10.23}$$

$$C_{BSS,t} = C_{BSS,t-1} + \eta_{BSS}^{ch} P_{BSS,t}^{ch} - \frac{P_{BSS,t}^{dis}}{\eta_{BSS}^{dis}} \tag{10.24}$$

Constraint (10.19) shows the capacity of BSS at $t + 1th$ time, where, $P_{BSS,t}^{ch}$ and $P_{BSS,t}^{dis}$ are charged and discharged power of BSS, respectively. η_{BSS}^{ch} and η_{BSS}^{dis} are charge and discharge coefficients of BSS. The limits of the rate of charged and discharged power at t^{th} time are given in Eqs. (10.20) and (10.21), where $P_{BSS,t}^{ch, max}$ and $P_{BSS,t}^{dis, max}$ are maximum rate of charge and discharge power, respectively. It should be noted that, BSS can be charged and service to the EV's owner, simultaneous. Constraint (10.22) shows the reserve capacity of BSS. For this reason, the reserve capacity in all times when BSS participate in reserve process, power output of BSS should not exceed the BSS capacity, where $T_{reserve}$ is all times that BSS participate in reserve, $C_{BSS, t}$, $C_{BSS, min}$ are capacity of BSS at t^{th} time and minimum level, respectively. Constraint (10.23) represents the battery constraint, means that sum of the full batteries ($N_{bat, ful}$), the empty batteries ($N_{bat, emp}$) and number of charged and discharged batteries ($N_{bat}^{ch} + N_{bat}^{dis}$) at t^{th} time must be equal to total number of BSS batteries. Constraint (10.24) shows the state of charge constraint of BSS batteries.

10.3.5 Determination of Power Exchanged Price Between MG and BSS Based on Real-Time Pricing Mechanism

We describe the mechanism of real-time pricing, and then the formulation of upper and lower-levels is presented. To consider the real-time pricing mechanism based on demand response, two principles are considered as follows:

- When the BSS participates in reserve, there is not any exchanging between MG and BSS. In such situation, the MG owner should pay a certain wage to BSS.
- When the power exchanged between BSS and MG through charging and discharging process, power price is determined based on the amount of exchanged power between BSS and MG, and MG's loads. If we consider the sum of the power exchanged amount and MG's loads as total demand level, two modes are imaginable. If total demand is greater than the load of MG's loads the power price is greater than reference price, otherwise, power price is lower than it.

To determine the exchange price that is denoted by λ_t, we consider the reference price to separate the two above principles:

$$\lambda_t = \begin{cases} \dfrac{P_{BSS,t} + P_t^{TOTAL}}{P_{rf}^{TOTAL}} \times \rho_{rf} & \text{if } P_{BSS,t} + P_t^{TOTAL} \geq P_{rf}^{TOTAL} \\ \dfrac{-P_{BSS,t} + P_t^{TOTAL}}{P_{rf}^{TOTAL}} \times \rho_{rf} & \text{if } P_{BSS,t} + P_t^{TOTAL} \leq P_{rf}^{TOTAL} \end{cases} \quad (10.25)$$

Where ρ_{rf} shows the reference price and P_{rf}^{TOTAL} is the reference power of the equivalent load of MG.

Based on real-time pricing mechanism, in the lower-level problem, the schedule of charging/discharging process of BSS is determined, then, it will be feedback to upper-level problem. Then, the real-time electricity price is determined by upper-level.

10.4 Simulation and Results

10.4.1 Case Study

The proposed scheduling in the previous section is implemented on 10-bus microgrid test system with BSS that can exchange the power with upstream network. The schematic of the case study is shown in Fig. 10.6, where SS allocated on bus 6. The characteristics of MG components consist of MTs, wind turbine, PV panel and BSS are given in Tables 10.1, 10.2, 10.3, and 10.4.

The price of swapping is 1.4 $/kWh and δ is 3 $, ρ_{rf} is 1.2 $/kWh and reserve price for all three MTs is 0.04 $/kW, as [32]. The number of batteries in the BSS is 30 batteries. The curve of exchanged power price between MG and upstream network is given in Table 10.5. Computer simulations and required coding are carried out in MATLAB software and using CPLEX 11.2 solver.

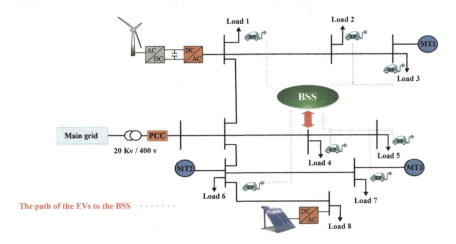

Fig. 10.6 The schematic of 10-bus MG test system in presence of BSS

Table 10.1 The characteristic of MTs

Unit	Min power generation (kW)	Max power generation (kW)	Startup/shut down cost ($)	a ($)	b ($/kW)	c ($/kw^2)
MT1	15	75	4.5	1.3	0.0304	0.000104
MT2	25	80	2.5	0.38	0.0267	0.00024
MT3	25	80	2.5	0.38	0.0267	0.00024

Table 10.2 The characteristics of PV panels

	T_c	P_{STC}	$G_{STC} W/m^2$	K	$P^{rated}(kW)$
PV	25^oC	250 kW	1000	0.001	400

Table 10.3 The characteristics of wind turbine

	V_{cut-in}	$V_{cut-out}$	V_{rated}	P^{min}_{wind}	P^{max}_{wind}	k_1	k_2	k_3	$P^{rated}(kW)$
WT	3	25	12	0	200	0.123	−0.096	0.0184	250

Table 10.4 The characteristics of BSS [6]

	η^{ch}	η^{dis}	$C_{BSS,max}(kWh)$	$C_{BSS,min}(kWh)$	$P^{ch}_{BSS,max}(kW)$	$P^{dis}_{BSS,max}(kW)$	$P^{dis}_{BSS,min}(kW)$
BSS	80%	80%	19	2	7.5	12	2

Table 10.5 Hourly exchanged power price [33]

Hour	Price ($/kW)	Hour	Price ($/kW)
1	1.1	13	1.7
2	1.1	14	1.7
3	1.1	15	1.95
4	1.1	16	1.8
5	1.1	17	1.8
6	1.1	18	1.6
7	1.1	19	1.3
8	1.1	20	1.3
9	1.3	21	1.25
10	1.3	22	1.3
11	1.3	23	1.2
12	1.4	24	1.1

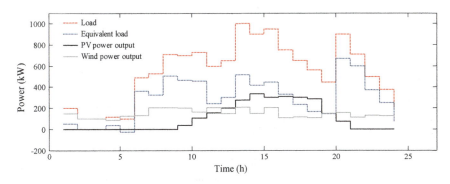

Fig. 10.7 The curve of power generation of renewable resources, load and equivalent load

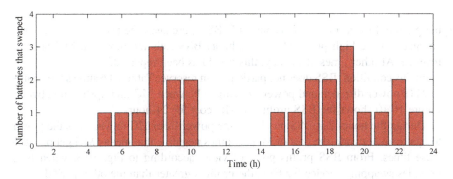

Fig. 10.8 The number of batteries that are replaced when the EVs arrived at the BSS

According to the uncertainty modeling that was presented in Sect. 10.2, based on the corresponding PDF of uncertain parameters, PV and wind power output, as well as, load demand are shown in Fig. 10.7. In addition, as discussed in Eq. (10.17), we consider the equivalent load as minus of load of MG and PV and WT power output to calculate the real-time exchanged power price, that is depicted in Fig. 10.7.

The number of EVs that arrived at the BSS, which is modeled in Eq. (10.6), is depicted in Fig. 10.8.

10.4.2 Results and Discussion

The proposed bi-level scheduling of MG in presence of the BSS, which proposed in Sects. 10.2 and 10.3, is implemented on 10-bus MG test system.

In this case, the price of power exchanged between MG and BSS follows the power exchanged price between MG and upstream network as given in Table 10.5. Figure 10.9 shows the charging/discharging scheme of BSS, where the negative values denote the charging mode. As can be seen, in the early hours of the day, when

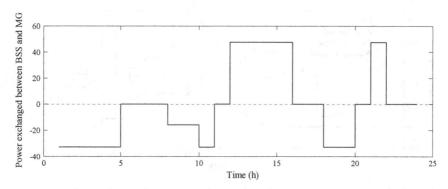

Fig. 10.9 Charging/discharging scheduling of BSS in

the price is low according to Table 10.5, BSS purchases the power from MG and charged. At the high prices (13- to 16-hour) BSS sells the power to MG and get revenue. At other times of the day, this trend has been repeated.

In this condition, BSS does not participate in reserve. Table 10.6 shows the results of MTs power dispatch and power exchanged between MG and upstream network. Figure 10.10 shows the BSS profit and MG cost for 24-hours.

During the hours that BSS purchases the power from MG or MG sells the power to upstream network, the cost of MG becomes negative that means MG get profit in these times. From BSS profits point of view, according to Fig. 10.8, when BSS provides swapping service for EVs, the profit is greater than the other periods.

Another striking point in Fig. 10.10, is the proportion of MG costs and equivalent load. According to Fig. 10.7, for hours 12–16 and hours 20- to 22, while equivalent load reaches its maximum value, MG needs to provide the balance between demand and supply, therefore, there are three options to provide this balance: (1) turn on all of the MTs, (2) purchase the power from upstream network and (3) purchases the power from BSS. It should be noted that we assumed that no load-shedding can occurred in the scheduling. Hence, the total cost of MG is greater than other periods. As can be seen, MT1 because of more expensive than the other two MTs, is committed only at peak hours (12- to 16-hour and 19- to 21 hour). In this case, the reserve capacity provided by MTs is 67.5 kW.

10.5 Conclusion

In this chapter, optimal scheduling of microgrid in presence of the battery swapping station was presented. Significant growth of utilization of EVs requires new method for exploitation. BSS in one of the suitable methods that can mitigate the integration of EVs with the power system by providing the swap service in a short time. Therefore, EV's driver can replace the empty battery with full one in shortest possible time and continue to travel. The problem of MG scheduling operation

10 Optimal Scheduling of Smart Microgrid in Presence of Battery Swapping...

Table 10.6 Results of MTs power dispatch and power exchanged between MG and upstream network in

Hour	Power dispatch (kW)			Power exchange (kW)
	MT1	MT2	MT3	
1	0	0	60	23
2	0	0	33	0
3	0	0	33	0
4	0	0	25	43
5	0	0	25	−50
6	0	80	80	200
7	0	80	80	165
8	0	80	80	361
9	0	80	80	321
10	15	80	80	314
11	75	80	80	7
12	75	80	80	17.5
13	75	80	80	235.5
14	75	80	80	134.5
15	75	80	80	163
16	75	80	80	96
17	15	80	80	60
18	15	80	80	24
19	75	80	80	−57
20	75	80	80	435
21	75	80	80	317.5
22	15	80	80	195
23	0	80	80	90
24	0	0	80	−5

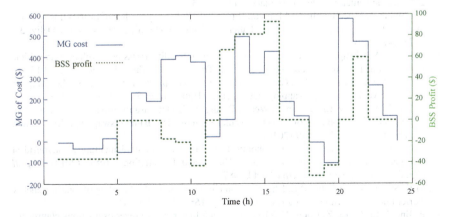

Fig. 10.10 The cost of MG and profit of BSS during the day

with the BSS, was formulated as a bi-level model, in the upper-level problem, MG operation cost minimized, while the lower-level problem maximizes the BSS profit. After simulation of the uncertain parameters consist of load, PV and wind power generation, as well as, arrival time of EVs to the BSS, real-time pricing mechanism as one of the demand response program is implemented to calculate the power exchange price between MG and BSS. The proposed model was implemented on the 10-bus MG test system. The presented results show the effectiveness of the proposed model. It should be noted that increasing the islanded operation time, causes the increase of the MG cost while guaranteeing the operation in the islanded mode, without any interruption.

References

1. J. Du, M. Ouyang, J. Chen, Prospects for Chinese electric vehicle technologies in 2016–2020: Ambition and rationality. Energy **120**, 584–596 (2017)
2. Z.-j. Pan, Y. Zhang, A novel centralized charging station planning strategy considering urban power network structure strength. Electr. Power Syst. Res. **136**, 100–109 (2016)
3. M. Hannan, M. Hoque, A. Mohamed, A. Ayob, Review of energy storage systems for electric vehicle applications: Issues and challenges. Renew. Sust. Energ. Rev. **69**, 771–789 (2017)
4. Y. Zheng, Z.Y. Dong, Y. Xu, K. Meng, J.H. Zhao, J. Qiu, Electric vehicle battery charging/ swap stations in distribution systems: Comparison study and optimal planning. IEEE Trans. Power Syst. **29**, 221–229 (2014)
5. Q. Kang, J. Wang, M. Zhou, A.C. Ammari, Centralized charging strategy and scheduling algorithm for electric vehicles under a battery swapping scenario. IEEE Trans. Intell. Transp. Syst. **17**, 659–669 (2016)
6. Y. Li, Z. Yang, G. Li, Y. Mu, D. Zhao, C. Chen, et al., Optimal scheduling of isolated microgrid with an electric vehicle battery swapping station in multi-stakeholder scenarios: A bi-level programming approach via real-time pricing. Appl. Energy **232**, 54–68 (2018)
7. P. You, Z. Yang, Y. Zhang, S.H. Low, Y. Sun, Optimal charging schedule for a battery switching station serving electric buses. IEEE Trans. Power Syst. **31**, 3473–3483 (2016)
8. S. Yang, J. Yao, T. Kang, X. Zhu, Dynamic operation model of the battery swapping station for EV (electric vehicle) in electricity market. Energy **65**, 544–549 (2014)
9. H. Farzin, M. Moeini-Aghtaie, M. Fotuhi-Firuzabad, Reliability studies of distribution systems integrated with electric vehicles under battery-exchange mode. IEEE Trans. Power Deliv. **31**, 2473–2482 (2016)
10. L. Cheng, Y. Chang, J. Lin, C. Singh, Power system reliability assessment with electric vehicle integration using battery exchange mode. IEEE Trans. Sustain. Ener. **4**, 1034–1042 (2013)
11. M.R. Sarker, H. Pandžić, M.A. Ortega-Vazquez, Optimal operation and services scheduling for an electric vehicle battery swapping station. IEEE Trans. Power Syst. **30**, 901–910 (2015)
12. L. Sun, X. Wang, W. Liu, Z. Lin, F. Wen, S.P. Ang, et al., Optimisation model for power system restoration with support from electric vehicles employing battery swapping. IET Gener. Transm. Distrib. **10**, 771–779 (2016)
13. S. P. Matthew, A technical and economic feasibility study of implementing a microgrid at Georgia Southern University. Electronic Theses and Dissertations. **1087**, (2014). https:// digitalcommons.georgiasouthern.edu/etd/1087
14. M.H. Shams, M. Shahabi, M.E. Khodayar, Stochastic day-ahead scheduling of multiple energy carrier microgrids with demand response. Energy **155**, 326–338 (2018)
15. P. Jin, Y. Li, G. Li, Z. Chen, X. Zhai, Optimized hierarchical power oscillations control for distributed generation under unbalanced conditions. Appl. Energy **194**, 343–352 (2017)

16. Y. Li, Z. Yang, G. Li, D. Zhao, W. Tian, Optimal scheduling of an isolated microgrid with battery storage considering load and renewable generation uncertainties. IEEE Trans. Ind. Electron. **66**, 1565–1575 (2019)
17. P. Xie, Y. Li, L. Zhu, D. Shi, X. Duan, Supplementary automatic generation control using controllable energy storage in electric vehicle battery swapping stations. IET Gener. Transm. Distrib. **10**, 1107–1116 (2016)
18. Hemmati, Mohammad, Saeid Ghasemzadeh, Behnam Mohammadi-Ivatloo, Optimal scheduling of smart reconfigurable neighbour micro-grids. IET Gener. Transm. Distrib.**13**(3), 380–389 (2018)
19. Z. Shi, H. Liang, S. Huang, V. Dinavahi, Distributionally robust chance-constrained energy management for islanded microgrids. IEEE Trans. Smart Grid **10**, 2234–2244 (2018)
20. A. Khodaei, Microgrid optimal scheduling with multi-period islanding constraints. IEEE Trans. Power Syst. **29**, 1383–1392 (2014)
21. B. Zhao, Y. Shi, X. Dong, W. Luan, J. Bornemann, Short-term operation scheduling in renewable-powered microgrids: A duality-based approach. IEEE Trans. Sustain. Ener. **5**, 209–217 (2014)
22. M. Vahedipour-Dahraie, H. Reza Najafi, A. Anvari-Moghaddam, J.M. Guerrero, Optimal scheduling of distributed energy resources and responsive loads in islanded microgrids considering voltage and frequency security constraints. J. Renew. Sustain. Ener. **10**, 025903 (2018)
23. G. Liu, M. Starke, B. Xiao, X. Zhang, K. Tomsovic, Microgrid optimal scheduling with chance-constrained islanding capability. Electr. Power Syst. Res. **145**, 197–206 (2017)
24. M. Vahedipour-Dahraie, A. Anvari-Moghaddam, J. Guerrero, Evaluation of reliability in risk-constrained scheduling of autonomous microgrids with demand response and renewable resources. IET Renew. Power Gener. **12**, 657–667 (2018)
25. G. Liu, M. Starke, B. Xiao, K. Tomsovic, Robust optimisation-based microgrid scheduling with islanding constraints. IET Gener. Transm. Distrib. **11**, 1820–1828 (2017)
26. J. Wang, M. Shahidehpour, Z. Li, Security-constrained unit commitment with volatile wind power generation. IEEE Trans. Power Syst. **23**, 1319–1327 (2008)
27. M. Hemmati, B. Mohammadi–Ivatloo, S. Ghasemzadeh, E. Reihani, Risk–based optimal scheduling of reconfigurable smart renewable energy based microgrids. Int. J. Electr. Power Energy Syst. **101**, 415–428 (2018)
28. N. Nikmehr and S. N. Ravadanegh, Reliability evaluation of multi–microgrids considering optimal operation of small scale energy zones under load–generation uncertainties. Int. J. Electr. Power Energy Syst. **78**, 80–87 (2016)
29. N. Nikmehr, S. Najafi–Ravadanegh, A. Khodaei, Probabilistic optimal scheduling of networked microgrids considering time–based demand response programs under uncertainty. Appl. Energy **198**, 267–279 (2017)
30. R.–C. Leou, J.–H. Teng, C.–L. Su, Modelling and verifying the load behaviour of electric vehicle charging stations based on field measurements. IET Gener. Transm. Distrib. **9**, 1112–1119 (2015)
31. T. Namerikawa, N. Okubo, R. Sato, Y. Okawa, M. Ono, Real–time pricing mechanism for electricity market with built–in incentive for participation. IEEE Trans. Smart Grid **6**, 2714–2724 (2015)
32. D. Zafirakis, K. J. Chalvatzis, G. Baiocchi, G. Daskalakis, Modeling of financial incentives for investments in energy storage systems that promote the large–scale integration of wind energy. Appl. Energy **105**, 138–154 (2013)
33. M. Honarmand, A. Zakariazadeh, and S. Jadid, Integrated scheduling of renewable generation and electric vehicles parking lot in a smart microgrid. Energy Convers. Manag. **86**, 745–755 (2014)

Chapter 11
Risk-Based Long Term Integration of PEV Charge Stations and CHP Units Concerning Demand Response Participation of Customers in an Equilibrium Constrained Modeling Framework

Pouya Salyani, Mehdi Abapour, and Kazem Zare

11.1 Introduction

Difficulties with delivering power from generation and transmission sector to the end users and the associated pollution problems encourage the distribution companies (DISCOs) to utilize the distributed generators (DGs) like wind turbines (WTs), photovoltaic cells (PVs), fuel cells (FCs) or combined heat and power units (CHP). In addition to this, voltage profile improve-ment and loss reduction can be accounted as the other goals of the DISCO owners. By this way, a portion of required energy in distribution sector can be supplied by DG units and the remainder is provided by the upstream network.

On the other hand, air pollution problems stem from existing vehicle cars as one of the important factors, environmental sensitivities and their dependence on the price of fossil fuels, directs the authorities and organizations to expand the usage of plug-in electric vehicles (PEVs) by preparing various kinds of incentives. However, the increase in the number of these PEVs implies the placement of PEV charge station to give service to the PEV owners. In addition to grid to vehicle (G2V) capability, their recent vehicle to grid (V2G) technology gives the possibility to the charge station owners to incorporate in the improvement of network electric indices and act as an energy delivery source.

The PEVs enter into the charge station usually tend to charge their PEVs in off-peak hours of the day in which the energy price is lower enough and acceptable for PEV owners. Also they can use from discharge facilities provided for them in order to sell their extra energy of battery to the grid in the peak hours of the day that DISCO deals with power shortage. This pattern of charge/discharge is economical

---

P. Salyani (✉) · M. Abapour · K. Zare
Faculty of Electrical Engineering, Tabriz University, Tabriz, Iran
e-mail: p.salyani@tabrizu.ac.ir; abapour@tabrizu.ac.ir; kazem.zare@tabrizu.ac.ir

© Springer Nature Switzerland AG 2020
A. Ahmadian et al. (eds.), *Electric Vehicles in Energy Systems*,
https://doi.org/10.1007/978-3-030-34448-1_11

for the PEV owners, however for many reasons some of them may be not able to follow this pattern and choose other hours for getting service from the charge station.

Long term integration of CHP units and PEV charge stations is the main purpose of this chapter. However, the size and location of these sources are directly affected by the total demand of network and the hourly price of energy that is purchased from the upstream grid. Hence participation of a number of customers in demand response (DR) can be cost-saving for the DISCO. Nevertheless, these customers should be economically justified and DISCO has to provide a long term contract which highly ensures that they benefit by participating in this program. The demand response program that is implemented herein is incentive-based. So what is done by the DISCO is to determine the optimal 24-hour incentives, penalties and the consumption profile for these customers through the predefined horizon.

But the main problem here arises from the contract between the profit of DISCO and customers eager to participate in demand response. Since it is desirable for DISCO to sell the energy with the highest price as it is possible for the consumption reduction of customers interested in demand response, these two variables should be determined in a way that is satisfactory for both the DISCO and each of the interested customers. Therefore the problem is modeled in a leader-follower Stackelberg framework in which the leader objective function as the upper level is the DISCO's payoff and the followers are the DR customers that their objective function is formulated in equilibrium constrained model with concerning the Kurash-Kuhn Tucker (KKT) condition. Furthermore, due to the inherent uncertainties in the consumption of customers and charge/discharge pattern of PEV charge stations, all the follower players of this game can offer their preferable risk strategy in order to model their profit in the form of CVaR function.

11.2 Literature Review

To have a brief review of previous literature, microgrid planning is studied in [1–4] in which optimal placement of distributed generators (DGs) and their sizing is the main goal. In the context of PEV parking lot planning in the distribution network, authors in [5] have first determined the uncertain parameters of PEVs arrival time, departure time and the mean traveled distance. Then the optimal location of parking lots is achieved in a reliability-constraint optimization. In [6], distribution network is partitioned into several zones candidate to the allocation of parking lot. After that, the optimal site and size of each parking lot are found by determining the probabilistic behavior of plug-in hybrid electric vehicle (PHEV) and their aforementioned uncertain parameters.

In [7], using Monte Carlo simulation to deal with uncertainties in PEV behavior and load pattern, the problem of optimal siting and placement of charge stations under demand response program is solved. The aim here is to maximize the profit of distribution system manager considering the time of use (TOU) demand response, optimal charge and discharge procedure and reliability issues. Optimal placement and sizing of renewable resources and PEV charging station are discussed in [8] under a deterministic environment with the objective of reducing power loss, voltage fluctuations, car battery maintenance costs and energy purchasing cost. Energy tariffs are specified for both the day-ahead market and the real time pricing. In [9], simultaneous placement of parking lot and renewable energy resources is studied within a two- stage stochastic programming considering the benefit of both the utility decision-maker and parking lot owner. In first stage candidate buses for parking lot allocation are introduced by the investor based on induces reliability, bus attraction and land price. Then considering the distributed resources capacity and loss minimization, optimal number of parking lots and their probabilistic model are obtained. Authors in [10] have implemented a two-step screening method and the modified primal-dual interior point algorithm and [11] has applied analytical hierarchical process to solve the charge station planning.

However it must be reminded that the energy is sold to the customers with different tariffs. This tariff or energy price is almost fixed in each hour for customers not accepted the demand response program while it is variable for customers participated in the demand response program. There exists several mechanisms for demand response program such as real-time pricing (RTP) which is used in this paper, incentive-based or TOU as another one [12–19]. What is important is that the tariff in any mechanism can be defined by DISCO within a contract however its optimal value along with the related demand can be determined through the optimization.

In [20], Stackelberg game theory is implemented to model the existing interactions in two level. The interaction between the electricity market and demand response aggregator (DRA) and the interaction among the DRA and its dependent consumers. DRA as the leader of the game reduces the demand bid and generators as the followers try to maximize their profit due to the bid reduction. Then in the second level, customers inconvenience have to be minimized by DRA while lowering of demand bid must also be served.

Virtual electricity-trading process is used in [21] to determine the optimal 24 hour consumption of customers participated in RTP demand response program. Virtual retailer that is energy management center acts as the leader player and offers the real time prices to the devices as the followers of the Stackelberg game. Authors in [22] have proposed the Vickrey-Clarke-Groves mechanism in order to maximize the social welfare in which consumers seek to maximize their total utility by receiving the optimal real-time price and company has the aim of minimizing the total generation cost. In [23], the best trading between the existing DRAs in the network and DISCO is found through the cooperative game based on bargaining Nash equilibrium, so optimal real-time tariffs that each DRA offers to its customers and also their consumption is achieved.

In [24], smart DISCO aims to optimally offer its bidding strategy with high penetration of wind farm and PEVs. PEVs are coordinated based on the determined real-time prices of customers so that the wind power imbalances be reduced. Considering the at-home electric vehicle charging, [25] introduces a Stackelberg game to maximize the benefit amount of both the retailer as the leader of the game and customers as the followers by setting the optimal real-time tariffs. Probabilistic distribution network and renewable energy expansion planning under demand response is proposed in [26] wherein by modeling the behavior of demands relative to the selling energy price, the problem is solved within a bi-level framework.

What is aimed in this chapter is the planning of distribution network under uncertainty in order to optimally siting and sizing of MTs and parking lots. In addition a number of customers that are eager to participate in DRPs can declare it to the DISCO, so the planning of distribution network is carried out under demand response of customers. The demand response that is applied is based on a long term contract to adjust the 24 hour selling energy price in a manner that becomes beneficial for these customers. In other word customers participate in the demand response program should get different and affordable tariffs rather than other customers having their normal consumption. However, the important point that should be noted is that reducing the selling energy price is not desirable for DISCO, especially when the number of requestor customers to participate in DRP is high enough. This is due to the fact that selling the energy to the customers with much higher price is profitable for DISCO.

Therefore since each of these customers and also the DISCO seeks to enhance their profit, interaction takes places between their strategies, the optimal planning decision of DsiCo and optimal demand scheduling decision of customers adopted DRP. One of the efficient ways to cope with this difficulty is to modeling the problem based on the game theory framework. To this end, leader-follower Stackelberg as the non-cooperative game is implemented in this chapter to give the equilibrium point in which both the DISCO and customers beneficial objectives are served. The leader of this game is DISCO that seeks to maximize its payoff. The followers are the customers that are interested in adopting DRP and each of them desires to achieve maximum payoff by implementing demand response.

After giving a general view about the abovementioned novel approach, the following section represents the mathematical formulation of the distribution planning and demand response of customers. Section 11.4 explains the Stackelberg game and its application in the planning problem, then elaborates problem modeling and the iterative distributed algorithm used to reach the equilibrium point for the DISCO and demand response customers. Section 11.5 discusses about the results obtained from simulation and the last section is the conclusion of the chapter about this novel approach.

11.3 Problem Formulation

With respect to the load growth rate in the distribution sector and considering the incremental rate of electric vehicles, it is aimed to implement a long-term integration of CHPs as dispatchable units and PEV charge stations. This long-term integration can lead to a reduction in the cost of buying the energy from the upstream network for DISCO besides the reduction in voltage anomalies and network loss. Starting

11 Risk-Based Long Term Integration of PEV Charge Stations and CHP Units...

with CHP and PEV charge station, their associated cost functions are given in Eqs. (11.1) and (11.2). One important point should be mentioned that is linearity of the problem and its model which is mixed-integer linear programming, thus binary variables that determine the location of CHP units and PEV charge stations are not modeled in below cost functions but appear in the related constraints.

$$
\begin{aligned}
\text{Cost}^{CHP} = &\sum_{i \in \Omega_L} CC^{CHP}.P_i^{CHP} \\
&+ \sum_{y=1}^{N_y} \left(\sum_{i \in \Omega_L} MC^{CHP}.P_i^{CHP}.T_h + \sum_{i \in \Omega_L} FC^{CHP}.P_i^{CHP}.T_h \right).PW_y
\end{aligned}
\tag{11.1}
$$

$$
\begin{aligned}
E_\omega \{\text{Cost}_\omega^{CS}\} = &\sum_{i \in \Omega_L} CC^{CS}.\Theta_i^{CS} \\
&+ \sum_{y=1}^{N_Y} \left(\sum_{i \in \Omega_L} MC^{CS}.\Theta_i^{CS}.T_d \right).PW_y \\
&+ \sum_{\omega \in \Omega_S} \pi_\omega \left\{ \sum_{y=1}^{N_Y} \sum_{h=1}^{N_T} \left(\sum_{i \in \Omega_L} \left[\rho_h^{PEV}.P_{i,y,h,\omega}^{dch}.\tau_{j,y,h,\omega}^{dch} - \rho_h^{G2P}.\frac{P_{i,y,t,\omega}^{ch}}{\mu_{conv}}.\tau_{j,y,h,\omega}^{ch} \right] \right) \right).PW_y
\end{aligned}
\tag{11.2}
$$

Where

$$
P_{i,y,h,\omega}^{dch} = \alpha_{i,h}^{dch}.\Theta_i^{CS}.G_{y,h,\omega}.Pr_{y,h,\omega}
\tag{11.3}
$$

$$
P_{i,y,h,\omega}^{ch} = \alpha_{i,h}^{ch}.\Theta_i^{CS}.G_{y,h,\omega}.Pr_{y,h,\omega}
\tag{11.4}
$$

$$
\tau_{i,y,h,\omega}^{dch} = \frac{SOC_{i,y,h,\omega}.E_i^{BCap}}{Pr_{i,\omega}}
\tag{11.5}
$$

$$
\tau_{i,y,h,\omega}^{ch} = \frac{\left(1 - SOC_{i,y,h,\omega}\right).E_i^{BCap}}{Pr_{i,\omega}}
\tag{11.6}
$$

As it is shown in Eq. (11.1), three terms are considered for CHP cost that are capital cost, maintenance cost and operation cost. Energy provision in a reliable manner and with the least of uncertainty is a prominent feature of CHPs known as dispatchable units. Nevertheless, their operation cost is the main drawback in comparison with the renewable ones like PV and WT. Maintenance cost could be calculated based on an almost fix percentage, but here it is a function of annual energy production.

Apart from CHP cost, associated PEV charge station cost constitutes three terms of capital cost which includes buying land and establishment cost, repair and maintenance cost and the third one that represents the expected payment and benefit exposed to DISCO in the case of PEVs charging and discharging. These expected payment and benefit are dependent of charge/discharge output power of each charge station, required time for charge/discharge of PEVs and energy price for both G2V and V2G mode of operation in each scenario.

It must be mentioned that as an assumption, DISCO is the owner of overall charge stations located in any bus of the medium voltage distribution network. For time horizons that a part of PEVs within the charge station are in their V2G mode, DISCO has to pay an amount to those PEV owners based on the price ρ_h^{PEV}. In turn for the charging mode periods, DISCO has to provide the power to the charge station in grid energy price and receives a payment from those G2V PEV owners to charge their battery of vehicles. It should be noted that the energy trading among the grid and the PEV charge stations does not include DRP and take places within a certain amount.

Output power of a charge station (charge and discharge) can be obtained through equations Eqs. (11.3) and (11.4) which are based on the parameters vehicle capacity of charge station, battery power rate of PEVs and the charge/discharge pattern of PEVs defined by parameters $\alpha_{i,h}^{ch}$ and $\alpha_{i,h}^{dch}$. These parameters give the percentage of PEVs that are present in hour h for the purpose of either charge or discharge service. The required time for charge and discharge of PEVs are calculated through Eqs. (11.5) and (11.6), both depend on state of charge (SOC) and maximum capacity of battery for PEVs. Parameter G is the hourly presence percentage of vehicles in the charge station. Because the exact number of vehicles presented in the charge station in such a long term integration level is uncertain, the presence percentage is used instead to model the output power concerning its uncertain nature.

The expected energy purchasing cost Eq. (11.7) is another cost term that is exposed to the DISCO which includes the consumed energy in the charge station and the network demand. However, a portion of this energy can be supplied by the CHP units located in the network or by PEVs in their V2G mode via charge station. Note that the network demand is divided into those customers willing to participate in DRP and other ones that are not interested in this program.

$$
E_\omega\{Cost_\omega^{EP}\} = \sum_{\omega \in \Omega_S} \pi_\omega \left\{ \rho_h^g \sum_{y=1}^{Ny} \sum_{h=1}^{N_T} \sum_{i \in \Omega_L} \left(P_{i,y,h,\omega}^{ch} - P_{i,y,h,\omega}^{dch} + P_{y,h,\omega}^{Loss} \right. \right.
$$
$$
\left. \left. + v_i^{Dr} . P_{i,y,h,\omega}^{Dr} + v_i^{Du} . P_{i,y,h,\omega}^{Du} - P_i^{CHP} \right) \right\}
\tag{11.7}
$$

$$
Cost^{Dis} = Cost_\omega^{CHP} + E_\omega\{Cost_\omega^{PL}\} + E_\omega\{Cost_\omega^{EP}\}
\tag{11.8}
$$

Total expected cost that DISCO incurs over the planning period as the upper-level problem is given in Eq. (11.8). However to minimize the total expected cost, following constraints Eqs. (11.9, 11.10, 11.11, 11.12, 11.13, 11.14, and 11.15) must be satisfied too. Applying an approximate linear power flow through Eq. (11.9), the per-unit value of voltages in every bus of the network can be obtained, so by Eq. (11.10) the network voltage is ensured to be in the determined permissible range in each scenario. Considering the radial structure of distribution network, the binary matrix $\phi_{i,j,n}$ is 1 if bus i covers the bus n up, (current can flow from bus i to bus n). Since the voltage of the first bus as the slack bus is 1 pu, the per-unit voltage of other buses can be known by Eq. (11.9).

Inequalities Eqs. (11.11) and (11.12) state that the output power of any CHP or PEV capacity of any charge station is limited between a minimum and a maximum value if it is selected in bus i. The constraint Eq. (11.13) indicates that every bus of the network can be equipped with just one of the CHP unit or PEV charge station. Also, the two last constraints refer to the fact that a limited number of CHP or PEV charge station can be sited in the network.

$$
\begin{aligned}
V_{i,t,y,\omega} = {}& V_{j,t,y,\omega} \\
& -r_{ij}\sum_{n\in\Omega_L}\phi_{i,j,n}\left(v_n^{Dr}.P_{n,t,y,\omega}^{Dr} + v_n^{Du}.P_{n,t,y,\omega}^{Du} + P_{n,t,y,\omega}^{ch} - P_{n,t,y,\omega}^{dch} - P_n^{CHP}\right) \\
& -x_{ij}\sum_{n\in\Omega_L}\phi_{i,j,n}\left(v_n^{Dr}.Q_{n,t,y,\omega}^{Dr} + v_n^{Du}.Q_{n,t,y,\omega}^{Du} - pf.P_n^{CHP}\right)
\end{aligned} \tag{11.9}
$$

$$
V^{\min} \le V_{i,t,y,\omega} \le V^{\max} \tag{11.10}
$$

$$
\xi_i^{CHP}.P^{\min} \le P_i^{CHP} \le \xi_i^{CHP}.P^{\max} \tag{11.11}
$$

$$
\xi_i^{CS}.\Theta^{\min} \le \Theta_i^{CS} \le \xi_i^{CS}.\Theta^{\max} \tag{11.12}
$$

$$
\xi_i^{CHP} + \xi_i^{CS} \le 1 \tag{11.13}
$$

$$
\sum_{i\in\Omega_L}\xi_i^{CHP} \le N_{\max}^{CHP} \tag{11.14}
$$

$$
\sum_{i\in\Omega_L}\xi_i^{CS} \le N_{\max}^{CS} \tag{11.15}
$$

In the lower-level problem, there are several load points who are interested in DRP and their motivation for participating in such a long term program is to increase the net gained profit by electricity consumption. Each customer has its own benefit function that has a general quadratic form shown in Eq. (11.16). The perceived benefit in hour h depends on the normal consumption in the absence of DR d_h^0, demand in the presence of DR d_h and the elasticity of the customer ε_h. Therefor the net profit that each participated customer in DRP obtains is calculated through Eq. (11.17) where ρ_h as a variable is the energy selling price determined by DISCO.

Since one of the main goals herein is to model the overall problem in a linear form, this nonlinear profit function is reformulated into Eq. (11.18) by its linearization. The benefit function is linearized through the piecewise method Eq. (11.19). But the nonlinear term, the bought energy from DISCO is replaced with bought energy in network tariff plus a linear penalty/incentive function defined in Eq. (11.20). This term has to be added to the cost function of DISCO.

$$B(d_h) = B_h^0 + \rho_h^0 \cdot (d_h - d_h^0) \cdot \left(1 + \frac{d_h - d_h^0}{2\varepsilon_h \cdot d_h^0}\right) \tag{11.16}$$

$$\Psi = B_h^0 + \rho_h^0 \cdot (d_h - d_h^0) \cdot \left(1 + \frac{d_h - d_h^0}{2\varepsilon_h \cdot d_h^0}\right) - \rho_h \cdot d_h \tag{11.17}$$

$$\Psi = B_h^0 + \rho_h^0 \cdot \left(\frac{1}{2\varepsilon_h} - 1\right) \cdot d_h^0 + \frac{\rho_h^0}{d_h^0} \sum_{b=1}^{N_B} S_{b,h} \cdot X_{b,h} - \frac{\rho_h^0}{d_h^0 \cdot \varepsilon_h} \sum_{b=1}^{N_B} X_{b,h} + \Phi_h \tag{11.18}$$

$$d_h = \sum_{b=1}^{N_B} X_{b,h} \tag{11.19}$$

$$\Phi_h = A_h \cdot (d_h^0 - d_h) \tag{11.20}$$

The above-mentioned equation Eq. (11.18) gives us the expected profit that any of DR customer wants to maximize it. Indeed there is a DISCO that seeks to minimize its related long-term integration cost and a number of customers that seek to maximize their own profit. This issue results in a competition among the DISCO and these customers, hence the bi-level leader-follower Stackelberg game can help to modeling the problem in which the DISCO acts as the leader and the DR customers act as the followers of the game.

The KKT condition method and using the Lagrangian method can be the best option to convert this bi-level problem into a single-level problem. From another aspect, these customers are willing to participate in a long term DRP which can contribute to getting the preferable outcome with a high degree of uncertainty. Thus in addition to the gained expected profit, the risk function is added to the objective function of DR customer k which is based on CVaR assessment Eq. (11.21) with the related constraints given in Eqs. (11.22, 11.23, 11.24, 11.25, 11.26, 11.27, and 11.28).

$$\begin{aligned}\Gamma_k &= (1 - \beta_k)E_\omega\{\Psi_{k,y,h,\omega}\} + \beta_k\Re_\omega\{\Psi_{k,y,h,\omega}\}\\ &= (1 - \beta_k)E_\omega\{\Psi_{k,y,h,\omega}\} + \beta_k \cdot \left(\eta_k - \frac{1}{1 - \gamma_k}\sum_{\omega \in \Omega_s} \pi_\omega \cdot \zeta_{k,\omega}\right)\end{aligned} \tag{11.21}$$

$$\eta_k - \sum_{y=1}^{N_y} \sum_{h=1}^{N_T} \Psi_{k,y,h,\omega} \leq \zeta_{k,\omega}, \quad \mu_{k,\omega}^R \tag{11.22}$$

$$\zeta_{k,\omega} \geq 0 \quad \mu_\omega^\zeta \tag{11.23}$$

$$X_{k,b,y,h,\omega} \geq z_{k,b,y,h,\omega}\left(P_{k,y,h,\omega}^{\min} + (b-1)\left[\frac{P_{k,y,h,\omega}^{\max} - P_{k,y,h,\omega}^{\min}}{N_B}\right]\right), \quad \mu_{k,b,y,h,\omega}^{x\min} \quad (11.24)$$

$$X_{k,b,y,h,\omega} \leq z_{k,b,y,h,\omega}\left(P_{k,y,h,\omega}^{\min} + b\left[\frac{P_{k,y,h,\omega}^{\max} - P_{k,y,h,\omega}^{\min}}{N_B}\right]\right), \quad \mu_{k,b,y,h,\omega}^{x\max} \quad (11.25)$$

$$E_{k,y}^{\min} \leq \sum_{\omega \in \Omega_S} \pi_\omega \sum_{h=1}^{N_T} \sum_{b=1}^{N_B} X_{k,b,y,h,\omega} \leq E_{k,y}^{\max}, \quad \mu_{k,y}^{E\min}, \mu_{k,y}^{E\max} \quad (11.26)$$

$$P_{k,y,h,\omega}^{\min} + b\left[\frac{P_{k,y,h,\omega}^{\max} - P_{k,y,h,\omega}^{\min}}{N_B}\right] - X_{k,b,y,h,\omega} \leq z_{k,b+1,y,h,\omega}H, \quad \mu_{k,b,y,h,\omega}^Z \quad (11.27)$$

$$\Phi_{k,y,h,\omega} = A_h \cdot \left(P_{k,y,h,\omega}^L - \sum_{b=1}^{N_B} X_{k,b,y,h,\omega}\right), \quad \lambda_{k,y,h,\omega}^\Phi \quad (11.28)$$

Inequality Eq. (11.22) refers to the CVaR constraint and Eq. (11.23) emphasizes that ζ is a positive variable. The constraints Eq. (11.24) and (11.25) give the demand limitation of each block and Eq. (11.26) states that the daily expected energy consumption must be lied within a specific range. Consecutive selection of blocks for the consumption of customer k is ensured by Eq. (11.27) and the penalty/incentive function is represented in Eq. (11.28). It is noteworthy that the binary variable z is 1 for the first block and can take 0 or 1 for the other blocks. H is a large enough number to set the binary variable in 1.

There exist several risk assessment methods but CVaR has a main and useful feature that is its ability to keep the convexity of the function. Convexity is of importance because it accounts as one of the essential and necessary factors for the implementation of KKT condition. After constructing the Lagrangian function L and applying the KKT condition, the following equations are obtained.

$$\nabla_{\mathbf{x}}L = (1 - \beta_k)T_d \cdot \pi_\omega \cdot \left[\frac{\rho_h^0}{P_{k,y,h,\omega}^L}S_{k,b,y,h,\omega} - \frac{\rho_h^0}{P_{k,y,h,\omega}^L \cdot \varepsilon_{k,h}} + A_h\right]$$

$$+ \mu_{k,\omega}^R T_d\left[\frac{\rho_h^0}{P_{k,y,h,\omega}^L}S_{k,b,y,h,\omega} - \frac{\rho_h^0}{P_{k,y,h,\omega}^L \cdot \varepsilon_{k,h}} + A_h\right] \quad (11.29)$$

$$+ \mu_{k,b,y,h,\omega}^{X\min} - \mu_{k,b,y,h,\omega}^{X\max} + \pi_\omega\left(\mu_{k,y}^{E\min} - \mu_{k,y}^{E\max}\right) + A_h \cdot \lambda_{k,y,h,\omega}^\Phi = 0$$

$$\nabla_{\mathbf{z}}L = \mu_{k,b,y,h,\omega}^{X\max} - \mu_{k,b,y,h,\omega}^{X\min} + \mu_{k,b,y,h,\omega}^z = 0 \quad (11.30)$$

$$\nabla_{\mathbf{\eta}}L = \beta_k - \sum_{\omega \in \Omega_S} \mu_{k,\omega}^R = 0 \quad (11.31)$$

$$\nabla_\zeta L = -\frac{\beta_k}{1-\gamma_k}\pi_\omega + \mu_{k,\omega}^R + \mu_{k,\omega}^\zeta = 0 \tag{11.32}$$

However the inequality constraints impose nonlinearity to the overall problem within the KKT framework but these terms can be linearized through the bigM method for a linear example constraint $\Im(x) \geq 0$ that is as follows. Remind that M should be large enough.

$$\mu.\Im(x) = 0, \quad \mu \geq 0 \tag{11.33}$$

$$\Im(x) \leq u.M, \quad u = \{0,1\} \tag{11.34}$$

$$\mu \leq (1-u).M, \quad u = \{0,1\} \tag{11.35}$$

As it was discussed above, the parameters and variables i.e. the presence percentage of PEVs in the parking lot, initial SOC of PEVs arrived at the parking lot, their power rate of batteries and the demand of load points either interested in DRP or not are inherent to uncertainty. In this chapter it is assumed that all the uncertain parameters are generated as presented in Eqs. (11.21, 11.22, 11.23, and 11.24). As it is stated, for these parameters a mean value μ and a standard deviation σ is considered. Then by a random number e between $[-1, 1]$ and using the Monte Carlo process, at each iteration an amount is assigned to these parameters.

$$G_{t,y,\omega} = \mu_G + e_{SOC,\omega}.\sigma_G \tag{11.36}$$

$$SOC_{t,y,\omega} = \mu_{SOC} + e_{SOC,\omega}.\sigma_{SOC} \tag{11.37}$$

$$Pr_{t,y,\omega} = \mu_{Pr} + e_{Pr,\omega}.\sigma_{Pr} \tag{11.38}$$

$$P_{i,t,y,\omega}^L = \mu_p + e_{p,\omega}.\sigma_p \tag{11.39}$$

Afterwards, the scenario generation is accomplished, due to the high number of scenarios using fast forward selection and using the Kontarovich distance scenario reduction technique the number of scenarios is reduced and the associated probabilities are obtained.

11.4 Numerical Results

The problem modeling is performed in the GAMS software and the CPLEX solver is selected due to its MIP model. To have a better and efficient evaluation on the proposed method, IEEE 33 bus distribution test system [27] in Fig. 11.1 is chosen for the long term integration problem. The load points {7, 14, 24, 31} are the medium voltage customers who are interested in adopting long term DRP within about 10 years. Also five scenarios after scenario reduction is obtained with probabilities of {0.30, 0.35, 0.25, 0.06, 0.04}.

11 Risk-Based Long Term Integration of PEV Charge Stations and CHP Units...

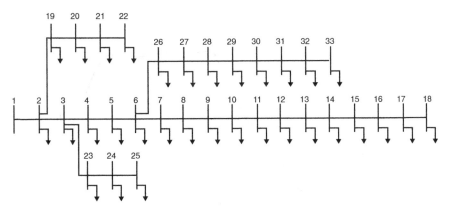

Fig. 11.1 IEEE 33 bus test system schematic

Table 11.1 Required data

Parameter	Value	Parameter	Value
MT investment cost ($)	900,000	Parking lot investment cost ($/EV)	304
MT operation cost ($/kWh)	80	Minimum voltage (pu)	0.95
MT maintenance cost ($/kWh)	15	Maximum voltage (pu)	1.05
Power factor of MT	0.9	Inflation rate	9%
Maximum number of MT	2	Interest rate	12%
Minimum power of MT (kW)	100	Load growth rate	7%
Maximum power of MT (kW)	1000	Efficiency of inverter	90%

Required data to solve this problem are given in Table 11.1. Figures 11.2 and 11.3 depict the 24-hour energy price bought from the upstream subtransmission substation and demand factor of entire customers in the network respectively. The standard deviation for all these uncertain parameters and also their mean value for this chapter are extracted from [7, 28]. Furthermore the total installed capacity of PEV charge station must be adequate to give charge/discharge service to PEV number equal to 500. It must be mentioned that for customers that have not participated in DRP, the energy selling price is 0.095 ($/kWh).

It is assumed that each PEV of charge station has a battery with a capacity of 80 kW and a mean charge and discharge rate of 50 kW/h. Table 11.2 gives the charge and discharge pattern of PEVs throughout the day in percentage that enter the charge station. Also the PEVs enter to the charge station have to pay the price equal to energy price represented in Fig. 11.2 for the purpose of 1 kWh charge of the battery. However when the EVs are in the mode of V2G in order to deliver their battery power to the grid, they can sell their energy with a price equal to 0.108 $/kWh.

Table 11.3 represents the risk strategy and the CVaR risk level of all the four customers willing to participate in DRP. For the DISCO, the total expected cost of 31.571 million ($) is obtained. Four CHP units with optimal size and location in the

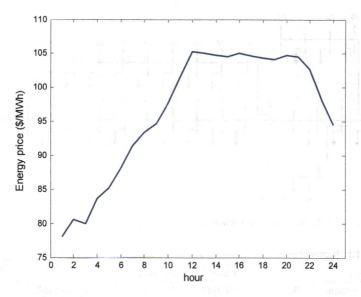

Fig. 11.2 24 hour energy price

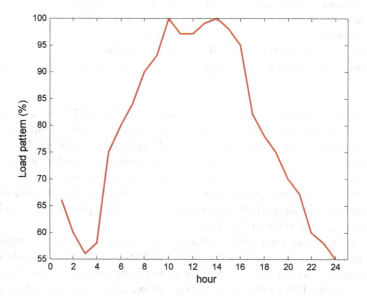

Fig. 11.3 Demand factor curve throughout the day

Table 11.2 Pattern of operation mode for PEVs in charge the stations

PEV mode of operation	Hours		
	0–6	6–15	15–24
V2G	0	5%	90%
G2V	100%	50%	10%

Table 11.3 Risk strategy and risk level of DR customers

DR customer	7	14	27
β_k	0.01	0.99	0.01
γ_k	0.99	0.98	0.93

Table 11.4 Optimal CHP size and location

Parameter	Values		
CHP location	5	11	14
CHP size (kW)	980	1000	720

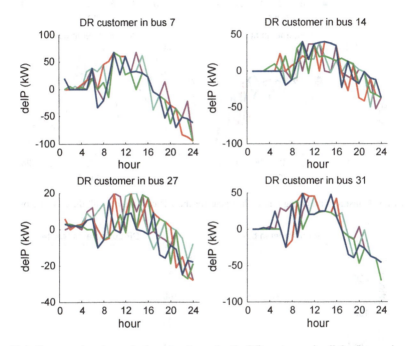

Fig. 11.4 Consumption change in the second year for the DR customers in all the 5 scenarios

network given in Table 11.4 are selected. Also, two PEV charge station with the capacities of 214 and 286 PEVs must be located in the busses 12 and 20 respectively.

Figures 11.4 and 11.5 demonstrate the amount of responded demand (consumption change) for every four load points participated in DRP. This change in consumption represented by 'delP' is depicted for two sample year of second and last and for all the five scenarios.

As it can be seen, for most of the scenarios in the off-peak hours of the day, the change in the consumption is negative and in the mid-hours of the day that are almost the peak hours, these changes are positive and considerable. In other words in the off-peak hours there is negative response while in the peak-hours DR customers have captured a positive response for DRP in both of the second and last years.

Furthermore, the expected energy consumption of these DR customers in the second and the last year are depicted in Figs. 11.6 and 11.7. Compared to the

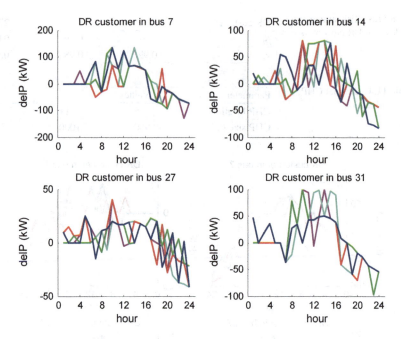

Fig. 11.5 Consumption change in the last year for the DR customers in all the 5 scenarios

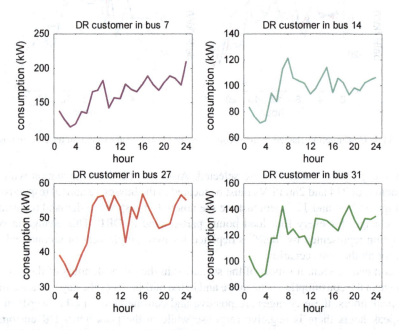

Fig. 11.6 Expected demand level of DR customers in the second year

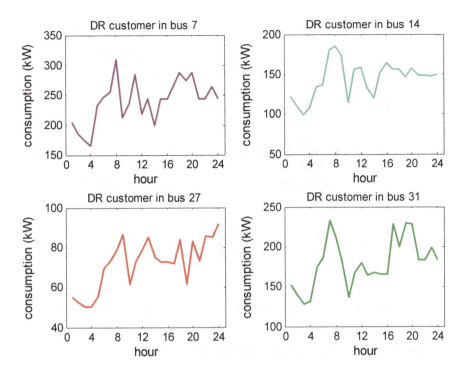

Fig. 11.7 Expected demand level of DR customers in the last year

expected consumption of these load points in the absence of DRP (do not participate in the long term DRP), the main part of the load is shifted from the peak-hours to the peak hours either the morning or mid-night in which the price for energy purchasing from the upstream network is lower.

It should be noticed that the energy selling price has a fixed value for all the load points within the distribution network, but the load shifting has taken placed in a proper manner. This is due to the fact that for any amount of demand reduction, DISCO has to pay an incentive to the customers participated in DRP. This incentive absolutely depends on the hours of the day, means that in the peak hours its value is high, unlike the off-peak hours that is low.

Thus it is beneficial for DR customers to reduce their consumption in the peak hours in order to get a higher incentive from DISCO. One another important point which may be hidden should be mentioned here is that DISCO seeks to lessen the investment and operational related to this long term integration. Surely demand reduction and peak shaving can help to reduce this considerable cost in the asset management framework, especially such these customers that have a high and considerable amount of consumption in comparison with other customers of the network.

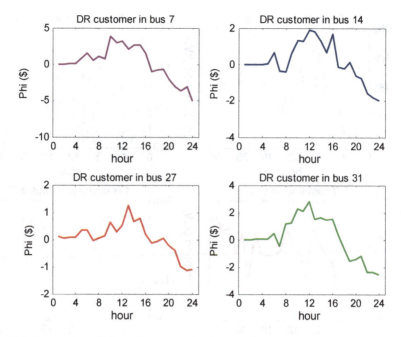

Fig. 11.8 Expected penalty/incentive amount for the second year

On the other hand, since the penalty imposed to the DR customers is in direction with the grid energy purchase price as well as the incentive price, it is preferable for these customers to shift their consumption to the off-peak times of the day. For this reason the negative response to the DRP (negative change in consumption) incurs a less enough penalty amount to these customers. However it should be reminded that the expected 24 hour energy consumption must not violate from its predefined range. Figures 11.8 and 11.9 respectively represent the amount of penalty/incentiveΦ, for the second and last year that verifies this issue.

11.5 Conclusion

This chapter discussed about the long term integration of PEVs and optimal siting and sizing of CHP units and charge stations in the distribution network. With consideration of a long term incentive-based DRP, a leader-follower Stackelberg problem was constructed in which the leader was the DISCO and the customers interested in DR were the followers of this game. By linearization and applying the KKT condition along with CVAR modeling of profit for DR customers, the bi-level problem was converted into a single-level problem. The two sited PEV charge station had the capacity of supporting about 500 PEV. In addition, these charge stations were located in busses that from DISCO viewpoint can help to maintain the

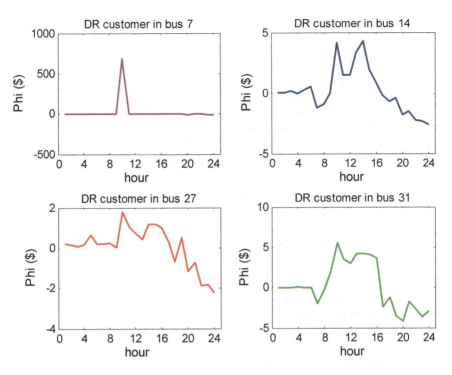

Fig. 11.9 Expected penalty/incentive amount for the last year

voltage profile of network within an acceptable range for all the scenarios. What was important is that a successful DRP was planned for the customers willing to participate. This was in a way that in the peak times of the day the DR customers capture a positive response to DRP benefit from the provided incentive for them and to help in long term integration cost reduction. While in the off-peak times they should try to increase their consumption to compensate for the reduction in their expected energy consumed throughout the day.

Appendix A

The nomenclature is shown below.

Ω_L	Set of buses
Ω_S	Set of scenarios
N_T	Number of hours in a day
N_y	Integration horizon
v_i^{Dr}	Binary parameter that is 1 if ith customer is participated in DRP
v_i^{Du}	Binary parameter that is 1 if ith customer is not interested in DRP

(continued)

$P^{Dr}_{i,y,h,\omega}$	Demand of ith DR customer in year y, hour h and scenario ω
$P^{Du}_{i,y,h,\omega}$	Demand of ith DU customer in year y, hour h and scenario ω
CC^{CHP}	Capital cost of CHP (\$/MW)
MC^{CHP}	Maintenance cost of CHP (\$/MWh)
FC^{CHP}	Fuel cost of CHP (\$/MWh)
CC^{CS}	Capital cost of charge station (\$/PEV number)
MC^{CS}	Maintenance cost of charge station (\$/PEV number)
Θ^{CS}_i	PEV capacity of each charge station selected to be installed in bus i
ξ^{CHP}_i	Binary decision variable that is 1 if a CHP is installed in bus i
ξ^{CS}_i	Binary decision variable that is 1 if a WT is installed in bus i
P^{CHP}_i	Rated power of CHP in bus i (MW)
$V_{i,\,y,\,h,\,\omega}$	Voltage of bus i in year y, hour h and scenario ω (pu)
ρ^0_h	Selling energy price to the entire customers in hour h
ρ^g_h	Energy price bought from upstream network in hour h
$E^{Min}_{i,y}$	Minimum energy consumption of ith DR customer in year t
$E^{Max}_{i,y}$	Maximum energy consumption of ith DR customer in year t
π_ω	Probability of scenario ω
PW_y	Present worth factor in year y
inf _ r	Inflation rate
int _ r	Interest rate
r_{ij}	Resistance between bus i and j
x_{ij}	Reactance between bus i and j
T_d	Total number of days in a year
pf_i	Power factor of CHP unit connected to bus i
γ_k	CVaR risk level of DR customer k
β_k	Risk aversion strategy of DR customer k
$\Phi_{k,\,y,\,h,\,\omega}$	Penalty/incentive function of DR customer k
η_k	Value at Risk of DR customer k
ζ_k	CVaR auxiliary variable of DR customer k
$X_{k,\,b,\,y,\,h,\,\omega}$	Piecewise demand of DR customer k, in block b, in year y, in hour h and in scenario ω
$S_{k,\,b,\,y,\,h,\,\omega}$	Piecewise benefit curve slope of DR customer k, in block b, in year y, in hour h and in scenario ω
$P^L_{i,y,h,\omega}$	Active power demand of load point in bus i, in year y, in hour h and in scenario ω

References

1. R. Hemmati, S. Hedayat, S. Pierluigi, Coordinated short-term scheduling and long-term expansion planning in microgrids incorporating renewable energy resources and energy storage systems. Energy **134**, 699–708 (2017)
2. S.S. Tanwar, D.K. Khatod, Techno-economic and environmental approach for optimal placement and sizing of renewable DGs in distribution system. Energy **127**, 52–67 (2017)

3. M. Kumar, P. Nallagownden, I. Elamvazuthi, Optimal placement and sizing of renewable distributed generations and capacitor banks into radial distribution systems. Energies **10**(6), 811 (2017)
4. J. Jung, M. Villaran, Optimal planning and design of hybrid renewable energy systems for microgrids. Renew. Sust. Energ. Rev. **75**, 180–191 (2017)
5. M.H. Amini, A. Islam, Allocation of electric vehicles' parking lots in distribution network, in *ISGT2014*, (IEEE, Washington, 2014)
6. M.J. Mirzaei, A. Kazemi, O. Homaee, A probabilistic approach to determine optimal capacity and location of electric vehicles parking lots in distribution networks. IEEE Trans. Ind. Inform. **12**(5), 1963–1972 (2016)
7. S. Shojaabadi, S. Abapour, M. Abapour, A. Nahavandi, Optimal planning of plug-in hybrid electric vehicle charging station in distribution network considering demand response programs and uncertainties. IET Gener. Transm. Distr. **10**(13), 3330–3340 (2016)
8. M.M. Rezaei, M.H. Moradi, M.H. Amini, A simultaneous approach for optimal allocation of renewable energy sources and electric vehicle charging stations in smart grids based on improved GA-PSO algorithm. Sustain. Cities Soc. **32**, 627–637 (2017)
9. M.H. Amini, M.P. Moghaddam, O. Karabasoglu, Simultaneous allocation of electric vehicles' parking lots and distributed renewable resources in smart power distribution networks. Sustain. Cities Soc. **28**, 332–342 (2017)
10. L. Zhipeng, F. Wen, G. Ledwich, Optimal planning of electric-vehicle charging stations in distribution systems. IEEE Trans. Power Deliv. **28**(1), 102–110 (2013)
11. X. Lin, J. Sun, Y. Wan, D. Yang, Distribution network planning integrating charging stations of electric vehicle with V2G. Int. J. Electr. Power Energy Syst. **63**, 507–512 (2014)
12. F. Wang, L. Zhou, H. Ren, X. Liu, S. Talari, Multi-objective optimization model of source–load–storage synergetic dispatch for a building energy management system based on TOU price demand response. IEEE Trans. Ind. Appl. **54**(2), 1017–1028 (2018)
13. A. Asadinejad, K. Tomsovic, Optimal use of incentive and price based demand response to reduce costs and price volatility. Electr. Power Syst. Res. **144**, 215–223 (2017)
14. A.S.O. Ogunjuyigbe, C.G. Monyei, T.R. Ayodele, Price based demand side management: A persuasive smart energy management system for low/medium income earners. Sustain. Cities Soc. **17**, 80–94 (2015)
15. A.H. Sharif, P. Maghouli, Energy management of smart homes equipped with energy storage systems considering the PAR index based on real-time pricing. Sustain. Cities Soc. **45**, 579–587 (2019)
16. K. Saberi, H. Pashaei-Didani, R. Nourollahi, K. Zare, S. Nojavan, Optimal performance of CCHP based microgrid considering environmental issue in the presence of real time demand response. Sustain. Cities Soc. **45**, 596–606 (2019)
17. M.H. Imani, P. Niknejad, M.R. Barzegaran, The impact of customers' participation level and various incentive values on implementing emergency demand response program in microgrid operation. Int. J. Electr. Power Energy Syst. **96**, 114–125 (2018)
18. Q. Yang, X. Fang, Demand response under real-time pricing for domestic households with renewable DGs and storage. IET Gener. Transm. Distr. **11**(8), 1910–1918 (2017)
19. A. Asadinejad, A. Rahimpour, K. Tomsovic, H. Qi, Evaluation of residential customer elasticity for incentive based demand response programs. Electr. Power Syst. Res. **158**, 26–36 (2018)
20. E. Nekouei, T. Alpcan, D. Chattopadhyay, Game-theoretic frameworks for demand response in electricity markets. IEEE Trans. Smart Grid **6**(2), 748–758 (2015)
21. M. Yu, S.H. Hong, A real-time demand-response algorithm for smart grids: A Stackelberg game approach. IEEE Trans. Smart Grid **7**(2), 879–888 (2016)
22. P. Samadi, A.H.M. Rad, R. Schober, V.W.S. Wong, Advanced demand side management for the future smart grid using mechanism design. IEEE Trans. Smart Grid **3**(3), 1170–1180 (2012)
23. S. Fan, Q. Ai, L. Piao, Bargaining-based cooperative energy trading for distribution company and demand response. Appl. Energy **226**, 469–482 (2018)

24. A. Ghasemi, S.S. Mortazavi, E. Mashhour, Hourly demand response and battery energy storage for imbalance reduction of smart distribution company embedded with electric vehicles and wind farms. Renew. Energy **85**, 124–136 (2016)
25. S.G. Yoon, Y.J. Choi, J.K. Park, Stackelberg-game-based demand response for at-home electric vehicle charging. IEEE Trans. Veh. Technol. **65**(6), 4172–4184 (2016)
26. M. Asensio, G. Munoz-Delgado, J. Contreras, A bi-level approach to distribution network and renewable energy expansion planning considering demand response. IEEE Trans. Power Syst. **99**, 885–895 (2017)
27. N. Acharya, P. Mahat, N. Mithulananthan, An analytical approach for DG allocation in primary distribution network. Int. J. Electr. Power Energy Syst. **28**(10), 669–678 (2006)
28. M. Moradijoz, M.P. Moghaddam, M.R. Haghifam, A multi-objective optimization problem for allocating parking lots in a distribution network. Int. J. Electr. Power Energy Syst. **46**, 115–122 (2013)

Chapter 12
Modelling the Impact of Uncontrolled Electric Vehicles Charging Demand on the Optimal Operation of Residential Energy Hubs

Azadeh Maroufmashat, Q. Kong, Ali Elkamel, and Michael Fowler

12.1 Introduction

Global attention to sustainability in energy use and reduction of greenhouse gas (GHG) emission has become a major driving force for the development and adoption of renewable and low-emission energy technologies. In the on-going development of the existing power grid towards a more sustainable energy future, the adoption of such novel technologies introduces the opportunity to shift towards more advanced energy networks. Under the smart energy network concept, the integration of distributed energy resources (DER) into existing communities provides the potential for more efficient and economic operation. These advantages may be achieved through the optimization of energy flows and through the coordinated operation of various distributed energy technology components within the network. With respect to its applicability to existing communities, there are near-term benefits for adopting smart energy network principles, particularly in consideration of the impacts of DER and mobility electrification on the residential sector.

12.1.1 Literature Review on Energy Hubs

The energy hub framework is an overarching concept for encapsulating the principles of smart energy networks for optimized energy vector dispatch and for the

---

A. Maroufmashat (✉) · Q. Kong · M. Fowler
Department of Chemical Engineering, University of Waterloo, Waterloo, ON, Canada
e-mail: amaroufm@uwaterloo.ca; qhkong@uwaterloo.ca; mfowler@uwaterloo.ca

A. Elkamel
Department of Chemical Engineering, University of Waterloo, Waterloo, ON, Canada

College of Engineering, Khalifa University of Science and Technology, The Petroleum Institute, Abu Dhabi, UAE
e-mail: aelkamel@uwaterloo.ca

© Springer Nature Switzerland AG 2020
A. Ahmadian et al. (eds.), *Electric Vehicles in Energy Systems*,
https://doi.org/10.1007/978-3-030-34448-1_12

coordinated utility of DERs, which have generally been studied via mathematical modelling techniques. Most notably, the formulation of the energy hub model as a mixed integer linear programming (MILP) problem was proposed by Geidl in [1]. This model has been further developed in [2] by Evins et al. to more accurately account for realistic operating characteristics of energy systems. The use of probabilistic considerations to account for uncertainty is presented in [3] by Alipour et al., which is implemented as a mixed integer non-linear programming (MINLP) model. Meanwhile, an iterative approach is discussed in [4] by Batic et al., which was aimed at addressing non-linearity in objective functions for energy vector dispatch within the model. Multi-objective optimization have also been considered, an example of which has been presented in [5] by Beigvand et al. for economic and energy utility criteria. In consideration of the flexibility of the energy hub model, it has been used as the basis for a number of energy hub simulation studies, most of which have been investigative works for unique energy systems or evaluative efforts that applied the energy hub model to examine various operating and optimization strategies. For example in [6], Vahid-Pakdel et al. applies the energy hub model to investigate a multi-energy vector system considering the presence of both thermal and electrical energy storage systems (ESS), demand response programs, and markets, as well as wind-based renewable energy resource (RES) adoption. In a study presented in [7], Moghaddam et al. applies an adaptation of the model using a MINLP approach for a system containing combined heat and power (CHP), electrical heat pump, boiler, absorption chiller, and electrical and thermal ESS technologies. Lastly in [8], Maroufmashat et al. consider an expansion of the energy hub to a network of interconnected hubs, in order to study the potential for more optimized energy vector dispatch resulting from diversity in energy consumption behavior and network size.

While the literature on energy hubs is fairly populated, there are particular topic areas that are of significant relevance to the content of this work. Specifically, the viability of RES integration for adoption into energy systems is critical for their consideration as DER. This characteristic has been investigated in a number of previous works, which have effectively concluded on their emission-reduction and economic potentials within existing energy systems. In [9], a study was conducted by Perera et al. to examine the potential for optimal integration of non-dispatchable renewable resources into electrical energy hubs. The study shows that optimal operation of the electrical energy hub can support RES integration to satisfy more than 60% of the annual electrical demand of the energy hub, under a Sri Lanka context. In another study, Sharma et al. [10] evaluated a centralized energy management system for residential energy hubs considering solar PV availability. The study shows that their energy dispatch strategy can potentially reduce energy consumption and costs by up to 8% and 17%, respectively. In [11], Ha et al. investigated the optimal operation of a residential energy hub implementing solar PV, solar-thermal, and battery ESS under a time-of-use electricity pricing scheme. Zhang et al. [12] present a multi-energy vector energy hub model implementing wind- and solar-based generation with hydrogen as the core energy vector. Both studies investigate the applicability of distributed

RES within the energy hub framework, while noting the need to address the intermittent nature of renewable energy technologies for significant integration into energy hub systems.

Also of significance to this work is the deploy-ability of CHP and ESS technologies, which have been studied in existing literature under the contexts of various unique systems. These studies, however, were aimed at justifying the deployment of such technologies and, as such, did not consider the relevance of EV adoption within complex energy hub systems. In [13], Mohsenzadeh et al. evaluate the operational and cost benefits of CHP implementation within energy hubs. Their study used a simulated energy hub system with electricity and gas energy vectors to demonstrate the potential total and operational cost savings of CHP implementation of up to 9.4% and 10.8%, respectively, as well as improved network reliability and reduced power losses of up to 15.4% and 16.8%, respectively. Similarly, Biglia et al. [14] examined the applicability of CHP implementation in a hospital energy hub based on energy and economic evaluation, under a Sardinia, Italy context. Wang et al. [15] explored the implementation of CHP technology in an integrated energy hub system containing heat pumps and electric boilers. The study investigated the effect of CHP implementation on both heat and electricity networks and the optimal operation of CHP technologies within the energy hub framework. Shams et al. [16] investigated the optimization of a multi-energy vector energy hub model with the presence of CHP technology, distributed renewable generation, and energy storage technologies. The presence of CHP technology in the multi-energy vector system was noted to affect the impact of electricity prices on the demand imposed on the natural gas network.

Meanwhile, the role of ESS technologies in energy hubs have been evaluated in [17] by Thang et al., who notes the advantages of ESS implementation within competitive electricity markets, highlighting the improvement in operational efficiency and flexibility due to inclusion of an energy storage system. Gabrielli et al. [18] discusses the role of both short-term and seasonal ESS technologies in maintaining system efficiency and flexibility in an energy hub subject to significant RES integration. Their study presents an optimized energy hub model incorporating thermal, battery, and hydrogen ESS, along with solar-based generation technologies, heat pumps, and power-to-gas systems. Another study, conducted by Maroufmashat et al. in [19], also considered the potential of hydrogen as a core energy vector in an energy hub containing renewable solar-based generation, hydrogen storage capabilities, and power-to-gas systems. The role of energy storage within a network of interconnected energy hubs has also been explored by Maroufmashat et al. in [20]. Their work illustrates that consideration of energy storage capabilities in combination with a variety of distributed generation technologies in large energy hub networks provides yields lower overall system costs and increased opportunities for integration of distributed generation resources into the network. In [21], Brahman et al. investigates the roles of electrical and thermal ESS within a multi-energy vector energy hub considering demand response programs in the energy vector dispatch optimization problem. Similarly, Javadi et al. [22] presents a study on the optimal operation of a multi-energy vector energy hub with the presence of battery ESS, while accounting for cycling degradation costs of the ESS in the optimization. In [23], Ye et al. incorporates both demand response programs and ESS functionality into an energy hub model and simulated the optimal dispatch of energy vectors within the energy hub based on a cost objective function. The study indicates the cost-cutting benefits of storage technologies in energy hubs that are subject to a time-of-use electricity pricing scheme.

Across these studies, the applicability of various DER technologies have been considered under a number of unique energy systems and conditions, which has established the viability and benefits of different DER technologies within energy hubs. However, recent market trends in electric mobility introduces EVs as another potentially disruptive energy technology that should be considered in the context of smart energy systems.

12.1.2 Plug-in Electric Vehicles in Energy Hubs

As an emerging technology, EVs have been developing at a rapid rate and has been projected to make up to 47% of the total light duty vehicle fleet by 2050 [24]. In comparison to traditional fossil fuel-based vehicles, EVs rely on grid-generated electricity and battery energy storage technologies for fuel. This allows EVs to incur significantly less GHG emissions during operation, particularly in energy systems that can meet their charging demand with electricity derived from renewable or low-emission energy resources. However, significant penetration of EVs into the automotive market will consequently result in tremendous increases in electricity consumption demand due to the charging behavior necessary to fuel EVs. This poses a major challenge to the power grid, which must allocate appropriate generation capacity to accommodate the additional demand. Realistically, much of the charging demand of EV fleets will originate from the residential sector, which provides the context for the adoption of EVs into residential energy systems as manageable components. Most importantly, significant EV charging demands can negatively impact the flexibility of the local energy system and, as such, must be appropriately managed to maintain energy reliability.

Currently, several levels of EV charging rates are available for EV charging, which can affect the shape of the electricity demand imposed on the energy hub by uncontrolled EV charging behavior. In level 1 charging, the low charging rate generally results in long charge durations, as well as in a flat charging profile. This contributes to increasing the base load of the energy hub during EV fleet charging periods. Meanwhile, the relatively higher rate of charging provided by level 2 charging will result in higher peaks in power demand, with a shorter charge period compared to a level 1 charging scenario. Finally, DC fast charging provides a significantly faster charge rate as compared to the other options. Thus, the charging profiles imposed by uncontrolled DC fast charging will be composed of short but significant power peaks during uncontrolled EV charging periods. However, EV charging stations with DC fast charging capabilities are unlikely to be implemented within residential energy hubs due to their high cost, and are therefore not considered in this work.

Within the existing literature, several forecasting efforts have been made to evaluate the relative impact of large-scale EV integration on the power grid. These works are often set in the context of unique power grid systems and broadly estimate the effects of uncontrolled EV fleet charging via total annual and peak charging demand criteria. For example, Clement et al. present in [25] a forecasting study on the impacts of uncontrolled EV charting at the residential level, based upon historic data of EV charging behaviors. A more recent evaluation of these impacts has been conducted by Fischer et al. in [26]. Both studies, however, evaluate scenarios of EV adoption within existing energy system conditions and do not consider how the energy hub concept may be leveraged to mitigate uncontrolled EV charging behaviors. Other notable developments in literature include the work of Dias et al. in [27], who compare impact scenarios between uncontrolled and controlled EV charging strategies within the residential sector. This study, again, is set in the context of conventional power systems and do not account for the role of DERs or for the energy hub concept. Meanwhile, further research has been conducted by Ul-Haq et al. in [28] to provide more realistic estimations of uncontrolled EV loads via stochastic methods. Concisely, there is a gap in the literature in evaluating scenarios of EV adoption into residential energy systems with uncontrolled charging behavior under an energy hub context, which may prove to be the most effective means of regulating volatile EV charging demands under medium to high market penetrations scenarios of EV fleets into the transportation sector.

In response to the significant impacts of uncontrolled EV charging behaviors, several strategies have been proposed to regulate EV charging. In one case, the controlled or smart EV charging mode has been considered for managing EV fleets as flexible loads via advanced communication and information technology. Similarly, the vehicle-to-grid (V2G) charging mode considers the adoption of bi-directional power flow infrastructure and intelligent centralized controls, in order to integrate EV fleets into energy systems as mobile BESS grid components. These two alternative charging modes have been discussed in a number of studies, which have aimed to justify their operational or economic feasibility. For instance, notable contributions to the feasibility evaluation of the V2G concept has been made by Kempton et al. in [29, 30], who concluded that V2G may contribute significantly to battery degradation in EVs and is consequently only economically justified for the provision of high-value services such as peak shaving. In another study, conducted by Locment et al. in [31], the coordinated dispatch of power is studied for an EV charging station system, which aimed to leverage controlled EV charging to improve the energy utility of local solar PV generation components. Anastasiadis et al. proposed a harmony search algorithm in [32] for controlling EV charging behavior in a microgrid containing mixed commercial and residential loads, as well as various DER components. Yao et al. [33] considered a particle swarm optimization approach for economic dispatch of power to a EV fleet with V2G enabled. The energy hub considered for this study contained both renewable and conventional energy technologies. In [34], Moeini-Aghtaei et al. presents a framework for scheduling the charging demands of a EV fleet considering charging patterns. The coordination of EV fleet charging demands is addressed using a particle swarm optimization approach for multi-objective optimization, considering financial factors, RES utilization, and a convenience criterion for EV usage. Alkahafaji et al. [35] considered the optimization of energy vector dispatch within a system containing EV fleets, using a mixed integer quadratic programming approach for the multi-objective optimization of financial and environmental criterion. The study indicates the cost-cutting potential of discharging EV fleets to maintain stability and reliability of the energy hub system. A scheme of integrating EV fleets into smart buildings is simulated and discussed in [36] by Wang et al. In [37], Liu et al. considers the economic and environmental optimization of an energy hub containing a EV fleet operating under both grid-connected and islanded modes. The study presents a comprehensive learning particle swarm optimization model for the coordinated dispatch of energy vectors within the energy hub. Similarly, Khederzadeh et al. [38] investigates the effects of EV fleet penetration in an energy hub operating between grid-connected and islanded modes, with a focus on the roles of the EV fleet, ESS, and responsive loads for maintaining islanded operation of the energy hub. In [39], Munkhammar et al. examine the potential of home-charging of EVs considering solar PV implementation at the household level, using a case study of Westminister London. The study notes the compatibility of solar PV generation and EV charging behavior, both at the single household level as well as at the grid level.

While these advanced charging modes have been considered in detail in research, they have yet to be successfully adopted in a real, large-scale energy system. Meanwhile, current trends of increasing EV penetration into the automotive market are likely to manifest in significant uncontrolled charging demands on existing power grids. As such, there is an immediate research need to evaluate the realistic impacts of uncontrolled EV charging behavior within energy hub systems, particularly for high impact areas such as the residential sector. Furthermore, an understanding of how these uncontrollable charging demands interact with grid components will provide insight into how best to implement available DER and technologies to mitigate their impact on the grid.

12.1.3 Contributions of This Chapter

In this chapter, we aim to address the research need of evaluating the potential of energy hubs for mitigating and regulating probable uncontrolled EV fleet charging demands considering systems with complex DER technology configurations. Specifically, we consider a case study of residential energy networks with solar PV arrays, CHP, electrical and thermal ESS, and conventional boiler heating technologies. This work employs an energy hub model based on [1] to simulate MILP-optimized system operation via a multi-objective approach based on economic and environmental criteria. The novelty of this study is the evaluation of realistic, near-term impacts of EV adoption into residential energy systems under than energy hub context. Furthermore, consideration of various scenarios of DER technology configurations provides insights into the planning and design of DER integration into the residential sector in consideration of disruptive EV integration into existing communities.

The contents of this chapter are structured as follows: the modelling approach and the simulation scenarios are discussed in Sect. 12.2, followed by a description of the examined energy hub system in Sect. 12.3. In Sect. 12.4, the results of the simulated scenarios are shown, and environmental and economic analysis of the results are presented. Lastly, concluding remarks for this work are made in Sect. 12.5.

12.2 EV Fleet Demand and Energy Hub Modelling Approach

12.2.1 Energy Hub Model

The operation of the residential energy hub considered in this work is formulated as a mixed-integer linear programming (MILP) problem and was modelled using the GAMS software, which is a mathematical modelling tool designed for linear, nonlinear, and mixed-integer optimization problems. The formulation of this model is based on the energy balance concept and is as shown in (12.1). In this approach, the operational flows of energy vectors within the residential energy hub is modelled as a process of energy vector transformation, conversion, and storage, beginning with grid feed and ending with consumption at the end-user. A holistic diagram of this energy hub model is as shown in Fig. 12.1.

Using this model, the aim of this work is to simulate the performance of the energy hub system under various energy technology configurations and loads, which is set by specifying the coupling matrix and the outflow energy vector set,

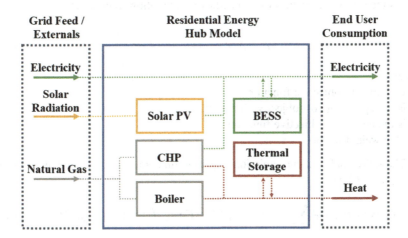

Fig. 12.1 Holistic diagram of the residential energy hub model

respectively. The coupling matrix is representative of conversion efficiencies of implemented energy technologies while the outflow energy vector set represents the various demand loads of end-users within the energy hub system. Furthermore, technology constraints such as flow and capacity constraints, as well as operational constraints, are accounted for by limiting the range of energy vector flows of the inflow energy vectors, as shown in (12.2). Optimization of energy vector flows is done using a weighted multi-objective MILP approach based on economic and environmental criteria, which are evaluated in correlation to the inflow energy vector set. The form of the overall objective function is as shown in (12.3). On the basis of optimizing the objective function, simulation of the energy hub under various scenarios is intended to reflect the optimal performance of the system under the specified conditions of each scenario.

$$O(t) + \frac{Q_{EV}(t)}{\varepsilon_{EV}} = C_{ij}I(t) + \dot{E}(t) \tag{12.1}$$

Where:
$O(t)$ is the energy demand load set of the energy hub
$Q_{EV}(t)$ is the charging required for the EV fleet
ε_{EV} is the efficiency of EV charging
C_{ij} is the coupling matrix for input energy vector i to load j
$I(t)$ is the inflow energy vector set
$\dot{E}(t)$ is the flow of energy into storage system

$$I_{min} \leq I(t) \leq I_{max} \tag{12.2}$$

Where:
I_{min} is the set of minimum flow capacities for the inflow energy set
I_{max} is the set of maximum flow capacities for the inflow energy set

$$Z = \mu \cdot Z_1 + (1 - \mu) \cdot Z_2 \tag{12.3}$$

Where:
Z is the overall objective function
Z_1 is the operating cost objective function
Z_2 is the emissions objective function
μ is the weight factor

The individual cost and emission objective functions are evaluated as shown in (12.4) and (12.5).

$$Z_1 = Cost_{fixed} + \sum_t \left(Cost_{oper,conv}(t) + Cost_{fuel}(t) + Cost_{oper,stor}(t) \right) \tag{12.4}$$

Where:

12 Modelling the Impact of Uncontrolled Electric Vehicles Charging... 297

$Cost_{fixed}$ is the fixed cost of the energy technology systems
$Cost_{oper,\ conv}(t)$ is the operating cost of energy conversion systems
$Cost_{fuel}(t)$ is the cost of fuels consumed in the energy hub during operation
$Cost_{oper,stor}(t)$ is the operating cost of energy storage systems

$$Z_2 = \sum_t EF \cdot I(t) \tag{12.5}$$

Where:
EF is the set of emission factors associated with inflow energy vector set $I(t)$.

12.2.2 Energy Storage Model

Energy storage technologies are incorporated into the mathematical model differently as compared to the energy conversion technologies, which are represented by the coupling matrix. Instead, energy storage technologies are constrained not only by their energy conversion efficiencies and power flow limitations, but also by their storage capacities and their temporal state-of-charge, which represents the current amount of energy stored. As such, these technologies are incorporated into the model as discrete temporal systems, where their performance are additionally constrained by a steady-state energy balance, as shown in (12.6). The state-of-charge of the technology is as calculated using a discrete temporal method as shown in (12.7). Further constraints were specified to limit the storage capacity of the energy storage systems, as shown in (12.8). Additionally, due to the bidirectional power flow of energy storage technologies, a further operational constraint is placed such that inflow and outflow of power cannot occur simultaneously.

$$\dot{E}_k(t) = Q_{charge,k}(t) \cdot \varepsilon_{charge,k} - \frac{Q_{discharge,k}(t)}{\varepsilon_{discharge,k}} - \dot{E}_{loss}(t) \tag{12.6}$$

Where:
$\dot{E}_k(t)$ is the flow of energy into storage system for energy vector k
$Q_{charge,k}(t)$, $Q_{discharge,k}(t)$ are the power charged and discharged to storage system k, respectively
$\varepsilon_{charge,k}$, $\varepsilon_{discharge,k}$ are the charge and discharge efficiencies for storage system k, respectively
$E_{loss}(t)$ is the standby loss of energy from the storage system k

$$SoC_k(t) = SoC_k(t-1) + \frac{\dot{E}_k(t)}{E_{max,k}} \tag{12.7}$$

Where:

$SoC_k(t)$ is the state of charge of the storage system k at timestep t
$SoC_k(t-1)$ is the state of charge of the storage system k at timestep $t-1$
$E_{max, k}$ is the maximum storage capacity of storage system k

$$SoC_{k,min} \leq SoC_k(t) \leq SoC_{k,max} \tag{12.8}$$

Where:
$SoC_{k, min}$ is the minimum charge capacity of the storage system k
$SoC_{k, max}$ is the maximum charge capacity of the storage system k

12.2.3 Monte Carlo Simulation of EV Fleet Charging Demand

The fleet charging demand of the EV fleet used in this work is derived using a Monte Carlo simulation, which considers stochastic elements affecting individual EV charging behavior including arrival and departure times, daily travelled distance, EV battery capacities, the efficiencies of EV charging nodes, and the non-linear charging characteristics of EV batteries. The use of the Monte Carlo method in this work is for the generation of representative fleet charging behaviors of hypothetical vehicle fleets based upon realistic vehicle use behavior. As the basis of this approach, 2009 National Household Travel Survey (NHTS) data [40] was used to derive the driving requirements of a fleet of light-duty vehicles in a residential context. A flow diagram of the Monte Carlo simulation used in this study is as shown in Fig. 12.2.

Based on this Monte Carlo approach, the following EV fleet charging profiles were derived for an EV fleet composed of 50 vehicles considering both level 1 and level 2 uncontrolled charging behavior. Under the level 1 charging mode, EVs were assumed to be able to charge at a power flow rate of 1.44 kW, whereas the level 2 charging mode was assumed to operate with a power flow rate of 7.2 kW. Under each of these charging level scenarios, the EV charging impact was evaluated and incorporated into the electricity consumption demand of the energy hub system, later described in this chapter. The electricity demand profiles were assumed to be consistent on a daily basis across the annual simulation, which represented the average annual charging requirement of the EV fleet. These profiles are as shown in Fig. 12.3. In the case of the level 1 charging scenario, an aggregate charging demand of 436 kWh was consumed for EV fleet charging, whereas the level 2 charging scenario required 458 kWh.

12.2.4 Model Inputs

In this mathematical formulation, the inputs to the model consist of the end-user energy vector consumption demands of the energy hub system, as represented by the

12 Modelling the Impact of Uncontrolled Electric Vehicles Charging... 299

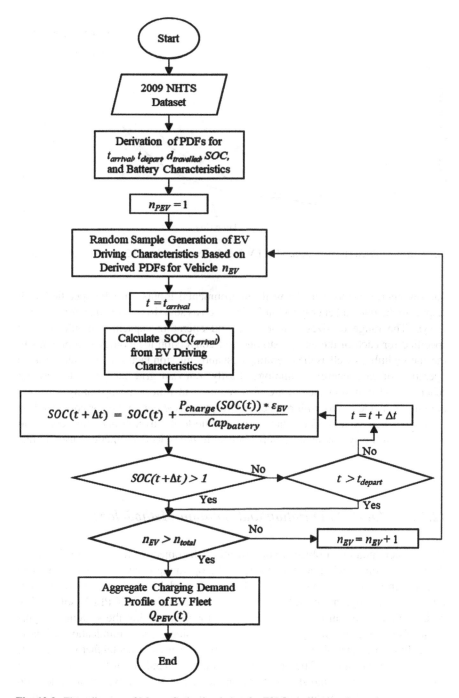

Fig. 12.2 Flow diagram of Monte Carlo simulation for EV fleet charging demand

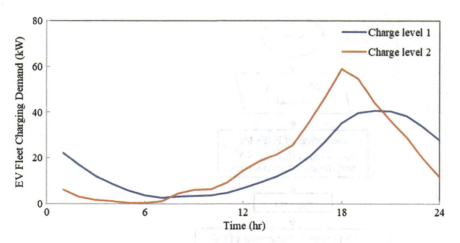

Fig. 12.3 Simulated charging demand of EV fleet for level 1 and level 2 uncontrolled charging

outflow energy vector set. As well, environmental inputs must be specified with respect to the relevant energy technologies, such as solar irradiation data for solar PV arrays. The range of sizes, efficiencies, and operating capacities should also be specified for each of the energy storage and conversion technology components in the energy hub, as well as the operating cost and emission factors associated with the operation of each energy technology. Lastly, relevant grid energy vector pricing schemes and emission factors are also required to reflect the operating cost and emission considerations with respect to grid-purchased energy vector consumed by the energy hub system. These factors contribute to the overall operating costs and emissions of the system and are thus relevant to the objective functions used in this model.

12.2.5 Model Optimization and Solution Methodology

Using a weighted multi-objective optimization, the objective function considered in the GAMS optimization account for both the operating costs and GHG emissions resulting from energy hub operation. The overall model is implemented as a mixed-integer linear programing problem, which is solved using the CPLEX solver. The optimization results in optimized energy vector flows within the system and purchases from the grid. The operating cost and emissions-related implications of these power flows are determined under the economic and environmental factors that were inputted into the model. The optimization is also dependent on the availability of energy transformation and storage technologies, as well as the type of load demand experienced by the energy hub. Thus, different optimized power flows will result under different simulation scenarios due to the conditions that the energy hub is subject to. A diagram illustrating the overall optimization process and optimization criterion, variables, and constraints is as shown in Fig. 12.4.

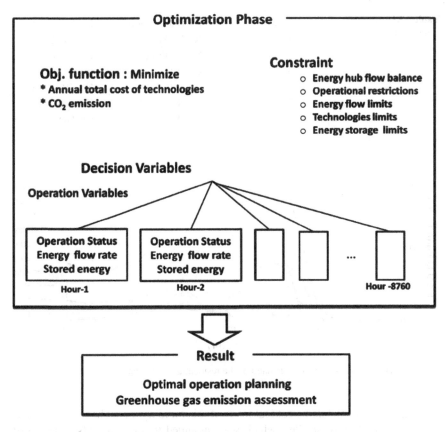

Fig. 12.4 Diagram of optimization process used in energy hub model

12.2.6 Simulation Scenarios

A total of 6 simulation scenarios were considered for the residential energy hub system. These scenarios were selected to evaluate the effect of different EV charging levels and the presence of distributed energy resources on the optimized operation of the energy hub, under the implemented optimization approach. A summary of these simulation scenarios is as shown in Table 12.1.

12.3 Residential Energy Hub System Case Study

In this study, the operational energy loads of a single residential complex were modelled and optimized under various simulated scenarios regarding EV fleet size and DER configurations. Within the energy hub model, the thermal and electrical loads of a 10-story residential complex was considered as the base load of the energy hub. The reference building model consists of 10 floors with a total floor area of

Table 12.1 Summary of simulation scenarios

Scenario	EV charging level	DER adoption
1 (Base Case)	No EV Fleet	Without DER
2	No EV Fleet	CHP and PV
3	Level 1	Without DER
4	Level 1	No CHP, only PV
5	Level 1	CHP and PV
6	Level 2	CHP and PV

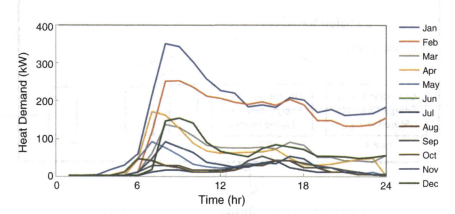

Fig. 12.5 Hourly profiles for heat demand of the residential energy hub system

7765 m^2. The thermal and electrical loads of this building model have been considered to follow hourly profiles, which were assumed to vary monthly. The hourly profiles for the thermal and electrical demands of the building are as shown in Figs. 12.5 and 12.6, respectively.

The costs of grid-purchased electricity are evaluated using a time-of-use pricing scheme, as reflective of Ontario, Canada conditions. Under this scheme, the cost of electricity is evaluated in tiers that consist of off-peak, mid-peak, and on-peak prices, which vary between summer and winter seasons and between weekdays and weekends. The values used for off-peak, mid-peak, and on-peak prices were 0.072 $CDN/kWh, 0.109 $CDN/kWh, and 0.129 $CDN/kWh, respectively. A summary of this pricing scheme for seasonal weekdays is as shown in Fig. 12.7 [41]. The price for weekends is valued consistently at off-peak prices. Meanwhile, the costs of natural gas were evaluated at a rate of 0.22 $CDN/m^3, based on Ontario conditions.

The emission factors used to evaluate the GHG emissions associated with energy hub operation are derived based on the fuels and grid-purchased electricity used to support energy hub operation. For grid-purchased electricity, a time-averaged emission factor of 0.187 kg CO_2/kWh was used to reflect Ontario, Canada conditions, which produces most of its electricity using a grid mix as shown in Fig. 12.8 [42]. Meanwhile, an emission factor of 1.9 kg CO_2/kWh was used for natural gas.

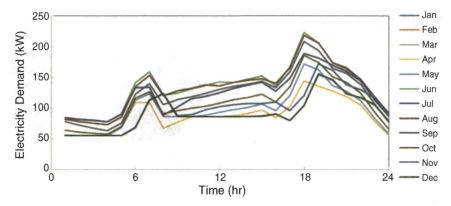

Fig. 12.6 Hourly profiles for electricity demand of the residential energy hub system

Fig. 12.7 Time-of-use pricing scheme for electricity costs in Ontario, Canada for: (**a**) summer weekdays (May 1st–October 31st) and (**b**) winter weekdays (November 1st–April 30th)

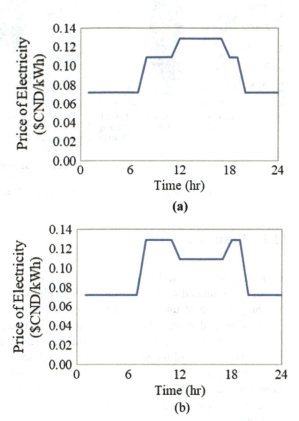

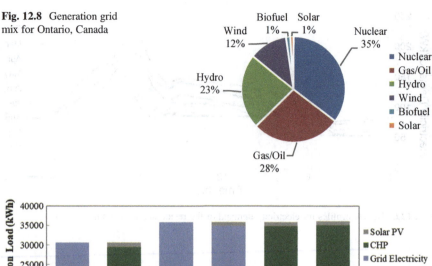

Fig. 12.8 Generation grid mix for Ontario, Canada

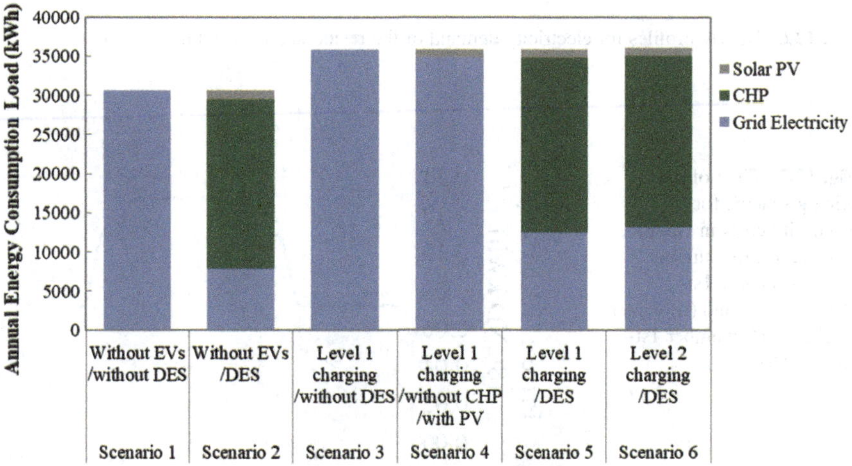

Fig. 12.9 Summary of energy consumption loads from simulated scenarios

12.4 Results and Discussion

A summary of the overall energy consumption behavior of the energy hub system in each of the simulated scenarios is as shown in Fig. 12.9. From these results, it is seen that the presence of uncontrolled EV fleet charging contributes to approximately 17.1% additional electricity consumption in the energy hub in comparison to a scenario in which a EV fleet was not considered. The significant increase in electrical power consumption of the residential energy hub indicates the potential for escalating power demand on the electrical grid as a result of EV penetration into the automotive market. Meanwhile, residential energy hubs must also adapt the necessary charging and power transfer infrastructure to accommodate the integration of EV fleets.

With respect to DER technology options in a residential context, the simulation results also indicate the distributed energy generation potential of CHP implementation in a residential context, which is shown to be able to supply up to 70% of the residential energy hub's overall consumption demand. This indicates the contribution of CHP implementation to improving energy security for residential energy hubs, as the overall system becomes significantly less reliant on grid generation for meeting its operational energy requirements. With respect to scenarios 5 and 6, in which CHP implementation was considered with EV fleet charging behavior, the results indicate the effects of significant DER implementation on alleviating the escalating demand of EV integration into residential energy hubs. This is due to the increased self-efficacy of the residential energy hub system, which in turn reduces the need for additional power transfer infrastructure and spinning reserve capacities at the grid level. PV adoption, however, is seen to play a minor role in meeting the energy hub's consumption demand, meeting only 3% of the total electricity requirements of the energy hub, considering the additional load of EV fleet charging. This is a result of the system's limitations for solar PV implementation in the target residential energy hub. Particularly, the lack of available rooftop surface area for solar PV array installation in residential high-rises limits the overall generation potential of solar PV technology, relative to the consumption needs of the building.

12.4.1 Operating Costs Analysis

A summary of the total operating costs derived for each of the scenarios is as shown in Fig. 12.10. As shown, adoption of fleet charging behavior into the residential energy hub results in an increase in operating costs. This increase in costs results from the additional electrical demand imposed onto grid generation and corresponds to an increase of 12.6% of the total operating costs of a scenario

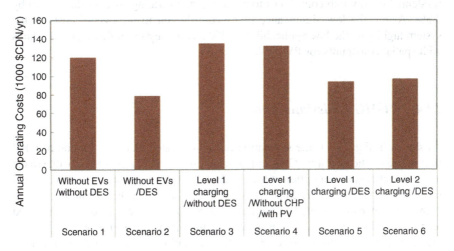

Fig. 12.10 Summary of cost analysis of simulated scenarios

without EV adoption. In comparison to the 17.1% increase in overall electricity demand determined from the energy analysis, the relatively lower increase in total operating costs results from the factoring of space heating costs as well as the time-of-use costs for EV fleet charging. These costs are incurred largely during mid-peak and off-peak periods, thus incurring a lesser impact on the operating costs of the residential energy hub as compared to its overall electricity consumption. A comparison between the two levels of EV fleet charging showed that level 1 charging results in lower operating costs for the system, due to the limitations in the rates of power purchase from the grid. These limitations extend the charging times of the EV fleet, thus constraining a larger portion of uncontrolled EV fleet charging behavior to occur during off-peak hours, thereby incurring lower charging costs during these periods. This indicates the advantages of controlled charging strategies, which can potentially schedule EV fleet charging to low-peak hours to minimize the costs of EV fleet charging, while still meeting the charging needs of the EV fleet.

With respect to DER implementation in the residential energy hub, scenarios considering DER implementation incur significantly lower operating costs. In comparison to scenarios not considering DER implementation, this corresponds to a reduction in operating costs of up to 34%. In particular, a significant portion of this cost reduction potential results from the adoption of CHP technology, due to the relatively cheaper costs of natural gas purchases in comparison to grid electricity costs. In these scenarios, optimization of the objective function for energy hub operation resulted in increased reliance on CHP operation, based on its economic advantage over mid- and on-peak costs of grid generation in Ontario's time-of-use pricing scheme. Based on this comparison, it is evident that significant DER implementation offers an economic advantage for residential energy hub systems, particularly in grids with high peaking prices for electricity. Lastly, a comparison between scenarios 3 and 4 indicates that solar PV implementation within the residential energy hub contributes to reducing the operating costs of the system by 2.5%. Again, the low significance of its contribution to the overall costs of the system highlights the low applicability of PV technology in residential energy hubs with spatial constraints for PV implementation.

12.4.2 GHG Emissions Analysis

As shown in Fig. 12.11, the simulation results indicated that the adoption of a EV fleet increases the operating GHG emissions of the residential energy hub, due to the additional energy consumption of the EV fleet. This represents an increase in annual GHG emissions of 11.3%. Meanwhile, it is also seen that the uncontrolled level 1 charging scenario resulted in higher emissions as compared to the uncontrolled level 2 charging scenario. This was because the optimized energy vector flows in the energy hub for the lower charging rate scenario satisfied more of its energy demand from CHP operation, which has a higher emission factor than compared to

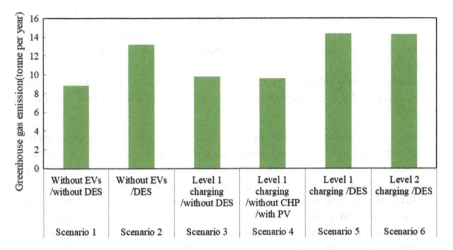

Fig. 12.11 Summary of operating emissions analysis of simulated scenarios

grid-generated electricity. In this case, the charging limitations of the level 1 charging scenario extended the charging demand of the EV fleet into a profile with a longer tail, with a less significant charging demand during on-peak periods. In this comparison, the results highlight the tradeoff between economic incentive offered by CHP operation and the environmental demerit incurred by natural gas consumption, relative to a power grid with a large portion of low emission generation.

Considering the implementation of DER technologies, CHP implementation was found to significantly increase GHG emissions resulting from energy hub operation. This was due to the effect of increased natural gas consumption resulting from CHP implementation, which has a significantly higher emission factor in comparison to Ontario's grid generation, which derives most of its generation capacity from low emission resources. This corresponds an increase in emissions of up to 49% of the scenario where DER implementation was not considered. These results indicate the negative environmental impacts of significant CHP implementation as a DER technology in a low emission power grid. Finally, comparison between scenarios 5 and 6 showed that solar PV implementation reduces the overall emissions of the energy hub system by 2.1%.

12.5 Conclusion

In this study, the following contributions to the literature has been made:

- Primarily, this work addresses the research gap in understanding the impact of realistic EV adoption scenarios into existing residential communities and the potential applicability of the energy hub concept in mitigating the volatile energy consumption behavior of uncontrolled EV fleet charging.

- This work provides insight into the compatibility of different DER technology configurations with uncontrolled EV fleet charging within residential energy hubs, which should aid in the planning and design of DER implementation within such systems.
- A case study of residential systems has been examined under an Ontario, Canada context, in order to evaluate the relevance of the study to real-world systems and conditions.

With respect to the results of the case study, analysis of the results based on energy, operating cost, and emissions criteria showed the impacts of EV adoption in escalating energy consumption, operating costs, and emissions at the residential level. Considering these effects, additional power transmission and distribution infrastructure, as well as spinning reserve capacities, may be necessary at a grid level to accommodate the additional demand. As well, sufficient charging infrastructure must also be adopted within residential energy hubs to accommodate the integration of EV fleets. Results concerning DER technology implementation indicated the benefits of significant DER implementation within residential energy hubs. Particularly, increased self-efficacy due to DER implementation allows the energy hub to address increasing EV fleet charging demand using its own DER generation resources. This could largely reduce the need for additional power transmission and distribution infrastructure, as well as the need for spinning reserve capacities at the grid level. The results also indicated the tradeoff between operating costs and emissions for the two levels of EV charging considered. The differences in EV fleet charging behavior indicate the potential benefits of controlled or scheduled charging behaviors, which could leverage time-of-use pricing schemes to provide economic and environmental benefits for the residential energy hub.

Appendix A

The nomenclature is shown below.

Nomenclature	
BESS	Battery energy storage system
CHP	Combined heat and power
DC	Direct current
DER	Distributed energy resource
GHG	Greenhouse gas
ESS	Energy storage system
MILP	Mixed integer linear programming
MILNP	Mixed integer non-linear programming

(continued)

NHTS	National Household Travel Survey
EV	Electric vehicle
PV	Photovoltaic
SOC	State of charge
V2G	Vehicle-to-grid
Variables	
$\varepsilon_{charge,\,k}$	Charge efficiency for storage system k
$\varepsilon_{discharge,\,k}$	Discharge efficiency for storage system k
ε_{EV}	Efficiency of EV charging
μ	Weight factor
C_{ij}	Coupling matrix
$Cap_{battery}$	Capacity of plug-in electric vehicle battery
$Cost_{fixed}$	Fixed cost of the energy technology systems
$Cost_{fuel}(t)$	Cost of fuels consumed in the energy hub during operation
$Cost_{oper,\,conv}(t)$	Operating cost of energy conversion systems
$Cost_{oper,\,stor}(t)$	Operating cost of energy storage systems
$d_{travelled}$	Distance travelled
$\dot{E}(t)$	Flow of energy into storage system
$\dot{E}_k(t)$	Flow of energy into storage system for energy vector k
$E_{loss}(t)$	Standby loss of energy from the storage system k
$E_{max,\,k}$	Maximum storage capacity of storage system k
EF	Emission factors associated with inflow energy vector set $I(t)$
$I(t)$	Inflow energy vector
I_{min}	Minimum flow capacities for the inflow energy set
I_{max}	Maximum flow capacities for the inflow energy set
i	Index for inflow energy vector set
j	Index for energy demand load set
k	Index for energy storage technologies
n_{PEV}	Index for plug-in electric vehicle in fleet
n_{total}	Total number of plug-in electric vehicles in fleet
$O(t)$	Energy demand load of the energy hub
$Q_{charge,\,k}(t)$	Power charged to storage system k
$Q_{discharge,\,k}(t)$	Power discharged to storage system k
$Q_{EV}(t)$	Charging required for the EV fleet
$SoC_k(t)$	State of charge of the storage system k at timestep t
$SoC_{k,\,min}$	Minimum charge capacity of the storage system k
$SoC_{k,\,max}$	Maximum charge capacity of the storage system k
t	Index for time
$t_{arrival}$	Time of arrival at energy hub
t_{depart}	Time of departure from energy hub
Z	Overall objective function
Z_1	Cost objective function
Z_2	Emissions objective function

References

1. M. Geidl, Integrated modeling and optimization of multi-carrier energy systems, Ph.D. thesis, ETH Zurich, 2007
2. R. Evins, K. Orehounig, V. Dorer, J. Carmeliet, New formulations of the 'energy hub' model to address operational constraints. Energy **73**, 387–398 (2014)
3. M. Alipour, K. Zare, M. Abapour, MINLP probabilistic scheduling model for demand response programs integrated energy hubs. IEEE Trans. Ind. Inf. **14**(1), 79–88 (2018)
4. M. Batić, N. Tomašević, G. Beccuti, T. Demiray, S. Vraneš, Combined energy hub optimisation and demand side management for buildings. Energ. Buildings **127**, 229–241 (2016)
5. S.D. Beigvand, H. Abdi, M. La Scala, A general model for energy hub economic dispatch. Appl. Energy **190**, 1090–1111 (2017)
6. M.J. Vahid-Pakdel, S. Nojavan, B. Mohammadi-ivatloo, K. Zare, Stochastic optimization of energy hub operation with consideration of thermal energy market and demand response. Energy Convers. Manag. **145**, 117–128 (2017)
7. I.G. Moghaddam, M. Saniei, E. Mashhour, A comprehensive model for self-scheduling an energy hub to supply cooling, heating and electrical demands of a building. Energy **94**, 157–170 (2016)
8. A. Maroufmashat, A. Elkamel, M. Fowler, S. Sattari, R. Roshandel, A. Hajimiragha, S. Walker, E. Entchev, Modeling and optimization of a network of energy hubs to improve economic and emission considerations. Energy **93**, 2546–2558 (2015)
9. A.T.D. Perera, V.M. Nik, D. Mauree, J.L. Scartezzini, Electrical hubs: An effective way to integrate non-dispatchable renewable energy sources with minimum impact to the grid. Appl. Energy **190**, 232–248 (2017)
10. I. Sharma, J. Dong, A.A. Malikopoulos, M. Street, J. Ostrowski, T. Kuruganti, R. Jackson, A modeling framework for optimal energy management of a residential building. Energ. Buildings **130**, 55–63 (2016)
11. T. Ha, Y. Zhang, V.V. Thang, J. Huang, Energy hub modeling to minimize residential energy costs considering solar energy and BESS. J. Mod. Power Syst. Clean Energy **5**(3), 389–399 (2017)
12. W. Zhang, D. Han, W. Sun, H. Li, Y. Tan, Z. Yan, X. Dong, Optimal operation of wind-solar-hydrogen storage system based on energy hub, in *2017 IEEE Conference on Energy Internet and Energy System Integration EI2 2017 – Proceedings*, vol. 2018, no. 1 (Jan 2018), pp. 1–5
13. A. Mohsenzadeh, S. Ardalan, M.-R. Haghifam, S. Pazouki, Optimal place, size, and operation of combined heat and power in multi carrier energy networks considering network reliability, power loss, and voltage profile. IET Gener. Transm. Distrib. **10**(7), 1615–1621 (2016)
14. A. Biglia, F.V. Caredda, E. Fabrizio, M. Filippi, N. Mandas, Technical-economic feasibility of CHP systems in large hospitals through the energy hub method: The case of Cagliari AOB. Energ. Buildings **147**, 101–112 (2017)
15. Y. Wang, K. Hou, H. Jia, Y. Mu, L. Zhu, H. Li, Q. Rao, Decoupled optimization of integrated energy system considering CHP plant based on energy hub model. Energy Procedia **142**, 2683–2688 (2017)
16. M.H. Shams, M. Shahabi, M.E. Khodayar, Stochastic day-ahead scheduling of multiple energy carrier microgrids with demand response. Energy **155**, 326–338 (2018)
17. V.V. Thang, Y. Zhang, T. Ha, S. Liu, Optimal operation of energy hub in competitive electricity market considering uncertainties. Int. J. Energy Environ. Eng. **9**(3), 1–12 (2018)
18. P. Gabrielli, M. Gazzani, E. Martelli, M. Mazzotti, Optimal design of multi-energy systems with seasonal storage. Appl. Energy **219**(2017), 408–424 (2018)
19. A. Maroufmashat, M. Fowler, S. Sattari Khavas, A. Elkamel, R. Roshandel, A. Hajimiragha, Mixed integer linear programing based approach for optimal planning and operation of a smart urban energy network to support the hydrogen economy. Int. J. Hydrog. Energy **41**(19), 7700–7716 (2016)

20. A. Maroufmashat, S. Sattari, R. Roshandel, M. Fowler, A. Elkamel, Multi-objective optimization for design and operation of distributed energy systems through the multi-energy hub network approach. Ind. Eng. Chem. Res. **55**(33), 8950–8966 (2016)
21. F. Brahman, M. Honarmand, S. Jadid, Optimal electrical and thermal energy management of a residential energy hub, integrating demand response and energy storage system. Energ. Buildings **90**, 65–75 (2015)
22. M.S. Javadi, A. Anvari-Moghaddam, J.M. Guerrero, Optimal scheduling of a multi-carrier energy hub supplemented by battery energy storage systems, in *Conference Proceedings - 2017 17th IEEE International Conference on Environment and Electrical Engineering 2017 1st IEEE Industrial and Commercial Power Systems Europe EEEIC/ICPS Europe 2017*, (2017)
23. L.H. Ye, B. Peng, J.B. Hao, Y.J. Zhang, The coordinated operation scheduling of distributed generation, demand response and storage based on the optimization energy hub for minimal energy usage costs, in *2017 2nd International Conference on Power and Renewable Energy, ICPRE 2017*, (2018), pp. 649–653
24. A. Keshavarzmohammadian, D.K. Henze, J.B. Milford, Emission impacts of electric vehicles in the US transportation sector following optimistic cost and efficiency projections. Environ. Sci. Technol. **51**(12), 6665–6673 (2017)
25. K. Clement, E. Haesen, J. Driesen, Stochastic analysis of the impact of plug-in hybrid electric vehicles on the distribution grid. IET Conf. Publ. **25**(1), 160–160 (2009)
26. D. Fischer, A. Harbrecht, A. Surmann, R. McKenna, Electric vehicles' impacts on residential electric local profiles – A stochastic modelling approach considering socio-economic, behavioural and spatial factors. *Appl. Energy* **233–234**(2018), 644–658 (2019)
27. F.G. Dias, D. Scoffield, M. Mohanpurkar, R. Hovsapian, A. Medam, Impact of controlled and uncontrolled charging of electrical vehicles on a residential distribution grid, in *2018 International Conference on Probabilistic Methods Applied to Power Systems PMAPS 2018 – Proceedings*, (2018), pp. 1–5
28. A. Ul-Haq, C. Cecati, E. El-Saadany, Probabilistic modeling of electric vehicle charging pattern in a residential distribution network. Electr. Power Syst. Res. **157**, 126–133 (2018)
29. W. Kempton, J. Tomić, Vehicle-to-grid power implementation: From stabilizing the grid to supporting large-scale renewable energy. J. Power Sources **144**(1), 280–294 (2005)
30. W. Kempton, J. Tomić, Vehicle-to-grid power fundamentals: Calculating capacity and net revenue. J. Power Sources **144**(1), 268–279 (2005)
31. F. Locment, M. Sechilariu, Modeling and simulation of DC microgrids for electric vehicle charging stations. Energies **8**(5), 4335–4356 (2015)
32. A.G. Anastasiadis, S. Konstantinopoulos, G.P. Kondylis, G.A. Vokas, Electric vehicle charging in stochastic smart microgrid operation with fuel cell and RES units. Int. J. Hydrog. Energy **42** (12), 8242–8254 (2017)
33. Y. Yao, W. Gao, J. Momoh, E. Muljadi, N. Carolina, Economic dispatch for microgrid containing electric vehicles via probabilistic modeling: preprint (2016), vol. 1, no. February
34. M. Moeini-Aghtaie, A. Abbaspour, M. Fotuhi-Firuzabad, P. Dehghanian, Optimized probabilistic PHEVs demand management in the context of energy hubs. IEEE Trans. Power Deliv. **30** (2), 996–1006 (2015)
35. M. Alkhafaji, P. Luk, J. Economou, Optimal design and planning of electric vehicles within microgrid. Commun. Comput. Inf. Sci. **763**, 677–690 (2017)
36. Z. Wang, L. Wang, A.I. Dounis, R. Yang, Integration of plug-in hybrid electric vehicles into energy and comfort management for smart building. Energ. Buildings **47**, 260–266 (2012)
37. T. Liu, Q. Zhang, J. Ding, T. Rui, Q. Wang, Grid-connected/island optimal operation of PV-diesel-battery microgrid with plug-in electric vehicles, in *Proceedings - 2017 32nd Youth Academic Annual Conference of Chinese Association of Automation (YAC) 2017*, (2017), pp. 419–423
38. M. Khederzadeh, H. Maleki, Coordinating storage devices, distributed energy sources, responsive loads and electric vehicles for microgrid autonomous operation. Int. Trans. Electr. Energy Syst. **25**, 2482–2498 (2015)

39. J. Munkhammar, J.D.K. Bishop, J. Jose, Household electricity use, electric vehicle home-charging and distributed photovoltaic power production in the city of Westminster. Energ. Buildings **86**, 439–448 (2015)
40. National Household Travel Survey [Online], http://nhts.ornl.gov
41. Ontario Hydro Rates, http://www.ontario-hydro.com/current-rates
42. Ontario's Energy Capacity, http://www.ieso.ca/learn/ontario-supply-mix/ontario-energy-capacity

Chapter 13
Optimal Operation of Electric Vehicle's Battery Replacement Stations with Taking into Account Uncertainties

Babak Mardan, Sahar Seyyedeh Barhagh, Behnam Mohammadi-ivatloo, Ali Ahmadian, and Ali Elkamel

13.1 Introduction

Increasing the concern about climate change and global warming have conviced the governments to solve these problems. In order to solve the mentioned problems, they try to reduce pollutant emissions like CO_2. One of the most consumed sections is the transportation industry. Vehicles have a critical role in environmental pollution [1], since the main fuel of the vehicles is fossil fuel. According to recent studies, 28% of CO_2 emissions in the United States and 20% in Korea are produced by transportation systems. A lot of efforts have been made to provide a suitable way to deal with these problems and despite many obstacles to the production of EVs. Advancement of technology and propagation of EVs are progressing quickly [2–4]. It is aimed that EVs will have a significant impact on the reduction of pollutants from the transportation sector. In 2015, more than

---

B. Mardan (✉)
Department of Electrical Engineering, Roshdiyeh Higher Education Institute, Tabriz, Iran

S. Seyyedeh Barhagh
Faculty of Electrical Engineering, University of Tabriz, Tabriz, Iran
e-mail: sbarhagh95@ms.tabrizu.ac.ir

B. Mohammadi-ivatloo
Faculty of Electrical Engineering, University of Tabriz, Tabriz, Iran

Department of Energy Technology, Aalborg University, Aalborg, Denmark
e-mail: bmohammadi@tabrizu.ac.ir

A. Ahmadian
Department of Electrical Engineering, University of Bonab, Bonab, Iran
e-mail: ahmadian@bonabu.ac.ir

A. Elkamel
Department of Chemical Engineering, University of Waterloo, Waterloo, ON, Canada

College of Engineering, Khalifa University of Science and Technology, The Petroleum Institute, Abu Dhabi, UAE
e-mail: aelkamel@uwaterloo.ca

© Springer Nature Switzerland AG 2020
A. Ahmadian et al. (eds.), *Electric Vehicles in Energy Systems*,
https://doi.org/10.1007/978-3-030-34448-1_13

550,000 EVs were sold worldwide, so the number of EV reached to 1.26 million [5]. In the future years, many European countries will decide to increase the number of EVs, such as German government, which plans to increase the number of EVs to 1 million by 2020 and reach 5 million by 2025. Accordingly, France has two scenarios. In the first scenario, it plans to increase its number of EVs to 1 million by 2020 and to 2.7 million by the year 2025, and in the second scenario it plans to increase the number of EVs to 3.5 million by 2020, and 7.6 million by 2025 [6]. The main challenges for buyers of these EVs are battery pricing, charging duration, and driving range limitation so that 45% of the price of EVs is related to their battery [7]. For solving the problem of battery replacement costs for consumers, some companies increase sales, by replacing old batteries with new ones with only a small amount of money and change their batteries. For example, in Europe, Renault Company uses this scenario [7]. Also, these vehicles are a kind of load on the network, each charging of the batteries of EV has many impacts on the daily load curve of the network, and the uncertainty of consumer behavior can have unpredictable effects on normal network performance [8]. The EVBRSs idea is posed to solve EVs problems in (13.3, 13.4). These stations help to reduce concerns about EV's limit range and their long duration for charging batteries of EV [9]. From the power grid point of view, the performance of these stations are as a flexible load and from consumers point of view, these stations are the service centers that can receive money and replace the batteries of EV. These services are like gas stations for cars which are internal combustion engines [10]. One of the EVBRSs targets is to increase profits by attending the market and providing services such as demand response and energy storage. These stations increase their profits by purchasing energy and storing them by charging EV's batteries when energy prices are low and discharge EV's batteries and sell energy when the price of energy in the electricity market is high [11]. Also, if renewable energy such as solar energy is added to the station's system, it can increase profits. In Ref. [12], the utilization of battery replacement stations for buses and taxis has been considered. This article focuses on the effects of driver behavior and load effects on the power network. In Ref. [13], the optimal charging was modeled by using the Marcov method and also dynamic programming was performed by taking various assumptions such as the unification of batteries, chargers and discharging battery's uncurtains and other assumptions into account. In [14], the scenario based on randomized optimization was used. The Monte Carlo method is used to create scenarios and investigate uncertainties. The amount of load and wind resources output were considered as uncertain variables. In Ref. [15], it was aimed to motivate EVs users to go to battery replacement stations at desired times. This was intended to provide discounts to consumers. In this chapter, consumer behavior has been studied in various scenarios. One of the goals of this chapter is optimal planning for increasing the benefit of EVBRSs. For proper decision making, the station's uncertainties are also modeled by (IGDT) method.

13.2 Electric Vehicle Battery Replacement Stations (EVBRSs)

13.2.1 Station Structure

Battery replacement stations are considered to be centers who provide services for end-users. One of the major problems in purchasing EVs is their higher prices. EVBRSs are responsible for the costs associated with batteries (costs of repair, protection, and battery replacement), and this reduces the final price for the end-users (approximately 20–35%) [16]. On the other hand, the costs associated with procurement of chargers and ancillary facilities would not be paid by the end-users, and the time for energizing the EV would be reduced [17]. To have a fast charge, the end-users should refer to EVBRSs, and it takes approximately 20–30 minutes to have the EVs charged [18] while it takes 5 minutes or less to charge the battery. From the electric netwoks viewpoint, EVs are extra loads that are imposed on the system. EVBRSs can solve such problems. High-tech facilities are installed in EVBRSs in which EV's charge processes are optimally scheduled to stabilize the network by discharging the batteries which gain profit for EVBRSs.

13.2.2 Station Performance

The overall duty of EVBRSs is to replace the EVs go to the station with empty or half empty batteries of EV with fully charged ones and to provide service from the station. It took about 5 minutes to change batteries in EVBRSs. Tesla Company has made an effort to reduce this time up to 90 second, which is less than the refueling process of traditional vehicles [19].

During batteries replacement, the amount of charge in each battery is one of the major variables in station modeling. The amount of charged battery and the amount of battery charge during replacement should be considered in planning and modeling. In Ref. [13], for charging less than 20%, the discount is considered for customers, but for more charging, in the other word when the battery is less than 80%, the replacement is not done. It should be kept in mind that the charge levels of batteries which delivered by customers to the EVBRSs, is a variable and can have different values. In EVBRSs modeling, for planning the batteries should have the maximum possible charge and also the EVBRSs should be able to respond to all customers. In references [11], planning is carried out in such a way that the EVBRSs will be fined if the EVBRSs is not able to supply EVs battery. According to the description in the planning and modeling of the EVBRSs, all details for the EVBRS's performance and customer satisfaction should be considered.

13.2.3 Combining the PV System with a Station

The presence of renewable resources, such as the PV system, will help us to reduce the number of pollutants in the main target of EVs. If the EVBRSs need power for charging EV, these resources will charge the battery at the EVBRSs. Otherwise, the product will be sold directly to the network. This keeps the network stable in peak times.

13.3 Uncertainties

Uncertainty in the parameters and decision variables is one of the main challenges in any system that can remove the system from its optimal point. Therefore, it is necessary to study these uncertainties for better decision making and proper planning. The market price causes the uncertainties, the amount of demand, the initial charge rate of batteries, and the amount of PV production in the EVBRSs. Various methods have been used to solve these uncertainties in different articles. In this chapter, the IGDT method has been employed to investigate existing uncertainties. Also, unlike the robust optimization method, this method does not need to determine the maximum uncertainty limit for non-deterministic parameters in this regard, it is more flexible.

13.3.1 Information Gap Decision Theory

IGDT method is one of the most powerful methods in describing the uncertainty [20]. Unlike some methods such as Monte Carlo and random planning (scenario based), this method does not require the probability density function of the uncertain parameters in the problem; robust decision-making is used against the drastic uncertainties. IGDT method is seeking to determine the maximum uncertainty permissive limitation for non-deterministic parameters, till the objective function doesn't fall outside the scope which determined by the decision maker [19, 21]. If the uncertain parameter value in the IGDT is different from what is anticipated, there are two strategies for designers: Risk-taking strategy and risk-averse strategy, in this chapter risk-taking strategy has been used. This strategy is about the state in which the uncertainty of the non-deterministic parameter has an undesirable effect on the objective function of the problem. In other words, the true realization of the uncertain parameter causes to increase the objective function amount from its base state. Therefore, this strategy seeks to find, which amount is specified and predetermined for getting the worst objective function of its base state; the maximum uncertainty limit for a non-deterministic parameter. The decision variables should be

13 Optimal Operation of Electric Vehicle's Battery Replacement...

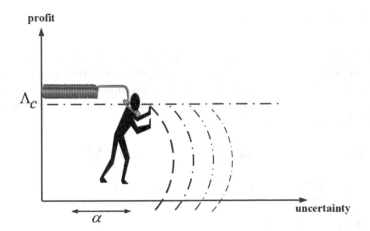

Fig. 13.1 Performance of IGDT

obtained in such a way that the actual objective function (f) is against the deviation of non-deterministic parameter γ than the predicted amount $\bar{\gamma}$ is calculated optimally when the objective function is rousted against the maximum uncertainty limitation, robust decisions are gained. In other words, the decision maker will be sure that the non-deterministic parameter changes in the uncertainty range, the amount of the objective function won't exceed from the limitation, which is intended to determine the uncertainty range of the non-deterministic parameter. The mathematical formulation of this strategy is as follows (13.1, 13.2, 13.3, and 13.4) (Fig. 13.1).

$$R_c = Max_x \alpha \tag{13.1}$$

$$H_i(X, \bar{\gamma}) \leq 0 \tag{13.2}$$

$$G_i(X, \bar{\gamma}) = 0 \tag{13.3}$$

$$\begin{cases} \left|\dfrac{\gamma - \bar{\gamma}}{\bar{\gamma}}\right| \leq \alpha \\ f(x, \gamma) \leq \Lambda_c \\ \Lambda_c = f_b(X, \gamma) \times (1 + \beta), \gamma \in \Gamma \end{cases} \tag{13.4}$$

13.4 Problem Formulation

Different methods have been used to simulate EVBRSs in various chapters. In this chapter, by using reference [10] and modifying it, EVBRSs is simulated. The formulation and explanation of these are discussed below.

13.4.1 Objective Function

The goal of this modeling is to maximize the benefits of EVBRSs as well as buying and selling energy. It should be noted that this model is intended to penalize the EVBRSs for inability to provide the demand of consumers. Also, in the case of the lack of charging at the time of replacement, according to the price of the replacement is reduced to the percentage of the lack.

$$Maxobj = profit = BSR \times \sum_{i \in I} \sum_{t \in T} X_{i,t} - VOCD \times \sum_{t \in T} bat_t^{short} -$$

$$BSR \times \sum_{i \in I} \sum_{t \in T} SOC_{i,t}^{short,\%} - \sum_{t \in T} \lambda_t \times \left(em_t^{buy} - PV_t^1 \right) + \qquad (13.5)$$

$$\sum_{t \in T} \lambda_t \times \left(em_t^{sell} + PV_t^2 \right)$$

13.4.2 Constraints

The level of SOC in each gap time is given in Eq. (13.6). If the battery is replaced at the desired time, the SOC value is equal to the initial charge of the battery, and in case of non-replacement $x = 0$, the amount of battery charge is obtained from the charge in the previous hour and the charge or discharge rate of the battery.

$$Soc_{i,t} = Soc_{i,t}^0 \times (1 - X_{i,t}) + \left(Soc_{i,t-1} + bat_{i,t}^{ch} \times \eta^{ch} - \frac{bat_{i,t}^{dis}}{\eta^{dis}} \right)$$

$$\times (1 - X_{i,t}) + Soc_{i,t}^{init} \times X_{i,t} \qquad (13.6)$$

$$0 \leq Soc_{i,t} \leq Soc_{i,t}^{max} \qquad (13.7)$$

$$F \times Soc_{i,t}^{max} \leq Soc_{i,t}^{init} \leq D \times Soc_{i,t}^{max} \qquad (13.8)$$

To meet the demand of the first hour, some fully charged batteries are considered, which is shown with $Soc_{i,t}^0$. The initial charge rate of batteries delivered by consumers to the EVBRSs for recharging is shown by $Soc_{i,t}^{init}$.

The following formula is used to obtain the value of bat_t^{short} in each time

$$\sum_i X_{i,t} + bat_t^{short} = N_t \qquad (13.9)$$

$$N_t^{min} \leq N_t \leq N_t^{max} \qquad (13.10)$$

The amount of lack of charge $Soc_{i,t}^{short}$ is gained from Eq. (13.11):

$$Soc_{i,t-1} + Soc_{i,t}^{short} \geq BC^{max} \times X_{i,t} \tag{13.11}$$

This formula has maximum nature, and it causes to increase the amount of charge and decrease $Soc_{i,t}^{short}$ of the battery during replacement.

To limit the charging and to discharge operation, the charging and discharging capacity of the existing chargers should be taken into account. There is no possibility of charging or discharging during the replacement operation.

$$bat_{i,t}^{dis} \leq \mu_t^{max} \times (1 - X_{i,t}) \tag{13.12}$$

$$P_{i,t}^{ch} \leq \mu_t^{max} \times (1 - X_{i,t}) \tag{13.13}$$

The amount of charging in each period of time should be between the minimum and maximum of charger capacity.

$$\mu_t^{min} \leq \mu_t \leq \mu_t^{max} \tag{13.14}$$

The amount of energy purchased and sold is modeled as follows:

$$em_t^{buy} - em_t^{sell} = \sum_i P_{i,t}^{ch} - P_{i,t}^{dis} \tag{13.15}$$

To avoid charging and discharging and selling and purchasing simultaneously, the following formulas are used:

$$bat_{i,t}^{dis} \leq \mu_t \times a_{i,t} \tag{13.16}$$

$$bat_{i,t}^{ch} \leq \mu_t \times (1 - a_{i,t}) \tag{13.17}$$

$$em_t^{buy} \leq M \times c_t \tag{13.18}$$

$$em_t^{sell} \leq M \times (1 - c_t) \tag{13.19}$$

13.5 Model Assumptions and Simulations

Simulation of the proposed model is done in GAMS software, and the DICOPT solver for the MINLP model and the CPLEX solver for the MIP model are used in this chapter. At the EVBRSs, there are about 500 batteries of 24 kWh and 130 chargers of 15 kWh. For each battery replacement, BSR \$70 is considered. The inability to respond to customer demand $VoCD$ a fine of \$200 is considered. The energy price of the PJM electricity market for the date of 2017/05/01 is given in Fig 13.2. The maximum $Soc_{i,t}^{short}$ is considered 25%. Five fully charged batteries are considered to meet the needs of the first hour in the model. In the EVBRS's

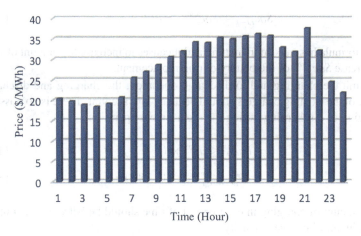

Fig. 13.2 Electricity Market price

uncertainty check section, a maximum of 10% reduction in the EVBRS's profit for the operator is considered (Fig. 13.2).

13.5.1 System Simulation Results

In this section, the base EVBRSs status is examined. The demand for different hours is shown in Fig. 13.3. The initial battery charge level between 15% to 35% of the battery capacity is altered. And also the maximum amount of energy in each battery during replacement is considered 25% of battery capacity. If the demand increases, the objective function will also increase. The results of the simulation are as follows. The amount of the objective function is $38805.270. The total power purchased during 24 hours is 15018.993 kWh, and the amount of power sold to the network is 3039.045 kWh. The amount of selling and purchasing power in each hour is depicted in Fig. 13.4. Figure 13.5 shows the power generation of the PV system. During the 24 hours a day, the battery replacement was done 555 times for different consumers. The amount of $Soc_{i,t}^{short}$ and $Soc_{i,t}^{short}$ in this case and in each gap, times are zero.

By maximizing the uncertainty, caused by the initial charge rate of the batteries in each period of time ($\alpha = 1$), the objective function will decrease by less than 1%. That reduction can't exclude the objective function from the specified range by the operator. In this case, the benefit of EVBRSs will be reached $38711.045. Also, by maximizing the uncertainty range resulting from the production of the PV system, the amount of EVBRSs profit will be reached $38,641,403. This reduction in the objective function will be less than 1%. In this chapter, the uncertainty of market price, decrease and increase in prices are investigated in two scenarios. By considering the scenario of market prices reduction, the profit of the EVBRSs increases. This increase in profit improves the objective function. But by examining the market

13 Optimal Operation of Electric Vehicle's Battery Replacement...

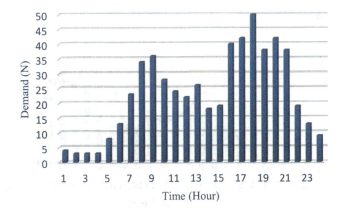

Fig. 13.3 Load Demand

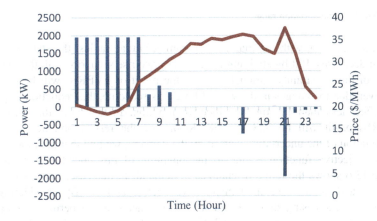

Fig. 13.4 Exchanged Power

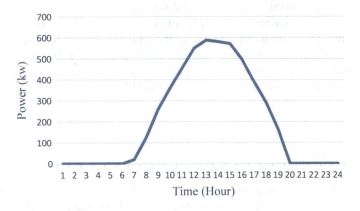

Fig. 13.5 The generation of PV

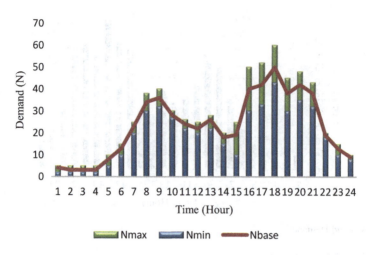

Fig. 13.6 Demand variation range

price rise scenario, with the maximum parameter range radius, the station's profit margin decreases. But by studying the market price rise scenario, with the maximization of the uncertainty parameter range, the profit of EVBRSs will be reduced. However, due to changes in the way of EVBRSs schedule, this reduction in profits will affect less than 1% on the objective function. In this case, the amount of EVBRSs profit will be $38,750,038. Regarding the results obtained from the investigation of the uncertainty caused by the demand, we find that the most impact on the objective function is due to the planning of this uncertainty parameter. Figure 13.6 shows the range of changes in demand. In this case, two scenarios of decrease and an increase in demand have been examined. By considering the scenario of decreasing demand and taking into account the permitted range of objective function reduction of 10% ($\beta = 0.1$), the maximum demand reduction range occurs from the amount of alpha at the same time interval on the system stability boundary. In this case, the demand is as shown in Fig. 13.7 and the amount of the objective function are $34,924,743 which decreased by 10% compared to base station status. This is the worst scenario for the EVBRSs. By contrast, by examining the scenario of increasing demand, we found that by maximizing the demand in each time interval ($\alpha = 1$), the revenue of EVBRSs increases significantly and equals $45045.78. This scenario is the best mode for the EVBRSs.

13.6 Conclusion

According to the content presented in this work, the creation of appropriate infrastructure for the development and creation of a suitable mark with the ability to compete with internal combustion engines is based as one of the research interest in this area. In this study, a suitable method for optimal operation of the EVBRSs is

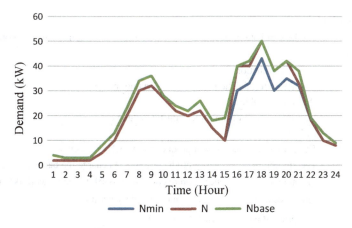

Fig. 13.7 Demand with the most uncertainty set

proposed, and the results indicate that planning can be profitable for the operator. The station profit in the base mode is $38,805,27. In addition, IGDT method is utilized to address the uncertainties. With this risk management method, the effect of each non-deterministic decision variable in the objective function is obtained, in order to achieve robust planning against the maximum range of uncertainties.

Appendix A

The nomenclature is shown below.

Indices	
T	Index of the time period
I	Index of batteries
$\psi_{eq|ineq}$	Total constraints of equal and unequal
Parameters	
γ	Non-deterministic parameter
Λ_c	The critical amount of the objective function
f_b	The deterministic amount of the objective function
β	The degree of bearing increases the objective function
M	A big integer number
BSR	Battery replacement cost
$VoCD$	The cost of customer dissatisfaction
η^{ch}	Charge efficiency
η^{dis}	Discharge efficiency
Binary Variables	
$X_{i,t}$	The binary variable of battery replacement
$a_{i,t}$	The binary variable of asynchronous charging and discharging of batteries
c_t	The binary variable of asynchronous purchasing and selling energy

(continued)

Integer Variable	
bat_t^{short}	The lack of battery numbers to meet the needs of customers in each hour
Variables	
of	Objective function
R_c	The non-deterministic parameter uncertainty range
X	Decision variable
α	Maximum range of non-deterministic parameter
Γ	Uncertainties
Pv_t	The amount of solar system production in each hour (kW)
PV_t^1	The amount of solar energy production used to charge the batteries of EV (kW)
PV_t^2	The amount of solar energy production used to sell to the network (kW)
$Soc_{i,t}^{short,\%}$	The percentage of energy loss in each battery per hour
λ_t	The price of energy in each hour ($/kWh)
em_t^{buy}	The amount of energy purchased in each hour (kWh)
em_t^{sell}	The amount of energy sold in each hour (kWh)
$Soc_{i,t}$	The amount of energy in each battery per hour (kWh)
$Soc_{i,t}^0$	The amount of energy in each battery in the first hour (kWh)
$bat_{i,t}^{ch}$	The amount of charge in each battery per hour (kWh)
$bat_{i,t}^{dis}$	The amount of discharge in each battery per hour (kWh)
$Soc_{i,t}^{init}$	Initial charge rate in each battery per hour (kWh)
μ_t	The amount of battery power per hour (kWh)

References

1. M. Kim, W. Won, J. Kim, Integration of carbon capture and sequestration and renewable resource technologies for sustainable energy supply in the transportation sector. Energy Convers. Manag. **143**, 227–240 (2017)
2. A.Y. Saber, G.K. Venayagamoorthy, Intelligent unit commitment with vehicle-to-grid—A cost-emission optimization. J. Power Sources **195**, 898–911 (2010)
3. D.B. Richardson, Electric vehicles and the electric grid: A review of modeling approaches, impacts, and renewable energy integration. Renew. Sustain. Energy Rev. **19**, 247–254 (2013)
4. A. Affanni, A. Bellini, G. Franceschini, P. Guglielmi, C. Tassoni, Battery choice and management for new-generation electric vehicles. IEEE Trans. Industr. Electron. **52**, 1343–1349 (2005)
5. N. Ito, K. Takeuchi, S. Managi, Do battery-switching systems accelerate the adoption of electric vehicles? A stated preference study. Econ. Anal. Policy **61**, 85–92 (2019)
6. M. Armstrong, C.E.H. Moussa, J. Adnot, A. Galli, P. Rivière, Optimal recharging strategy for battery-switch stations for electric vehicles in France. Energy Policy **60**, 569–582 (2013)
7. J. Yang, F. Guo, M. Zhang, Optimal planning of swapping/charging station network with customer satisfaction. Transp. Res. Part E **103**, 174–197 (2017)
8. Q. Kang, J. Wang, M. Zhou, A.C. Ammari, Centralized charging strategy and scheduling algorithm for electric vehicles under a battery swapping scenario. IEEE Trans. Intell. Transp. Syst. **17**, 659–669 (2016)

9. N. Liu, Q. Chen, X. Lu, J. Liu, J. Zhang, A charging strategy for PV-based battery switch stations considering service availability and self-consumption of PV energy. IEEE Trans. Ind. Electron. **62**, 4878–4889 (2015)
10. M.R. Sarker, H. Pandžić, M.A. Ortega-Vazquez, Optimal operation and services are scheduling for an electric vehicle battery swapping station. IEEE Trans. Power Syst. **30**, 901–910 (2015)
11. M.R. Sarker, H. Pandzic, M.A. Ortega-Vazquez, Electric vehicle battery swapping station: Business case and optimization model, in *2013 International Conference on Connected Vehicles and Expo (ICCVE)* (2013), pp. 289–294
12. R. Rao, X. Zhang, J. Xie, L. Ju, Optimizing electric vehicle users' charging behavior in battery swapping mode. Appl. Energy **155**, 547–559 (2015)
13. S. Bo, X. Tan, D. Tsang, Optimal charging operation of battery swapping and charging stations with QoS guarantee. IEEE Trans. Smart Grid **9**, 4689–4701 (2017)
14. Y. Sun, J. Zhong, Z. Li, W. Tian, M. Shahidehpour, Stochastic scheduling of battery-based energy storage transportation system with the penetration of wind power. IEEE Trans. Sustain. Energy **8**, 135–144 (2017)
15. W.F. Infante, J. Ma, Y. Chi, Operational strategy and load profile sensitivity analysis for an electric vehicle battery swapping station, in *2016 IEEE International Conference on Power System Technology (POWERCON)* (2016), pp. 1–6
16. B. Vedran, H. Pandzic, Lithium-ion batteries: Experimental research and 315 application to battery swapping stations, in *2018 IEEE International Energy Conference 316 (ENERGYCON)*, pp. 1–6. IEEE, (2018), https://doi.org/10.1109/ENERGYCON.2018.8398829
17. K. Morrow, D. Karner, J. Francfort, U.S. Department of energy vehicle technologies program – advanced vehicle testing activity: plug-in hybrid electric vehicle charging infrastructure review (2008)
18. D. Sara, A.S. Masoum, P.S. Moses, M.A.S. Masoum, Real-time coordination of plug-in electric vehicle charging in smart grids to minimize power losses and improve voltage profile. IEEE Transactions on Smart Grid **2**(3), 456–467 (2011), https://doi.org/10.1109/TSG.2011.2159816
19. B. Mohammadi-Ivatloo, H. Zareipour, N. Amjady, M. Ehsan, Application of information-gap decision theory to risk-constrained self-scheduling of GenCos. IEEE Trans. Power Syst. **28**, 1093–1102 (2013)
20. A. Soroudi, M. Ehsan, IGDT based robust decision making tool for DNOs in load procurement under severe uncertainty. IEEE Trans. Smart Grid **4**, 886–895 (2013)
21. A. Rabiee, A. Soroudi, A. Keane, Information gap decision theory based OPF with HVDC connected wind farms. IEEE Trans. Power Syst. **30**, 3396–3406 (2015)

Chapter 14
Participation of Aggregated Electric Vehicles in Demand Response Programs

Maedeh Yazdandoust and Masoud Aliakbar Golkar

14.1 Introduction

According to the report published by the international energy agency (IEA), about 30% of energy consumption and 61% of petroleum consumption is related to the transportation system [1]. This increase in energy consumption rises economic and political concerns, and by considering environmental concerns, EVs stand in the center of attention as an alternative option [2]. The main reasons and motivations for utilizing EVs are summarized as follow:

- their capability of increasing the energy security
- their capability of reducing the fossil fuels cost
- their capability of decreasing the greenhouse gases
- their capability of increasing the penetration of renewable resources [3]
- their capability of storing energy to compensate for the renewable energy uncertainties [4]
- their capability of reducing the cost of establishing new power plant [5]

On the other hand, implementing Demand Response Programs (DRPs) as one of the most important resources to mitigate the power losses by smoothing the load profile, becomes widespread. In addition, DRPs provide lots of advantages for consumers, aggregators, and utilities, that can be summarized as follow:

- Benefits from relative and absolute reductions in electricity demand
- Benefits resulting from short-run marginal cost savings from using demand response to shift peak demand
- Benefits in terms of displacing new plant investment from using demand response to shift peak demand or respond to emergencies;
- Reducing Network Losses

M. Yazdandoust (✉) · M. A. Golkar
Faculty of Electrical Engineering, K. N. Toosi University of Technology, Tehran, Iran
e-mail: golkar@kntu.ac.ir

© Springer Nature Switzerland AG 2020
A. Ahmadian et al. (eds.), *Electric Vehicles in Energy Systems*,
https://doi.org/10.1007/978-3-030-34448-1_14

- Increasing reliability of power systems
- Better integration of renewable energy sources in power systems
- Reducing the cost of electricity generation
- Reducing emissions of greenhouse gases [6]

Due to the high price of fossil fuels and high CO_2 emissions, electric vehicles could be widely used in future transportation systems to reduce both fossil fuels consumption and CO_2 emissions. Many studies have been done considering the impact of EVs on the power systems. Most of the research has expressed that the impact of EVs is closely related to the schedule of charging and discharging process and moreover the electrical tariffs [7, 8]. Soares et al. [9] evaluated the impacts of EVs on the power systems by using a simulator. Furthermore, the effect of charging process on bus voltage can be studied by the same method.

Besides the advantages of EVs mentioned above, the energy demand will be increased due to required energy for charging a large number of electric vehicles. This demand increase will lead to a significant new load on the existing energy distribution systems when vehicles are being charged. Therefore, due to the limitation of distribution system capacity and EVs that should be charged during the specific period, the process of charging should be controlled. This control should be done to prevent overload and increase the stability of the system. In [10], a demand response strategy is proposed for identifying overloaded transformers and remove the problem in case of using the electric vehicle with other loads. In this way, a demand response strategy is used to reduce the loading of distribution transformers. In [11] association of EVs in demand response program is proposed. The main aim of this paper is to shift the specific loads from peak hours to off-peak hours and to study the impact of V2G and G2V options.

To deal with this issue, smart charging and discharging of EVs is highly recommended. In this way, by managing the charging process by shifting the charging process to the off-peak hours, the load profile of the system will be smooth. As a result of this, power losses and operation costs of the system will be decreased [12]. However, coordinated charging behavior of EVs should be considered. In the presence of a large number of EVs, demand response flexibility of the whole loads should be evaluated to design the smart charging scheme and the maximum charging delay that each EV can stand should be calculated [13]. In [14], the natural charging behavior characteristics of the large number of EVs is proposed through stochastic simulation.

Another solution is the implementation of demand response programs. DRPs cause changing in consumers behavior due to the different electrical tariffs or incentives which is done during high wholesale market price and when the reliability problem occurs [15].

Integration of EVs in DRPs has many advantages for utility, EV owners, and PL operators. In [16], utility benefits consist of charging/discharging vehicle benefit, increasing the reliability of the system, and loss reduction benefit was studied.

Although the penetration of EVs in power system leads to some problem mentioned above, they are in the center of attention due to their capabilities which help

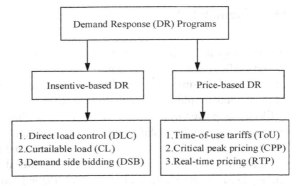

Fig. 14.1 the main categories of DRPs

the power system. For example, their capability to behave as a shiftable load in peak hours which can postpone the charging process and their energy storage capability that helps the power system in this period [17].

With proper charging and communication infrastructure, EVs may eventually either turn into Interruptible Loads (IL) when plugged in for charging or act as gridable storage responding to the pricing commands [18].

Since the generation of renewable resources has some uncertainties, the importance of using energy storage systems to manage these fluctuations becomes clear. EVs, as one of the energy storage resources, could take part in this process by the integration of EVs in DRPs [19].

Studies show that by the implementation of DRPs, greenhouse gases emission, and cost of the power system in the presence of EVs have been reduced. In this way, the two main goals of implementation of DRPs are fulfilled.

In this chapter, the main aim of implementing DRPs is described, and the benefits of cooperative participation of costumer and utility in DRPs displayed. In order to perform the DRPs in a different situation, various types of programs will be presented. The main categories of DRPs are shown in Fig. 14.1.

The demand response capability of EVs will be described and analyzed in Sect. 14.3. This section is categorized in two main groups as follow:

- Charging and discharging behavior of EVs
- EVs as responsible demand

In Sect. 14.4, the optimization problem of the smart distribution system and electric vehicle parking lots in the presence of EVs is formulated, and the constraints and uncertainties are described.

Finally, the conclusion will be presented in Sect. 14.5.

14.2 Demand Response Definition

Demand response (DR) has been widely documented as a potential solution for several challenges the electrical power system is facing. Some of these issues are the integration of intermittent renewable electricity generation and maintaining system reliability under rapid and global electrification.

Electrical power systems are currently in a phase of transition. Part of this transition is a paradigm shift towards more utilization of demand-side flexibility. Traditionally, the balance between supply and demand is ensured by using supply-side flexibility. The flexibility of consumers' power demands is traditionally only actively used for large industrial consumers at high power levels. With the advent of smart distribution systems; however, there has been increasing attention for the potential use of demand side flexibility for residential consumers as well. The usage of demand side flexibility sources is what is usually referred to as demand response (DR).

A demand response program tries to influence consumers to change their electricity consumption behavior in response to a signal such as dynamic prices or incentive payments. These different types of demand response implementations will be discussed [20].

As mentioned before, the implementation of DRPs has lots of advantages for consumers, environment, and utilities, which can be summarized in Table 14.1 [21].

14.2.1 Types of DR Programs

A demand response program can be implemented in several forms. They can be divided into two main groups based on how behavior changes are obtained: incentive-based and price-based programs. In price-based programs, consumers react to an electricity price signal, while in incentive-based programs, they receive incentive payments independent from electricity price.

Table 14.1 advantages of DRPs implementation

Consumers	Environment	Utilities
Supply electricity demand improvement	environmental degradation reduction	Operational cost reduction
Electricity bill reduction	Resources consumption reduction	System efficiency improvement
Service quality improvement	Environmental protection	Electricity generation cost reduction
Lifestyle improvement	Greenhouse gases reduction	System reliability improvement

14.2.1.1 Incentive-Based DR

Depends on the way in which the demand change is triggered, and the form of the incentives this type of demand response programs can be categorized into three main groups:

1. Direct load control (DLC): These programs usually involve an aggregator as a third party, who is given direct control over some of the consumers' appliances. The aggregator can then offer the flexibility of a group of consumers on the market and makes incentive payments to the consumers in turn.
2. Curtailable load (CL): Here, utilities also issue requests for decreasing or increasing demand, but the end-user remains in control over their own appliances. Consumers are rewarded with bill credit or participation fees for following these requests. Failing to do so will typically result in penalty fees.
3. Demand-side bidding (DSB): In these programs, consumers can bid on load reductions in a dedicated market. If their bid is cleared, they are obliged to change load accordingly.

14.2.1.2 Price-Based DR

This type of program tries to influence consumption behavior by providing an electricity price that varies over time. This price would be in contrary to a flat rate model where the price is the same at every point in time and has three main groups:

1. Time-of-use tariffs (TOU): This market model divides the day into different periods in which different electricity prices are applied. Typically, these periods and prices are fixed over a longer time. A typical example of TOU is a different tariff for day and night.
2. Critical peak pricing (CPP): This pricing is mostly used in the form of an extra component to a flat rate or time-of-use tariff. The CPP component adds an additional part during a limited number of peak hours per year.
3. Real-time pricing (RTP): This tariff scheme provides a price signal that varies hourly, reflecting the changes in the electricity spot price. Consumers can be notified of the price on a day-ahead or hour-ahead basis [22].

14.2.2 Customer Response: Electric Vehicle

The main purpose of the demand response program (DRP) implementation summarized in two sections: (1) energy efficiency (EE) and (2) load shifting, by the aim of load profile flattening.

In this way, EVs, as an electricity consumer, increase the total consumption of energy. However, by using the smart charging capability of them, they could be an excellent resource to achieve the second aim of DRP implementation.

In power system operation, minimizing the fluctuation of the demand and satisfying electricity demand during peak load period are the primary concern. On the other hand, by increasing the number of EVs in the system, these issues become more important.

By considering the flexibility of demand on the customer side and EVs as shiftable loads, implementation of DRP could be a wise approach to overcome these issues [23].

14.3 Demand Response Capability of Electric Vehicle

By increasing air pollution and global warming crisis, the usage of electric vehicles is growing all around the world. Therefore, the integration of EVs into the power system and its impacts on the grid is not negligible in recent years.

Loading demands are various in different hours of a day. This difference means the residential loads consume more power at nights, which cause peak demand in these hours and light demand right after that. On the other hand, power losses and overloading of transformers and line are the main concern in power system operation. In this way, demand response plays an important role in peak shaving and valley filling by shifting some loads from on-peak hours to off-peak hours. Moreover, by smoothing the load profile, it can reduce power losses.

Studies show participating EVs in DR programs can have a positive effect on load profile. This participation can occur during the charge and discharge process of EVs.

Cost reduction effects, the comfort of users, and power loss, as well as some standards of these smart charging, should be considered as summarized below:

- Charging EVs without considering its impacts on the power system causes higher peak demands. DR programs that focused on discharging during peak hours and start charging just after peak hours, effectively helped peak shaving and valley filling. As a result, the total electricity bill decreased in spite of the increase in some of the off-peak hours.
- Using the DR capability of the EVs by managing the charging process in both V2G and G2V modes is an effective way to reduce the operational costs of the system and satisfying the electricity demand as well.
- In spite of the advantages provided by smart charging, power losses due to the numerous charge and discharge process will be increased. Besides the required infrastructures made these process more complicated.
- EVs can also operate in vehicle to grid (V2G) mode, in which they send power back to the grid. In this situation, a central control system is needed to communicate between the utility demand and SOC of EVs battery. Some standards indicate the infrastructures and requirements for this kind of communication [24].

By considering the massive penetration of EVs and advantage of the implementation of DRP in smart grid, integration of EVs in DRPs become an essential issue nowadays. In this way, the role of EVs as a part of DRP in both smart grid and parking lots should be investigated. Besides this, the model of DR and types of DRP that can be performed for EVs must be extracted, and for the last step, the performance of EVs in each DRP should be analyzed.

For the implementation of DR programs for an electric vehicle, various information should be gathered. Some of these data are as follow:

- Price of charge/discharge
- Time of charge/discharge
- Operation of parking lots
- Traveling distance of the vehicle
- Vehicle traveling time features
- Real amount of generation (based on the number of vehicles, state of charge and owner's permission)

By considering the availability of these data, the next step is finding appropriate demand side management (DSM) programs for EVs participation. Among different kinds of DR programs that explained in the previous subsection, the DSM schemes that would be available for the electric vehicles will be discussed in the following [21].

14.3.1 Charging and Discharging Behaviour of EVs

Electric vehicles as shiftable loads in charging period and as DGs in the discharging period

Most of the times, vehicles are parked at homes, streets, parking lots, or garages; hence, EVs battery capacity can be fully utilized during such times. Therefore, EVs could serve as decentralized energy storage in a smart grid and can act as either a load or a generator as needed.

Considering EVs as shift able load, make this concern about the time of the second charging occurs. One of the technologies that can be used for this issue is basic timers. In this way, EV owners connect their cars as soon as they arrived home. However, the charging process could be postponed to the other time by considering the cost of electricity. Otherwise, EV owners may start charging process as soon as they arrive home since if they don't, they may forget charging their vehicle.

In vehicle to grid operation, it should be considered that this process causes battery degradation and decrease the efficiency of the system though, it provides advantages both for customers and utility. In order to overcome the battery degradation issue, advanced battery technology could be a reasonable candidate. Since EVs as an important part of the system could be more effective if they operate in both V2G and G2V mode and in the large numbers.

Beyond enabling technology, various rate structures can be used to provide incentives for EV owners to take advantage of the enabling technology.

Considering these capabilities of EVs mentioned above, and the benefits of implementing demand response programs in smart grid, integration of demand response using the electric vehicle in the smart grid would be a good idea.

Charging Scheduling Scheme in Parking lots

By increasing the number of EVs, establishing parking lots with enough poles to cover this number of requests is not negligible. In this way, parking lot owners introduced to manage the process of charging. In order to evaluate the operation of parking lot owners, the power sent to parking lots from the utility (G2V) and the power sent from parking lots to utility (V2G), should be calculated in both energy and reserve markets. Actually, in the demand response events, the time that EV owners parked their vehicle, and the expected SOC at the departure time, besides all constraints and uncertainties should be considered [25].

As mentioned, establishing parking lots in a smart microgrid (SMG) to manage and cover the large numbers of EVs is not negligible. By considering the energy storage capability of EV, in the situation of over generation, energy could be stored in EVs batteries and consumed in the peak hours. On the other hand, in the case of day-ahead market notice, EVs could participate in both energy and reserve markets.

In the following, the operation of EVs in a residential area is investigated; however, EVs in parking lots due to the two main reasons are more important. Actually, the main factor in parking lots is a large number of EVs that can be controlled at the same time. By considering a large number of flexible loads can participate in demand response programs, all the EVs could be charged, and all the curtailments will be satisfied.

In order to evaluate the operation of different parking lots in DR events, types of parking lots should be considered. There are three types of parking lot based on the location.

- Residential parking lot
- Commercial parking lot
- Industrial parking lot

Each of these types is suitable for the specific mode of EVs operation (V2G or G2V). For example, as the residential and industrial parking capacity is not available during daytime, they more tend to operate in G2V mode. On the other hand, as the commercial parking capacity is available during daytime, the EVs could operate in G2V mode in these parking lots. In addition, due to the traffic in the commercial parking lot, they are not a proper candidate for operating in V2G mode.

Studies show, participating in price-based demand response programs have cost reduction benefits for parking lot operators. However, participating in incentive-based demand response programs depends on the parking lot pattern, and they may not have an acceptable operation in both G2V and V2G mode [26]. This problem happens due to the different available capacity of parking lots.

14 Participation of Aggregated Electric Vehicles in Demand Response Programs

Charging Scheduling Scheme in Residential Area

Vehicle to building (V2B) is defined as exporting electrical power from a vehicle battery into a building. V2B considers EVs batteries as a generation resource for the buildings via bidirectional power transfer at certain periods, which could increase the flexibility of the electrical distribution system operation.

It is expected that V2B operation improves the reliability of the distribution system and provide extra economic benefits to the vehicle owners. Furthermore, V2B reduce the home or building electricity purchase cost based on DSM programs with customer incentives.

The objective of DSM is to improve the reliability of power supply for the building and create revenue. For demand side management, the peak load shifting strategy using EVs can reduce peak load demand and energy consumption. In turn, it will reduce the electricity purchase cost for the customer and vehicle owner [21].

Due to the increase of EVs penetration and limitation of residential given demand, consumers satisfaction become an important issue. Implementation of demand response programs in residential level can be performed in two layers: the neighborhood area network (NAN) and the home area network (HAN). Actually, the main purpose of DR implementation in a residential area is to supply EVs charging demand so that the satisfaction level of consumer do not disturb.

To evaluate EVs in a residential area, loads divided into two main categories. The first group is shiftable loads that can be shifted to the other time due to their flexibility and less importance. The second group is non-shiftable loads that cannot be shifted to the other time.

The traveling duration of the EVs, the place, and the duration of the EVs parked in that place are the important items in analyzing their charging profile. Since EVs plug-in time follows a normal distribution curve, the normal PDF use to present the EV plug-in time.

The main concern in the residential area is the peak hours that occur in the evening due to the charging of arrived EVs and the peak demand of the system.

The layers of DR implementation are as follow:

A. *Implementation of demand response through NAN strategy*

In this strategy, to determine the impact of the presence of EVs in the distribution system, the peak load of the system in the absence of EVs should be specified as the maximum demand. Then each house determines its maximum demand so that the summation of all the houses demands do not exceed the maximum demand level of the system. Figure 14.2 shows the way in which the amount of demand dedicated to each house is specified.

This strategy is explained as below:

All the houses demand sorted from maximum to minimum at first. Then the maximum demand of the household section (DL_i, dashed line) specified so that the summation of all the houses demand do not pass the peak load in the absence of EVs. As illustrated in Fig. 14.2 the houses which their consumption is more than the maximum demand of the household section, including in the demand limitation.

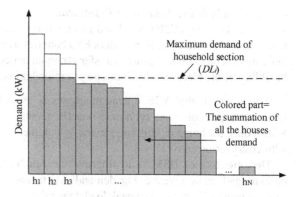

Fig. 14.2 Implementation of demand response in NAN strategy [27]

$$\text{maximize}(DL_i)$$

Subject to :

$$\sum_{m=1}^{N} D_{m,i} \leq DL_{total,i} \qquad (14.1)$$

$$D_{m,i} = \begin{cases} L_{m,i}, & L_{m,i} < Dl_i \\ DL_i, & L_{m,i} \geq Dl_i \end{cases} \quad m = 1, 2, \ldots, N$$

where $D_{m,i}$ and $L_{m,i}$ show demand of the mth house after DR (kW) and original demand of the mth house in time slot i; respectively, and $DL_{total,i}$ shows available supply (kW) of the distribution circuit of interest in time slot i.

B. *Implementation of demand response on HAN strategy*

Loads in the residential area divided into two main categories. The first group is shiftable loads that can be shifted to the other time due to their flexibility and less importance. The second group is non-shift able loads that cannot be shifted to the other time due to their importance. In this strategy, shift able loads are controlled by HAN based on their determined maximum demand. For the implementation of DR in a residential area, customers preset some of their loads as shift able loads that can be controlled by HAN control center. In this way, if the demand of the household is more than the determined limit, HAN control center do not support the specific shift able loads in that time. Actually, in this situation, DRPs applied in the system. The implementation of the DRPs in the HAN strategy has three steps that described as follow:

Step (1) In this way, customers based on the importance of the loads and their appliance, prioritize them so that some loads are the first priority and the other will be in the next priority.

Step (2) In this way, customers organized and schedule their loads. For example, the time of the washing machine job and temperature limitation of the room could be scheduled in advance.

14 Participation of Aggregated Electric Vehicles in Demand Response Programs

Step (3) In the last step, based on the priority allocated in step 1 and the preset schedule in step 2, the demand response program will be performed. Demand response program in the HAN strategy will be performed as follow:

- In the situation that demand exceeds the determined limit, EVs will be stopped charging
- If the HAN control center estimates that EVs need more time for charging, the EVs charging resumed

it should be noted that in HAN strategy, customers can apply changes in both step 1 and 2 in different time of the day [27].

14.3.2 EVs as Responsible Demand

Modeling the Demand Response Programs

The main purpose of DR implementation is to change the customer's behavior based on the price changes. Actually, these price changes could be as a result of price-based DRPs and their tariffs or incentive-based DRPs, and their incentive or penalty applied to customers.

By considering that as a result of price changes, customer's demand change from d_t^{ini} to d_t at hour t, so that the impact can be formulated as follow:

$$\Delta d_t = d_t^{ini} - d_t \tag{14.2}$$

The amount of incentive, ζ_t, based on the rate of the incentive of reducing the demand (Inc_t) is expressed as:

$$\zeta_t = Inc_t \Delta d_t \tag{14.3}$$

Similarly, the amount of penalty, ξ_t, based on the rate of the penalty of not reducing the demand (Pen_t) can be formulated as:

$$\xi_t = Pen_t \left(d_t^{cont} - \Delta d_t \right) \tag{14.4}$$

where d_t^{cont} denotes the contract level for hour t.

The customer's benefit, B, at hour t can be obtained as follows:

$$B_t = Revenue_t - d_t \lambda_t + Inc_t \Delta d_t - Pen_t \left(d_t^{cont} - \Delta d_t \right) \tag{14.5}$$

where, $Revenue_t$ is the customer's revenue from consuming d kWh energy in t-th hour. The second item of Eq. (14.5) refers to the amount of electrical energy consumed by the customer. The cost of energy at hour t. in addition, as mentioned above, the third and fourth items are related to the incentive of reducing the demand

and penalty of not reducing the demand at hour t, respectively. In this way, the total benefit can be obtained through Eq. (14.6) presented as below:

$$B_{tot} = \sum_{t=1}^{T} \left(Revenue_t - d_t\lambda_t + Inc_t\Delta d_t - Pen_t\left(d_t^{cont} - \Delta d_t\right) \right) \tag{14.6}$$

Actually, the total benefit is a good factor that helps customer to decide how to behave in case of price-based and incentive-based demand response [25].

EV in price-based DR Tariff

Time-of-use, critical peak pricing and real-time pricing are kinds of priced-based DR tariff. In the time of use tariff, customers based on the different tariff that noticed by utility companies can easily manage their consumption. In critical peak pricing, tariffs are a bit higher in specific hours of a day, and in this way, customers decide whether to participate in this type of DR or not. Actually, this type of DR is more complicated than TOU due to the time limitation of applying tariff. In real-time pricing events, electricity tariff announced by the utility in the day-ahead market and based on this, customer manage their consumption behavior. Due to the limitations and complication of CPP and RTP tariffs, TOU is more popular among residential customers [28].

TOU and CPP tariffs are the most effective DSM schemes that usually use for the participation of EVs in DR programs. In TOU tariff, EV owners and EV aggregators charge and discharge based on different tariffs announced by utility companies. Economy 7 and Economy 10 are differential tariffs available in the market. In Economy 7 tariff, energy consumers get a lower price between 0:00 AM – 7:00 AM, and in Economy 10 tariff energy consumers get a lower price between 8 PM – 8 AM. In CPP tariff energy consumers noticed for higher energy rates for selected hours/days.

The price changing event is a reasonable operator that encourage consumers to shift the EV charging process. It is difficult for the utility to make a tradeoff between the shift able loads and the incentive cost that should be paid to the customer in exchange for this behavior.

Performance of EV in price-based DR programs

In order to analyze the performance of EV in price-based DR programs, the price changing event and the motivation of EV owners to change their charging behavior should be considered.

TOU

First of all, the time of use tariff is presented, and its assumptions and constraint are formulated. In this way, by considering the evening as off-peak hours, T_s and T_e as the beginning and the end time of this period, respectively; the maximum amount of charging can be formulated as follow:

$$E = P_{EV}.(T_e - T_s).\eta \qquad (14.7)$$

where E is the maximum amount of charging in the evening, P_{EV} and η are the charging power provided by the utility and charging efficiency, respectively. The minimum SOC of EV at T_s should be as follow to guaranty that vehicle battery is charged up to the expected level at the departure time.

$$SOC_{min} = 1 - \frac{P_{EV}.(T_e - T_s).\eta}{E_{EV}} \qquad (14.8)$$

where SOC_{min} is the minimum SOC of EV at T_s and E_{EV} is the battery capacity of EV. Actually, in order to postpone the charging process, the SOC of EV should be more than SOC_{min}. Otherwise, the charging process should be started before T_s to guaranty that vehicle battery fully charged before the departure time. In this way, the appropriate time to start charging can be achieved as below:

$$T_s = T_e - \frac{E_{EV}.\left(1 - SOC_{(T_{current})}\right)}{P_{EV}.\eta} \qquad (14.9)$$

where $SOC\ (T_{current})$ is the current SOC of the vehicle battery.

In the case of discharging, the process could be started if $SOC(T_{current}) > SOC_{min}$. If it is so, the economic aspect of the discharging process can be formulated as follow:

$$C_{sale} = C_{current}.\eta_{cha}.\eta_{dis} \qquad (14.10)$$

where $C_{current}$ and η_{dis} are the electricity price at the current time and discharging efficiency, respectively. Moreover, C_{sale} is the sale price in the case of discharging by considering EV efficiency.

Actually, the discharging process of EV should be started whenever the utility couldn't supply a large amount of demand.

According to the above, the discharge is implemented by considering the following constraints:

$$\begin{cases} C_{eqsale} > C_{average} \\ P_{total} > P_{max} \end{cases} \qquad (14.11)$$

where P_{total} and P_{max} are the actual power demand and expected power demand by the utility, respectively.

The discharge operation following the procedures in the Fig. 14.3.

The above-mentioned formulations are available for TOU and RTP events as well [29].

Fig. 14.3 Flowchart of discharging process of EV

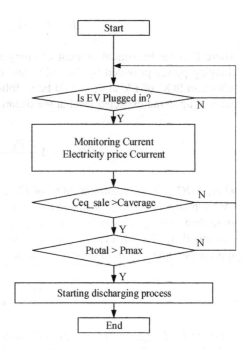

CPP

In this part, critical peak pricing tariff is presented, and its assumptions and constraint are formulated.

The pricing of CPP by considering assumptions and constraint can be formulated as below:

$$\min \left[\sum_{m=0}^{M}\left(P_{t+m}^{C} - P_{t+m}^{W}\right) + \sum_{n=m}^{M+N} P_{t+n}^{W} \times \Delta Q_{t+n}\right] \quad (14.12)$$

$$\Delta Q_{t+m} = Q_{t+m}^{EV} \times \frac{P_{t+m}^{C}}{\partial_{t+m}} \quad (14.13)$$

$$\sum_{m=0}^{M} \Delta Q_{t+m} = \sum_{n=m}^{M+N} \Delta Q_{t+n} \quad (14.14)$$

$$\frac{\Delta Q_{t+n}}{\sum_{m=0}^{M} \Delta Q_{t+m}} \leq \theta \quad (14.15)$$

$$P_{t+m}^{C} \geq 0 \quad (14.16)$$

$$\partial_{t+m} \geq P_{t+m}^{C} \quad (14.17)$$

The objective function in this tariff could be modeled as Eq. (14.12), where P_{t+m}^{C} refers to the price in the critical peak situation at time $t + m$ in which m illustrate the

14 Participation of Aggregated Electric Vehicles in Demand Response Programs 341

length of CPP event, P_{t+m}^W and P_{t+n}^W refer to the price of the wholesale market at time $t + m$ and $t + n$, respectively. ΔQ_{t+m} refers to the amount of EV energy that could be shifted to the other time and ΔQ_{t+n} refers to the amount of energy that could be shifted to the $t + n$ time period.

For this objective function, some constraints should be considered that described in Eqs. (14.13), (14.14), (14.15), (14.16), and (14.17). These constraints illustrate the connection between the price of CPP tariff and the amount of participation of EV owners in CPP events [30].

RTP

In RTP tariff, every EV computes its willingness to defer charging through a reservation price based on the remaining charge needed and the customer's desire to reach full charge and offers this price to the market operator. If the price is above an EV buy price, it will forgo charging for the next period. If the price is above an EV sell price, it will discharge energy back to the feeder during the next period.

The EV buy price is determined as:

$$P_B = P_M + P_D K \frac{\Delta t_R}{\Delta t_A} \tag{14.18}$$

where P_M and P_D are the mean and standard deviation of the expected electricity price over a time interval between the real-time and departure time. K is the consumer comfort control setting, which enables the consumer to control its charge/discharge behavior. The required time to fully charge is:

$$\Delta t_R = (1 - SOC) \frac{\beta}{\rho} \tag{14.19}$$

where SOC is the battery state of charge, β is the battery capacity in kWh, and ρ is the charging rate in kW. The available time until the departure is:

$$\Delta t_A = t_{dep} - \tau \tag{14.20}$$

where t_{dep} is the departure time, and τ is the real-time.

Electric vehicles can also potentially discharge electricity to the power grid in super peak times. Similar to the charge bidding strategy, every car can compute its willingness to sell electricity to the grid based on the potential opportunity price of recharging later. In the vehicle-to-grid case, an extra component is added to take the battery degradation account. The EV sell price is:

$$P_S = \frac{P_O}{\eta^2 \gamma} + P_C \tag{14.21}$$

where η is the round trip efficiency and γ accounts for battery aging. The cycling cost, P_C, accounts for the additional degradation costs of using V2G, which for $SOC < 80\%$ are estimated based on experimental data as:

$$P_C = \frac{0.001\, k}{(SOC + 0.4)^2} \tag{14.22}$$

where k is battery capital cost in \$/kWh; if $SOC > 80\%$ the cycle cost is zero. P_O is opportunity cost for discharging during the next time increment δt [31]:

$$P_O = P_M + P_D\, k\, \frac{\Delta t_R + \delta t}{\Delta t_A - \delta t} \tag{14.23}$$

EV in incentive-based DR tariff

Direct load control, curtailable load, and demand-side bidding are kinds of the incentive-based tariff. Among these three types of the tariff, DLC is more popular between customers. In the case of peak load and increase in wholesale market price, the utility companies have permission to control the specific load remotely by the preset preference of customers. Actually, in this way, customers convinced to allow the utilities to change their devices time of use or interrupt them [30].

Performance of EV in incentive-based DR programs

As mentioned before, DLC as a kind of incentive-based program has a basis of implementation of EVs. When peak load anticipated, the operator looking for interruptible loads and EVs as a flexible load can participate in DLC program.

For participation of EVs in DLC tariff some assumption should be considered that summarized as follow:

- It is assumed that all the EVs are plug-in in the case of DLC event are participated in DLC program
- All the EVs by more than 1-hour charging duration, which connected to the grid 1-hour before DLC events, are suitable for DLC program.
- All the EVs by more than 2-hour charging duration, which connected to the grid 2-hour before DLC events, are suitable for DLC program.
- All the EVs which connected to the gird 3-hour before DLC events are considered as uninterruptable loads [30].

14.4 Optimization

Optimization in the presence of EV investigated in the form of two subjects as follow:

1. Optimization of a smart distribution company in the presence of EV
2. Optimization of parking lots

Fig. 14.4 The Flowchart of solving optimization problems

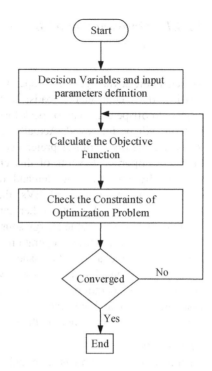

Each of the items has its objective function, uncertainties, and constraints that explained in following.

There are many optimization algorithms which are used in different papers such as genetic [16], ARIMA [19], MILP [25, 34], etc. in following optimization process presented in Sect. 14.4, MILP optimization algorithm is utilized.

To solving the optimization problem, the first step is defining the decision variable and input parameters. Based on the [16, 19, 25, 34], the primary decision variables in this phase are the number of EVs present in PLs, charging/discharging schedule, and types of DR programs. Furthermore, the input parameters are mostly SOC, arrival time, and departure time of EVs uncertainties, which expressed in Eqs. (14.24), (14.25), and (14.26), market uncertainties Eqs. (14.35) and (14.36), PV, and Wind generation uncertainties.

In the next step, objective function and its items are calculated, which illustrated in Eqs. (14.27) and (14.37).

For the final step, after checking the constraints of optimization problem Eqs. (14.28), (14.29), (14.30), (14.31), (14.32), (14.33), and (14.34), conversion investigated, and in this way, the optimization problem could be solved. The flowchart of solving the optimization problem is presented in Fig. 14.4.

14.4.1 Optimization of Smart Distribution System in Presence of EV

To benefit the advantage of EVs operation in smart distribution systems (SDSs), they considered as controllable loads. In this situation, by postponing the charging process to off-peak instead of peak load period, the efficiency of smart distribution systems will be improved. Actually, in this strategy, they consider a priority for loads, and the one with less preference will be connected in off-peak hours. One of the most important benefits of this charging mode is peak shaving that can be occurred by implementing demand response problems in SDSs operation. By increasing the penetration of EVs, the management and operation of the SDSs become more critical. On the other hand, uncertainties of EVs and other elements in SDSs such as renewable energy sources made it more complicated and should be considered. Thus, optimal operation of SDSs by considering uncertainties and constraint become an essential issue.

Both utility and EV aggregators try to reduce their cost. In the case of utility companies, the operational cost will be reduced by the optimal operation of conventional generation and renewable resources. Besides, in the case of EVs aggregators, this goal will be ascertained by the optimal charging of EVs based on DRPs [26].

Uncertainties
Each part of the system based on their generation and operation have uncertainties. In this section, the possibilities of EVs and RER generation are explained.

DGs Uncertainties
According to the topology of the system shown in Fig. 14.5, DGs are consist of PVs and Wind power. Each of them has uncertainty due to the generation that depends on

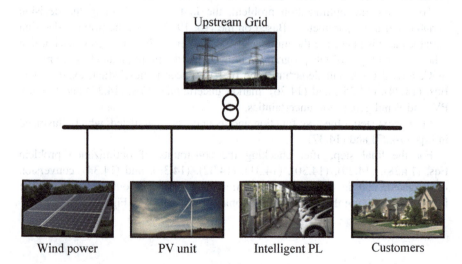

Fig. 14.5 Topology of the system [17]

14 Participation of Aggregated Electric Vehicles in Demand Response Programs   345

solar irradiance and wind speed, respectively. In order to model the impact of uncertainties for these DGs, there are some standard probability distribution functions (PDF) that can be used for each of them. These functions are summarized as follow:

For wind power:

- Weibull PDF
- Rayleigh PDF
- Rice
- Lognormal

For PV:

- Weibull PDF
- Lognormal
- Beta PDF

Each of these PDFs has its characteristic and be chosen based on the specific situation to approximate the nearest function to the real operation. Among the PDFs mentioned above, Weibull PDF for wind power and Beta PDF for PVs are the most common.

EVs uncertainty

As well as DGs, EVs have uncertainty due to some non-fixed items which are summarized as follow:

- Type of EV from battery capacity and rate of charge point of view
- Routine traveling pattern from time of plug in and timespan of charging possibility point of view
- State of charge of the battery at first and expected SOC when disconnected
- The schedule of charging/discharging process

There are some PDFs to estimate the timespan of charging possibility in PLs. One solution to considering these uncertainties is Monte Carlo Simulation (MCS) which presented in [32]. In the case of multi-variables functions that each of these random variables has their PDFs, MCS used to obtain the PDF of this function. In this method, the PDF of the function is estimated as a normal PDF with specific mean and standard deviation. Presence possibility of EVs in PL i, in demand level of h and each MCS, is estimated in Eq. (14.24):

$$M_{i,h}^{EV} = \mu_{i,h}^{EV} + 0.1 \times \mu_{i,h}^{EV} \times \lambda_{i,h}^{EV} \tag{14.24}$$

where $M_{i,h}^{EV}$ is presence possibility of EVs in PL i, in demand level of h, $\mu_{i,h}^{EV}$ is forecasted mean value of presence possibility of EVs in PL i, in demand level of h and $\lambda_{i,h}^{EV}$ is a random variable of EVs in PL i, in demand level of h.

In [16] to model the SOC uncertainty, the scenario-based approach is presented. The distribution curve related to the scenario-based method is divided into several areas and each area associated with the specific scenario. In this way, two central values should be obtained from this distribution curve:

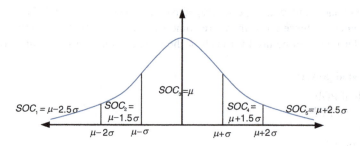

Fig. 14.6 Initial SOC [33]

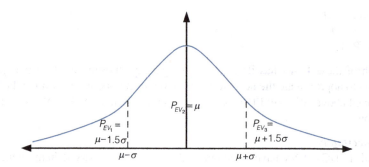

Fig. 14.7 Charging/discharging rates

1. Each scenario probability
2. Input variable average value

By considering five areas for initial SOC and three areas for charging/discharging rates in the distribution curve and the scenario-based approach, the initial SOC and charging/discharging rates can be calculated. The initial SOC and charging/discharging rates illustrated in Figs.14.6 and 14.7, respectively.

To model the uncertainty of the charging/discharging process which is dependent on the SOC, we have:

$$t_{charge} = \frac{(SOC_{max} - SOC) \times E_{EV}}{P_{EV}}$$
$$t_{discharge} = \frac{(SOC - SOC_{min}) \times E_{EV}}{P_{EV}} \quad (14.25)$$

where P_{EV} is charging power provided by the utility, E_{EV} is the battery capacity of EV, SOC_{max} and SOC_{min} are maximum and minimum SOC, respectively.

In this way, the expected power of PL i in demand level of h and each MCS obtained through Eq. (14.26):

14 Participation of Aggregated Electric Vehicles in Demand Response Programs

$$P_{i,h}^{PL} = \sum_{s=1} \sum_{j=1} CP_i \times P_{EV} \times M_{i,h}^{EV} \times SOC \qquad (14.26)$$

where CP_i is the capacity of parking station i and SOC is the initial state of charge of EV j in the scenario s.

Objective function

In smart distribution system consist of EVs, renewable energy resources (RERs), parking lots, and customers, to manage peak load periods, all capabilities of the component should be considered. RERs generation is one of the options that smart distribution owners can count on them in peak periods to reduce the energy bought from the wholesale electricity market. Customers, by participating in demand response programs, can manage their consumption and shift their controllable loads to off-peak periods. Besides this, intelligent PLs by scheduling charging and discharging of EVs can play an important role in peak load periods. In this situation, SDS should consider incentives for customers and EVs participating in DRP and costs for EVs battery degradation in the recharging process as well.

$$Max\ OF = Revenue_1 + Revenue_2 - Cost_3 - Cost_4 - Cost_5 - Cost_6 \qquad (14.27)$$

where:

$Revenue_1$: This part related to the revenue earned in exchange for charging EVs

$Revenue_2$: This part related to the revenue earned in exchange for customers consumptions

$Cost_1$: This part related to the cost paid for buying extra energy from the wholesale electricity market

$Cost_2$: This part related to the cost paid to EV owners in the mode of V2G

$Cost_3$: This part related to the cost paid to EV owners in exchange for battery degradation

$Cost_4$: This part related to the cost spent on DRP implementation

Constraints

For the above objective function, some constraints should be considered that explained as bellow:

1. PV and wind power generation: as mentioned before, the amount of generation of PV and wind power depended on solar irradiance and wind speed, respectively. Therefore their generation is bounded by this issue.
2. Permitted voltage limitation: in optimal power flow, voltage drop, and overvoltage of all buses should be in permitted limitation.
3. Permitted line capacity: loading of each line should be under the nominal line capacity and don't be overloaded.
4. Power generation and consumption balance: equilibrium should be existing between the power generated and the power consumed, and they should be equal.

5. Constraints of DR implementation: by implementing DR programs, controllable load during peak hours shift to another time. It should be considered that on off-peak hours, another peak load not occurred.
6. Constraints of EVs: As indicated in Eq. (14.28), the charging and discharging process cannot be implemented at the same time ($X_{n,t,s}^{ch}$ and $X_{n,t,s}^{dch}$ are binary variables show the status of charge and discharge, respectively). According to Eq. (14.29), the total SOC of the EVs ($SOC_{n,\,t,\,s}$) is limited to the minimum and maximum SOC ($SOC_{n,t,s}^{min}$, $SOC_{n,t,s}^{max}$). Also, according to Eqs. (14.30) and (14.31), the EVs hourly SOC is related to some factors like the SOC remains from the previous hour ($SOC_{n,\,t-1,\,s}$), the amount of power exchanged with the SDS and the PLs ($P_{n,t,s}^{ch}$, $P_{n,t,s}^{dch}$), the charge/discharge efficiency (η^{ch}, η^{dch}), and the EVs initial SOC ($SOC_{n,t,s}^{arv}$). The amount of EVs power buy/sell from/to the PLs (($P_{n,t,s}^{ch}$) and ($P_{n,t,s}^{dch}$)), are limited to their maximum value (see Eqs. (14.32) and (14.33)), respectively. Finally, according to Eq. (14.34), the charging and discharging procedure should be managed in such a way that the SOC reached the expected level.

$$X_{n,t,s}^{ch} + X_{n,t,s}^{dch} \leq 1 \quad \forall n, t, s \tag{14.28}$$

$$SOC_{n,t,s}^{min} \leq SOC_{n,t,s} \leq SOC_{n,t,s}^{max} \quad \forall n, t, s \tag{14.29}$$

$$SOC_{n,t,s} = SOC_{n,t-1,s} + \left(P_{n,t,s}^{ch} \times \Delta t \times \eta^{ch}\right) - \left(\frac{P_{n,t,s}^{dch} \times \Delta t}{\eta^{dch}}\right) \quad \forall n,$$
$$> t^{arv}, s \tag{14.30}$$

$$SOC_{n,t,s} = SOC_{n,t,s}^{arv} + \left(P_{n,t,s}^{ch} \times \Delta t \times \eta^{ch}\right) - \left(\frac{P_{n,t,s}^{dch} \times \Delta t}{\eta^{dch}}\right) \quad \forall n, t > t^{arv}, s \tag{14.31}$$

$$0 \leq P_{n,t,s}^{ch} \leq X_{n,t,s}^{ch} \times R_n^{ch} \quad \forall n, t, s \tag{14.32}$$

$$0 \leq P_{n,t,s}^{dch} \leq X_{n,t,s}^{dch} \times R_n^{dch} \quad \forall n, t, s \tag{14.33}$$

$$SOC_{n,t,s} = SOC_{n,t,s}^{dep} \quad \forall n, t^{dep}, s \tag{14.34}$$

The process of solving the problem

For solving this kind of problem, stochastic programming is usually used. Actually, for problems with various uncertainties this kind of programming recommended since Uncertainties, are considered as random variables which expressed in the collection of scenarios. As mentioned before, to considering these uncertainties, there are some PDFs which truncated Gaussian distribution, beta and Weibull are the most common. In Fig. 14.8. The flowchart shows the process of solving the problem.

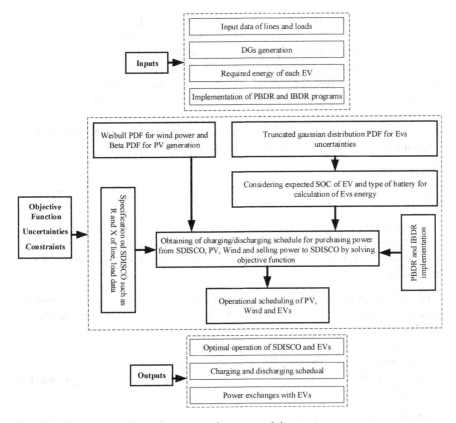

Fig. 14.8 The flowchart shows the process of program solving

According to Sadati et al. [34], integration of smart charging/discharging of EVs and implementation of DR programs have different results based on the kind of DRPs. In this way, results show that PBDR programs have better performance, and among PBDR programs, CPP programs give more profit to SDSs.

On the other hand, the best case for SDSs in terms of profit is the participation of customers in DRP and smart charging and discharging of EV. In this way, by discharging of EVs in peak hours, SDSs buy less power from the wholesale electricity market and sell more power to EVs.

This type of integration gives profit to SDSs, moreover, can reduce losses and have peak shaving affects.

14.4.2 Optimization of Parking Lots

In this part, EV parking lot optimization is investigated in which PL participate in DRPs. Based on the operating schedule and by considering the DR capability of

EVs, the PLs decided when to charge and discharge the EVs. By considering the battery capacity of the EVs, expected SOC of EV at the departure time and the period of EVs attend in PLs, the charging process managed by PL operators. The main purpose of PL operators is to gain more profit by the integration of EVs in DRPs, and the revenue resulted from customers. PLs participate in the energy market, moreover, based on the type of DRP participate in reserve market as well, so that, by considering the cost of buying energy from wholesale market and the revenue of selling energy back to the wholesale market due to the V2G capability of EVs, maximum profit achieved for PL operators.

Uncertainties

In this section, the uncertainties of EVs operation and the related market factors are explained.

EV uncertainty

In order to model the possibilities of EVs behavior, truncated Gaussian distribution and normal PDF is widely employed for arrival, and departure times and the SOC at arrival time.

In order to generate the scenarios of EVs, the behavior of each EV, and the amount of energy stored in it should be modeled and formulated. The capacity of each EV depends on the EV battery class.

One of the items in EVs uncertainties is the type of EVs due to its battery capacity, which can be modeled by using probability distribution.

The details about the modeling of EVs behavior uncertainties are presented in Sect. 14.4.1.

Electricity market uncertainty

The parking lots can be cooperated in the energy and reserve markets, to maximize the profits. This cooperation requires accurate and reliable information about the clearance level in the energy market and the activated spinning reserve in the reserve market.

Since, the energy market is a dynamic procedure, and prediction of the player's behavior and contingencies are so difficult. Therefore, these uncertainties are modeled probabilistically by PDFs like uniform distribution, Beta distribution, normal distribution, etc.

In [35] through the scenario-based approach, the price electricity bought from the grid in demand level of h, year t, and MCS is formulated in Eq. (14.35):

$$\rho_{t,h} = \rho_s \times PLF^e_{t,h} \tag{14.35}$$

where ρ_s is the base price in scenario s and $PLF^e_{t,h}$ is the price level factor in demand level of h, year t, and MCS. By considering normal PDF for price level factor, it can be calculated as following in Eq. (14.36):

14 Participation of Aggregated Electric Vehicles in Demand Response Programs 351

$$PLF^e_{t,h} = \mu^p_{t,h} + \sigma^p_{t,h} \times \lambda^p_{t,h} \tag{14.36}$$

where $\mu^p_{t,h}$ and $\sigma^p_{t,h}$ are the forecasted value of the price level factor and standard deviation in demand level h and year t, respectively. $\lambda^p_{t,h}$ is a random variable for electricity price, in demand level of h, year t, and MCS.

Objective function
The objective function is to maximize the parking lots profit, and as well as Sect. 14.4.2, to solve this problem, stochastic programming is used, and the objective function is formulated in Eq. (14.37).

$$Max\ OF = \sum_{i \in DRPs} \alpha_i \begin{pmatrix} Revenue_1 + Revenue_2 + Revenue_3 + \\ Revenue_4 + Revenue_5 - Cost_1 - \\ Cost_2 - Cost_3 - Cost_4 \end{pmatrix} + \tag{14.37}$$
$$\alpha_i(Revenue_6 - Cost_5 - Cost_6 - Cost_7)$$

where:
Revenue$_1$: The revenue of selling energy to the wholesale
Revenue$_2$: The revenue of participating EVs in the reserve market
Revenue$_3$: The revenue resulted from EV owners in exchange to charge of EVs
Revenue$_4$: The revenue received from the use of parking tariff
Revenue$_5$: The revenue of participating in IBDR programs
Revenue$_6$: The revenue resulted from supplying the amount of reserve
Cost$_1$: The cost regarding the EV owners share due to the participating in DRPs in the energy market
Cost$_2$: The cost of buying power from the wholesale market
Cost$_3$: The cost of battery degradation in the energy market
Cost$_4$: The cost of participating in IBDR programs in case of penalty
Cost$_5$: The cost of battery degradation in the reserve market
Cost$_6$: The cost of lack of supply the offered reserve
Cost$_7$: The cost regarding the EV owners share due to the participating in DRPs in reserve market
α_i: The participation level of the EV parking lot in each DRP

According to Shafe-khah et al. [25], the participation of PLs as an effective load in the main system in DR programs have different results based on the kind of DRPs. In this way, results show that PBDR programs have better performance, and among PBDR programs, TOU programs give more profit to PLs.

This type of integration gives profit to PLs, moreover, as PLs can supply spinning reserve of system, power system benefits too.

Optimal charging scheduling
Besides the cases studied above, one of the most important issue in the operation of parking lots are optimal charging scheduling of EVs. Although the number of

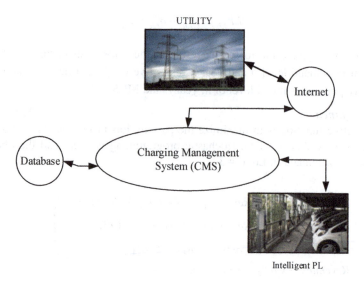

Fig. 14.9 Communication way of PLs in DR event and utility [36]

parking lots and their poles cover numerous EVs, not all the EVs could be charged at the same time. EVs that connected to the pole in this scale assumed as an effective demand; therefore, the limitation of the distribution system in supplying them should be considered.

In this way, the importance of charging schedule is determined. In spite of the limitation of the distribution system, EVs connected to the poles should be charged to the expected level. To achieve this goal, PL owners should manage to charge by optimal charging schedule.

As mentioned before, based on the type of EV, the battery capacity is different. Besides this, the time that EV owners park their cars and in consequence, the time that they leave the PLs are different. By considering these differences, PL operators, prioritize all EVs present in PLs to achieve the two main goals. The goals are EV owner's satisfaction and observing the limitation of the system.

In addition to the items discussed above, one more thing that could be concerned is the electricity bill. By participation of PLs in DRPs and considering the charging schedule for EVs, electricity bill could be minimized. The charging of EVs could be managed and based on the priority, shifted to the other time, which is one of the most important factors for participation in DRPs. Charging management system (CMS) as a center of communication can manage the connection of parking lots and utility. As shown in Fig. 14.9 to achieve the goals, an infrastructure in which two-way communication established between parking lots, the CMS and database is needed. In this way, the utility sends the DRPs, and PLs send load profile to the CMS. The utility observes the operation of parking lots in DRPs based on the received load profile.

14 Participation of Aggregated Electric Vehicles in Demand Response Programs

By considering the DRPs defined by utility and number of connected EVs in PLs and their priority, the CMS manages the charging schedule.

By considering many numbers of EVs in PLs and limitation of time caused by DRP, on the other hand, the best way to charge more EVs in minimum time is by charging with maximum power. The CMS monitors the charging poles and based on the priority decide which EV should be charged in this way.

In the case of the day-ahead market, utility request the DRP and PL owners bid, both through the CMS and based on internet communication. In this way and by considering DRP, the demand reduction schedule presented by parking lot owners.

According to Leether Yao et al. [36], all the EVs in parking lots cannot be charged at the same time. PLs assumed as a large load, and the limitation of the distribution system should be considered. On the other hand, PL owners can manage the charging process and based on the battery capacity of the EVs, and expected SOC at the departure time, shift charging to the other time. Based on this capability of PLs, they can participate in DRPs. DSB is one of the DR programs in which PL operators can bid on load reductions in a day-ahead market. If their bid is cleared, they are obliged to change load accordingly and in this way reduce electricity bill.

CMS as a connector, make communication between utility and PLs. Due to the large number of EV that should be charged and distribution system capacity limitation, CMS manages the charging pole. In this way, it connects the EVs based on the priority and constraints of demand based on the DR events. In order to charge more number of EVs, the maximum charging power method use to charge EVs in minimum time.

The main object of the system, charging a maximum number of the EVs and minimum electricity bill, could be achieved in this way.

14.5 Conclusion

In this chapter, the integration of EVs in demand response programs has been studied. By increasing environmental concerns, the idea of finding an alternative option for transportation systems receives special attentions. As a result, the electric vehicles were chosen because of their advantages for the utility, EV owners and PL operators. Furthermore, implementation of demand response programs due to their advantages such as reducing network losses, increasing reliability of the power system, and reducing the electricity generation costs, becomes widespread.

On the other hand, charging EVs without considering its impacts on the power system, causes higher peak demands. DR programs which focusing on discharging during peak hours and start charging just after peak hours, effectively help peak shaving and valley filling. Using the DR capability of the EVs by managing the charging process in both V2G and G2V modes is an effective way to reduce the operational costs of the system and satisfying the electricity demand. In vehicle to grid operation, it should be considered that this process causes battery degradation and to overcome this issue, advanced battery technology could be a reasonable candidate.

By increasing the number of EVs, establishing parking lots in a smart microgrid (SMG) is inevitable. The main factor in PLs is a large number of flexible loads that can be participated in demand response programs. All the EVs should be charged in a suitable time, and all the curtailments caused by DRPs should be satisfied. To evaluate the operation of PL, three types of PL based on the main occupied time considered: residential, commercial, and industrial parking lots.

Due to the increase of EVs penetration and limitation of residential given demand, consumers' satisfaction becomes an important issue. Implementation of demand response programs in residential level can be performed in two layers: the neighbourhood area network (NAN) and the home area network (HAN). The main purpose of DR implementation in a residential area is to supply EVs charging demand so that the satisfaction level of the consumer does not disturb. In NAN strategy, the houses which their consumption is more than the maximum demand of the household section, are included in demand limitation. In HAN strategy, shift able loads are controlled by HAN based on their determined maximum demand.

The main purpose of DR implementation is to change the customer's behavior based on the price changes. These price changes could be as a result of price-based DRPs, their tariffs or incentive-based DRPs, and their incentive or penalties which are applied to customers.

To achieve maximum profit through the integration of EVs in DRPs, optimization of the smart distribution system and PLs usually are studied. In the case of the smart distribution system in the presence of EVs, the operational cost will be reduced by the optimal operation of conventional generation and renewable resources. In the case of EVs aggregators, this goal will be ascertained by the optimal charging of EVs based on DRPs. Thus, optimal operation of SDSs by considering uncertainties and constraints becomes an essential issue. In order to solve the objective function, DGs uncertainties, EVs uncertainties, and constraints related to the voltage limitations, power balance and etc. should be considered. As results show, PBDR programs have a better performance, and among PBDR programs, CPP programs give more profit to SDSs. Actually, by participating of customers in DRPs and smart charging and discharging of EVs, SDSs achieve the most benefit. In this way, by discharging of EVs in peak hours, SDSs buy less power from the wholesale electricity market and sell more power to EVs.

In the case of PLs, The main purpose of PL operators is to gain more profit by the integration of EVs in DRPs, and the revenue resulted from customers. As results show, PBDR programs have a better performance, and among PBDR programs, TOU programs give more profit to PLs. Although the number of parking lots and their poles cover numerous EVs, not all the EVs could be charged at the same time. All the EVs connected to the poles should be charged to the expected level in spite of the limitation of the distribution system so that PL owners should manage to charge by optimal charging schedule. Charging management system (CMS) as a center of communication can manage the connection of parking lots and utility. By considering the DRPs defined by utility and number of connected EVs in PLs and their priority, the CMS manages the charging schedule.

As the knowledge of authors, utilizing EVs and cooperation of DRPs are in a transient phase from research to implementation. In this situation, providing confident electrical and communication infrastructure for the integration of EVs in DRPs is necessary. Based on the studies, EV modeling, operational constraints, optimization algorithm, PDFs and uncertainties modeling, and optimization problem solving have been the primary concerns.

In this situation, considering below items can show the road map in this field:

1. In most of the studies, uncertainty modeling is considered for PLs. However, more often, EVs have been parked in the city, not the PLs. On the other hand, individual behavior uncertainties are different, and using a single PDF for all the EV owners cause losing some of the EV owner's behaviors. Therefore, using a number of PDFs for modeling the EV owner's behavior to model PDF interactions is suggested.
2. Most of the customers do not participate in the spot market and ancillary services individually, which cause a reduction in the participation of EVs in DRPs. Therefore, investigating the effects of financial incentives for EVs participation in DRPs seems necessary.
3. Study the impacts of different types of DRPs in the amount of EVs participation can be the other issue.

References

1. International energy agency, 2006, available at: http://iea.org/subjectqueries
2. R.A. Waraich, M.D. Galus, C. Dobler, M. Balmer, G. Andersson, K.W. Axhausen, Plug-in hybrid electric vehicles and smart grids: Investigations based on a microsimulation. Transp. Res. C Emerg. Technol **28**, 74–86 (2013)
3. E. Pashajavid, M.A. Golkar, Multivariate stochastic modelling of plug-in electric vehicles demand profile within domestic grid, in *Reliability Modelling and Analysis of Smart Power Systems*, (Springer, New Delhi, 2014), pp. 101–116
4. C. Pang, P. Dutta, M. Kezunovic, BEVs/PHEVs as dispersed energy storage for V2B uses in the smart grid. IEEE Trans. Smart Grid **3**(1), 473–482 (2011)
5. H. Morais, T. Sousa, Z. Vale, P. Faria, Evaluation of the electric vehicle impact in the power demand curve in a smart grid environment. Energy Convers. Manag. **82**, 268–282 (2014)
6. P. Bradley, M. Leach, J. Torriti, A review of the costs and benefits of demand response for electricity in the UK. Energy Policy **52**, 312–327 (2013)
7. Z. Darabi, M. Ferdowsi, Aggregated impact of plug-in hybrid electric vehicles on electricity demand profile. IEEE Trans. Sustainable Energy **2**(4), 501–508 (2011)
8. I.I. Green, C. Robert, L. Wang, M. Alam, The impact of plug-in hybrid electric vehicles on distribution networks: A review and outlook. Renew. Sust. Energ. Rev. **15**(1), 544–553 (2011)
9. J. Soares, B. Canizes, C. Lobo, Z. Vale, H. Morais, Electric vehicle scenario simulator tool for smart grid operators. Energies **5**(6), 1881–1899 (2012)
10. R. Johal, D.K. Jain, Demand response as a load shaping tool integrating electric vehicles, in *2016 IEEE 6th International Conference on Power Systems (ICPS)*, (2016), pp. 1–6

11. S. Khemakhem, M. Rekik, L. Krichen, Impact of electric vehicles integration on residential demand response system to peak load minimizing in smart grid, in *2019 19th International Conference on Sciences and Techniques of Automatic Control and Computer Engineering (STA)*, (2019), pp. 572–577
12. H.K. Nguyen, J.B. Song, Optimal charging and discharging for multiple PHEVs with demand side management in vehicle-to-building. J. Commun. Netw. **14**(6), 662–671 (2012)
13. M. Alizadeh, A. Scaglione, A. Applebaum, G. Kesidis, K. Levitt, Reduced-order load models for large populations of flexible appliances. IEEE Trans. Power Syst. **30**(4), 1758–1774 (2014)
14. P. Liu, J. Yu, Charging behavior characteristic simulation of plug-in electric vehicles for demand response, in *2016 UKACC 11th International Conference on Control (CONTROL)*, (2016), pp. 1–6
15. A. Rabiee, A. Soroudi, B. Mohammadi-Ivatloo, M. Parniani, Corrective voltage control scheme considering demand response and stochastic wind power. IEEE Trans. Power Syst. **29**(6), 2965–2973 (2014)
16. S. Shojaabadi, S. Abapour, M. Abapour, A. Nahavandi, Optimal planning of plug-in hybrid electric vehicle charging station in distribution network considering demand response programs and uncertainties. IET Gener. Transm. Distrib. **10**(13), 3330–3340 (2016)
17. J. Jannati, D. Nazarpour, Multi-objective scheduling of electric vehicles intelligent parking lot in the presence of hydrogen storage system under peak load management. Energy **163**, 338–350 (2018)
18. E. Akhavan-Rezai, M.F. Shaaban, E.F. El-Saadany, F. Karray, Demand response through interactive incorporation of plug-in electric vehicles, in *2015 IEEE Power & Energy Society General Meeting*, (2015), pp. 1–5
19. P. Aliasghari, B. Mohammadi-Ivatloo, M. Alipour, M. Abapour, K. Zare, Optimal scheduling of plug-in electric vehicles and renewable micro-grid in energy and reserve markets considering demand response program. J. Clean. Prod. **186**, 293–303 (2018)
20. F. Boshell, O.P. Veloza, Review of developed demand side management programs including different concepts and their results, in *2008 IEEE/PES Transmission and Distribution Conference and Exposition: Latin America*, (2008), pp. 1–7
21. C. Pang, P. Dutta, M. Kezunovic, BEVs/PHEVs as dispersed energy storage for V2B uses in the smart grid. IEEE Trans. Smart Grid **3**(1), 473–482 (2011)
22. N.G. Paterakis, O. Erdinç, J.P.S. Catalão, An overview of demand response: Key-elements and international experience. Renew. Sust. Energ. Rev. **69**, 871–891 (2017)
23. J. Soares, Z. Vale, H. Morais, N. Borges, Demand response in electric vehicles management optimal use of end-user contracts, in *2015 Fourteenth Mexican International Conference on Artificial Intelligence (MICAI)*, (IEEE, 2015), pp. 122–128
24. J. Soares, M.A.F. Ghazvini, N. Borges, Z. Vale, Dynamic electricity pricing for electric vehicles using stochastic programming. Energy **122**, 111–127 (2017)
25. A. Rautiainen, C. Evens, S. Repo, P. Järventausta, Requirements for an interface between a plug-in vehicle and an energy system, in *2011 IEEE Trondheim PowerTech*, (IEEE, 2011), pp. 1–8
26. N. Nezamoddini, Y. Wang, Risk management and participation planning of electric vehicles in smart grids for demand response. Energy **116**, 836–850 (2016)
27. S. Shao, M. Pipattanasomporn, S. Rahman, Grid integration of electric vehicles and demand response with customer choice. IEEE Trans. Smart Grid **3**(1), 543–550 (2012)
28. M. Mallette, G. Venkataramanan, Financial incentives to encourage demand response participation by plug-in hybrid electric vehicle owners, in *2010 IEEE Energy Conversion Congress and Exposition*, (IEEE, 2010), pp. 4278–4284
29. Y. Liu, S. Gao, X. Zhao, S. Han, H. Wang, Q. Zhang, Demand response capability of V2G based electric vehicles in distribution networks, in *2017 IEEE PES Innovative Smart Grid Technologies Conference Europe (ISGT-Europe)*, (IEEE, 2017), pp. 1–6

30. K. Zhang, S. Zhou, Data-driven analysis of electric vehicle charging behavior and its potential for demand side management, in *IOP Conference Series: Earth and Environmental Science*, vol. 223, no. 1, (IOP Publishing, 2019), p. 012034
31. S. Behboodi, D.P. Chassin, C. Crawford, N. Djilali, Electric vehicle participation in transactive power systems using real-time retail prices, in *2016 49th Hawaii International Conference on System Sciences (HICSS)*, (IEEE, 2016), pp. 2400–2407
32. A. Soroudi, R. Caire, N. Hadjsaid, M. Ehsan, Probabilistic dynamic multi-objective model for renewable and non-renewable distributed generation planning. IET Gener. Transm. Distrib. **5** (11), 1173–1182 (2011)
33. S. Nojavan, K. Zare, B. Mohammadi-Ivatloo, Stochastic energy procurement management for electricity retailers considering the demand response programs under pool market price uncertainty. Majlesi J. Energy Manag. **4**(3), 49–58 (2015)
34. S.M.B. Sadati, J. Moshtagh, M. Shafie-khah, J.P.S. Catalão, Smart distribution system operational scheduling considering electric vehicle parking lot and demand response programs. Electr. Power Syst. Res. **160**, 404–418 (2018)
35. S. Abapour, K. Zare, B. Mohammadi-Ivatloo, Dynamic planning of distributed generation units in active distribution network. IET Gener. Transm. Distrib. **9**(12), 1455–1463 (2015)
36. L. Yao, W.H. Lim, T.S. Tsai, A real-time charging scheme for demand response in electric vehicle parking station. IEEE Trans. Smart Grid **8**(1), 52–62 (2016)

Chapter 15
Optimal Charge Scheduling of Electric Vehicles in Smart Homes

Arezoo Hasankhani and Seyed Mehdi Hakimi

15.1 Introduction

The electric vehicles (EVs) are developing in the today network, which has significant effect on the network demand. Defining suitable energy management system and demand response method is an inevitable issue in today networks. In the first step, present demand response considerations in the presence of electric vehicles should be addressed. It is additional important to review the EVs models and its behavior in different studies. Charging management of EVs has been done considering different objectives in objective functions including cost minimization, power loss minimization, frequency regulation, contamination minimization, flattening the demand and etc.

In [1], the demand response program have been developed considering the presence of residential EVs, and its effect on electricity market have been studied. The charging and discharging of EVs can have important role in real-time market and can diminish the diverse effect of intermittent resources by suitable demand response programs. In [2], the real-time pricing based demand response method has been proposed for the smart grid including EVs and photovoltaic units, which consider forecasting price and developing a dynamic price vector. The demand response program has been proposed for charging EVs, which has been solved by binary optimization problem [3]. The energy transaction through

A. Hasankhani
Department of Computer and Electrical Engineering and Computer Science, Florida Atlantic University, Boca Raton, FL, USA
e-mail: ahsankhani2019@fau.edu

S. M. Hakimi (✉)
Department of Electrical Engineering and Renewable Energy Research Center, Damavand Branch, Islamic Azad University, Damavand, Iran
e-mail: sm_hakimi@damavandiau.ac.ir

© Springer Nature Switzerland AG 2020
A. Ahmadian et al. (eds.), *Electric Vehicles in Energy Systems*,
https://doi.org/10.1007/978-3-030-34448-1_15

neighbor smart homes for charging EVs has been considered as a solution for efficient demand response in [4].

In [5], the EVs' charging stations have been considered as smart loads, which have been forecasted by neural network in demand response programs. In [6], the potential of EVs participation in demand response programs in United kingdom has been investigated, and different EVs' manufactures have been interviewed. The stochastic behavior of EVs as a load has been modelled by probabilistic modelling of EVs, and its behavior in demand response programs has been studied in [7].

The EVs model has been addressed in different studies. In [8], the charging demand of EVs has been modelled by an integrated dynamic method, which has also been tested in different case studies. Modelling the demand of EVs based on different criteria including driver behavior, location of charging station and electricity pricing has been done in [9]. In [10], the charging demand of EVs has been modelled in different locations and purposes. The charging demand of EVs has been studied in New Zealand base on high penetration of EVs [11]. The EVs variable demand has been supplied by wind-powered stations in [12]. In this study, the intermittent behavior of wind turbines has been forecasted by autoregressive integrated moving average method, so the accessible power for supplying EVs has been determined.

Modelling the behavior of EVs has been addressed in different studies. In [13], the stochastic behavior of EVs and its demand has been investigated. The EVs demand has been forecasted considering real-data traffic data [14]. The EVs demand has been predicted by mixed generalized extreme value model [15]. In [16], the EVs' demand has been forecasted considering the behavior of consumers. The transport and trip behavior of EVs have been studied in [17]. The EVs' travelling behavior has been forecasted by a novel artificial neural network, which improve the accuracy of forecasting by 10% [18]. In [19], the location of charging stations has been identified by studying the behavior of EVs and applying hybrid heuristic algorithm. In another study [20], the demand of EVs has been forecasted and controlled by using it as a storage unit. In [21], the day-ahead demand of EV has been forecasted at business level.

Charging management of EVs have been done by different objectives. The integration of EVs in the UK network has been investigated considering heating systems in order to maximize system profit and minimize carbon dioxide emission [22]. In [23], the produced power of photovoltaic units has been forecasted in order to charge EVs in the workplace, and the main objective in this problem is minimizing the charging cost. In [24], the controlled charging of EVs has been done in order to flatten the demand profile and reach the acceptable level of charge in charging timeout. In [25], the coordination between home energy management system and grid energy management system has been studied in order to address the EV considering photovoltaic unit, which has been tested by Japanese distribution system data. In another study [26], the effect of EV on grid by EV to grid view has been studied, which has been applied in two different energy management method.

In [27], the charging of EV has been controlled in order to regulate frequency. The stochastic optimization has been applied in order to increase the application of photovoltaic units and minimize the charging cost in [28]. In [29], the optimization problem in charging EVs has been defined by two objectives of cost and waiting time minimization. The decentralized charging plan, in which the consumer can decide locally about their charging plan, has been proposed in [30] based on augmented Lagrangian method. In another view, the distributed controlled charging has been applied in order to minimize the power loss [31]. The decentralized controlled charging of EVs has been addressed considering three different socio-technical issues including reliability, discomfort and fairness [32].

The demand response subject and its relation with EVs will be discussed in Sect. 15.2. In addition, the EV model will be presented in Sect. 15.3. Demand in smart home in the normal condition will be addressed in Sect. 15.4. The charging management method will be proposed in Sect. 15.5. The comparison between uncontrolled and controlled charging will be done in Sect. 15.6. Finally, the conclusion remarks will be presented and summarized in Sect. 15.7.

15.2 Demand Response

As defined by the US Department of Energy, demand response is the making the ability for industrial, commercial and residential customers to improve the pattern of electric energy consumption in order to achieve reasonable prices and improve network reliability. In other words, demand response can change the shape of the electric energy consumption in such a way that the peak of the system is reduced and expenses are transferred to off-peak hours. Figure 15.1 shows the effect of load response program on customer consumption curve.

Demand response programs are split into two main branches and several subclasses, which is shown in Fig. 15.2 [33].

According to the application of each demand response programs, they can be applied in different timescales (Fig. 15.3). For example, direct load control is directly applied for consumers. Whenever the network needs load decrease, the consumers reduce their load which should be done in the 1-h period. However, ancillary programs should be managed monthly in order to be efficient.

The presence of the customer in the market and using demand response programs will lead to more competitive markets and ultimately reduce the price of electricity on the market. On the other hand, using these resources will increase the reliability of the system. Advantages of presence of customers in the market can be divided into three major groups [34]:

- Customer benefits

These benefits include economic benefits such as not purchasing electricity in expensive times and buying at cheap prices, which is due to participating in demand response programs for participating loads.

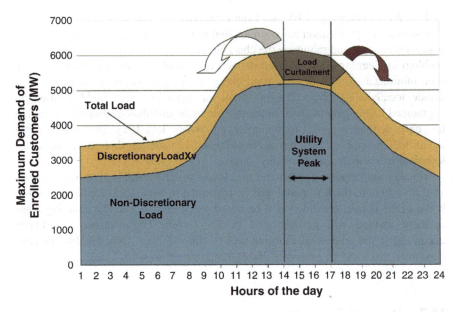

Fig. 15.1 The effect of demand response on residential demand

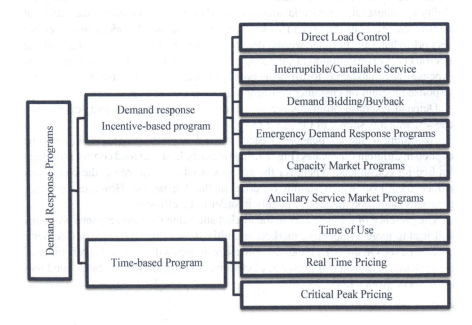

Fig. 15.2 Demand response categories [33]

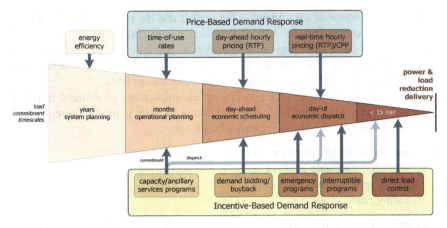

Fig. 15.3 Application timescale of demand response programs [33]

- Network benefits

Network benefits can be categorized into several general groups: short-term effects on the market, long-term effects on the market, savings in investment and operating costs and improving system reliability.

- Additional benefits

Additional benefits that result from the implementation of demand response programs include increasing the strength of retailers and the retail market, providing new tools for customer management, improving market operations, and market competitiveness and proper supply and demand interaction. The market also has an effective relationship between the retail and wholesale markets and the isolation of subscribers and retailers as well as the environmental benefits of reducing the generation of fossil generators, and the other benefits that customers have on the market.

15.2.1 Demand Response Effects on the Network

The results of the survey conducted by the Federal Energy Regulatory Committee of the United States from regional electricity companies show that the implementation of demand response programs puts the following effects on the power system [33]:

- Impact on reducing the costs of electricity companies (55%)
- Effect on improving network reliability (52%)
- Effect on the reduction of subscriber bill (36%)
- Impact on reducing network black-out (28%)
- Effect on increasing electricity sales (22%)
- Effect on reducing environmental pollution (18%)

The high presence of subscribers in the market and the use of demand response programs and optimizing the use of electrical energy improve the competitive market. On the other hand, the implementation of demand response programs has important implications for reducing electricity prices in the electricity markets, and also improves the reliability of the network to the optimum level.

A properly designed network considering the importance of demand response programs must have at least the following features in order to meet requirements of resources in the side of consumers and network side:

- Proper access of all customers to the network
- Establishing a reliable network
- Maximizing economic benefits
- Reducing costs

Demand response resources can be provided to the system operator as a virtual resource. Capacity created by demand response, in addition to increasing system reliability, can also be used as a replacement for supplying power at peak times. In addition, demand response resources in terms of technical characteristics are also superior to traditional ones, such as the absence of pollution and pollution, low running costs and a very high rate of increase or decrease in load.

There are some constraints for utilization of conventional generators, such as the minimum start-up time, the minimum shut-down time, and the minimum load. Other constraints exist on the use of demand response programs:

- The number of times a load can be interrupted within a specified period
- Maximum operating time
- Accuracy in response

15.3 Electric Vehicles

In general, electric vehicles are divided into three categories:

- Electric Vehicle (EV)
- Hybrid Electric Vehicle (HEV)
- Plug-in Hybrid Electric Vehicles (PHEV)

Electric Vehicles
These vehicles have an electric motor with batteries for electric power supply, and the energy of the batteries is used as a driving force for the electric motor and for the supply of energy for other equipment. The batteries can be charged by connecting to the power grid and the braking energy of the vehicle, and even from non-network electrical sources such as solar cells.

The main advantages of these cars are:

- Absolutely free of greenhouse gas emissions.
- Generating very low noise.
- The efficiency is much higher than the internal combustion engines.
- The price of their electric motors is low.

The main disadvantage of these cars is the full dependence on the battery (whose technology still does not have the same capacity and energy density as fossil fuels).

Hybrid Electric Vehicle (HEV)
These vehicles have a fuel engine and an electric motor with sufficient battery life (1–3 kWh) with energy saving power from the engine and the car's brakes. Batteries come at a time when they need help with the vehicle to produce auxiliary power, or at the low speeds, by turning off the fuel engine, to provide the vehicle's driving force.

Over the past decade, about 1.5 million hybrid EVs have been sold. In developed countries such as the United States, about 3% of existing cars are hybrid.

The disadvantages of these cars are:

- Failure to charge batteries from the power grid.
- Dependence on the fossil fuel consuming engine (inability to drive a car only with an electric motor).

Plug-in Hybrid Electric Vehicles (PHEV)
These cars are designed to eliminate the disadvantages of hybrid electric vehicles, and they can be charged from the network and hence require more batteries than HEV are. In these cars, there is a complete fossil fuel engine system.

PHEVs have more batteries than HEVs (about 5 times more than HEVs). The main difference between the batteries of these two types of EVs is that PHEV batteries should be capable of rapid discharging and fast charging, while HEV batteries operate in almost complete charge, and their discharge is rarely done.

The cost of the PHEV batteries is between 1.3 and 1.5 times the cost of EV batteries. However, because of the lower battery life, the total cost of batteries in PHEVs is less than EV.

For these cars, the following points can be mentioned:

- With massive battery production, it costs $ 750 per kilowatt hour, which for a mid-range car (40 km with a battery of 8 kWh), the total cost of batteries will be about $ 6000.
- If the car's lifespan is 200,000 km, the cost of the saved fuel will be about $ 4000, which is less than the cost of the battery.
- Reducing the battery cost to $ 500 per kilowatt-hour creates a competition between the PHEVs and conventional gasoline vehicles.

In this study, the EVs has been applied, and its characteristics is explained in this section. Since the development of electric vehicles in power grids is inevitable, their management in the grid and the use of electric vehicles as an element in the direction of load shading can play a significant role in accelerating the development process.

The power consumption of an electric vehicle depends on the battery discharge characteristic and the vehicle's driving pattern [35]. In this study, Nissan Altra is chosen as the sample electric vehicle, with its lithium-ion battery [36].

15.3.1 Effect of Electrical Vehicles on the Power System

EVs connected to the network have a significant impact on reducing environmental pollution and transportation costs, especially in large cities. Considering the current trends in the use of this device, it is a good future for this, so the study of the interactions between EVs and distribution networks is essential. With the widespread use of EVs, the pattern of consumption has changed over and over again, and many variables for planning, designing and operating power distribution networks will change. On the other hand, considering the changes in the structure of distribution networks for the use of smart equipment, the study of the effects of these two issues on each other seems necessary.

EVs with chargeable batteries must be connected to the power system to charge their batteries. Therefore, with the widespread influence of these cars, the performance of the power system will change especially in the distribution network. Uncontrolled battery charging can result in undesirable effects such as overload, overvoltage, power loss, unbalanced load, harmonic, and instability [37, 38]. By applying demand side management, not only can such problems be prevented, but the power curve is also flattened. As a result, the capacity of a network that is only used to respond to peak power and is only used in a very limited period of the year is better utilized [39].

Many studies have been conducted on the integration of electric vehicles with power grids. In [40], the effect of charging PHEVs on the distribution system has been evaluated. In this study, the optimal charging profile has been designed to minimize losses, and since precise prediction of residential loads is not possible, a statistical program is used to predict it. In [41], a new planning method for charging EVs, considering network, voltage and power constraints, while responding to the needs of individual consumers, has been proposed. The accuracy of the proposed method by simulating on an electric network has been confirmed. Another reference has investigated the effect of uncontrolled and controlled PHEV charging on the calculation of power losses and voltage deviations in three different scenarios. In this study, controlled charging has aimed to minimize losses, and dynamic programming techniques have been used to solve this optimization problem.

In the case of smaller networks, references [42, 43] have focused on aggregating EVs with a distributed transformer that supplies a small number of home energy, and has proposed new control methods to overcome the transformer overload problem. In [44], a method based on the response to the cost of electricity at the time of consumption has been proposed to control the charge of the EVs. Comparing the optimum charging results with uncontrolled charging indicates that controlled charging, in addition to minimizing charging costs, increases demand at off-peak times, so the power curve becomes more flattened.

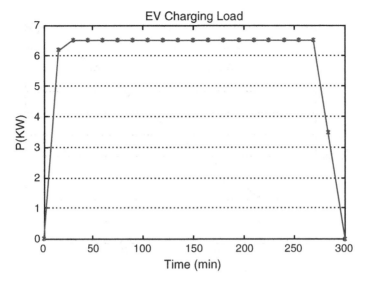

Fig. 15.4 Electric vehicle demand characteristics based on Li-Ion charging time (Nissan Altera) [45]

15.3.2 Electric Vehicle Charging Methods

Electric Vehicle Power Curve

In the figure below (Fig. 15.4), the electric vehicle demand is shown in terms of battery charge time [45]. As shown in the figure, the time required to charge the battery is 5 h. The Electric Vehicle demand is dependent on the starting point of charging and the initial SOC. The Eq. 15.1 defines the relation between charging point and the electric vehicle demand.

$$f(x) = \begin{cases} 0.413x, & 0 \leq x < 15 \\ 0.02x + 5.9, & 15 \leq x < 30 \\ 6.5, & 30 \leq x < 270 \\ -0.217x + 65 & 270 \leq x < 300 \end{cases} \quad (15.1)$$

Where, f(x) is electric vehicle demand (kW), and x shows charging point (minute). As it can be seen in the figure, EV charging characteristics include four different steps. At the first step, the electric vehicle demand sharply increases during 15 minutes, whereas this increase slow down in the next 15 minutes. The electric

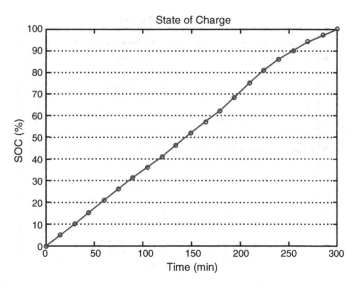

Fig. 15.5 Battery SOC based on charging time

vehicle demand remains constant 6.5 kW during next 4 h, and it decreases in the final 30 minutes.

Battery Charging Modelling

The Li-Ion battery SOC is shown in the figure below based on charging time. The relation between charging point and SOC defines by Eq. 15.2.

$$SOC = \begin{cases} 0.353x, & 0 \leq x < 255 \\ 0.22x + 34, & 255 \leq x \leq 300 \end{cases} \quad (15.2)$$

Where, SOC shows the battery state of charge (%), and x is charging time. As it can be seen in the Fig. 15.5, SOC has a linear relationship with x, so SOC increases from 0 to 100 by increasing x. Reversing Eq. 15.3 defines x based on SOC.

$$x = \begin{cases} 2.83 SOC, & 0 \leq SOC < 90 \\ 4.55 SOC - 154.55, & 90 \leq SOC \leq 100 \end{cases} \quad (15.3)$$

Equation 15.3 is the base of calculating x0 and SOC0, which show initial charging time and initial state of charge respectively.

The initial SOC is modelled by normal distribution function [46].

$$f(soc_0, \mu, \sigma) = \frac{1}{\sigma\sqrt{2\pi}} e^{-\frac{(soc_0 - \mu)^2}{2\sigma^2}} \quad (15.4)$$

15 Optimal Charge Scheduling of Electric Vehicles in Smart Homes 369

Where, μ and σ are average and standard deviation of initial SOC. It has been expressed in [47] that initial SOC has relation with the travelled distance by electric vehicle, which is explained in the next section.

Modelling the Behaviour of Electric Vehicle with Probability Distribution Functions

The behaviour of electric vehicle can be modelled by different distribution functions. The behaviour of electric vehicles' users should be analysed for two different travelled distance, arrival time and departure time. In [48], the electric vehicle behaviour has been studied for men and women separately. Modelling electric vehicle travelled distance can determine the amount of remained charge in battery when users receive to home. In addition, arrival time and departure time can determine when electric vehicle is at home, which can be used for charging time.

As it is expressed in the previous section, μ is average of initial state of charge which is related to the travelled distance of electric vehicle [47].

$$\mu = \frac{Range - AverageDailyDrivingRange}{Range} \qquad (15.5)$$

Where, Range specifies the travelled distance by electric vehicle which is fully charged, and it is equal to 140 mile [45]. Average daily driving range was equal to 32.72 in 2001, which is assumed 35 mile in this study. By replacing these values in (15.5), the value of μ is obtained at 75%. Therefore, the normal distribution SOC0 is determined with mean values of 0.75 and a standard deviation of 0.1.

The arrival time of electric vehicles is modelled by normal distribution function according to Eq. 15.6.

$$f\left(t_{\text{arrival}}, \mu, \sigma\right) = \frac{1}{\sigma\sqrt{2\pi}} e^{-\frac{\left(t_{\text{arrival}} - \mu\right)^2}{2\sigma^2}} \qquad (15.6)$$

Where t_{arrival} is the arrival time of electric vehicle to home. μ and σ for this distribution function are equal to 17 and 2.8 respectively. As a result, according to the number of electrical vehicles in the study system (1000 cars), the normal distribution of arrival time is modelled between 0 and 24. On the other hand, times between 0 and 10 are not acceptable due to the fact that they are not the usual time to get home, and if they exist, the data will be replaced. The arrival time for different 1000 electric vehicles is shown in Fig. 15.6.

The leaving time of electric vehicle is also modelled by normal distribution function as Eq. 15.7.

$$f\left(t_{\text{leaving}}, \mu, \sigma\right) = \frac{1}{\sigma\sqrt{2\pi}} e^{-\frac{\left(t_{\text{leaving}} - \mu\right)^2}{2\sigma^2}} \qquad (15.7)$$

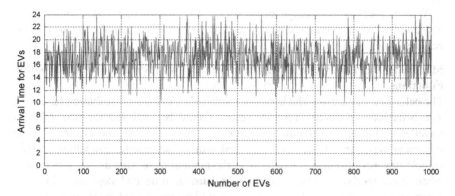

Fig. 15.6 Arrival time to home for different 1000 electric vehicles by normal distribution function with mean of 17 and standard deviation of 2.8

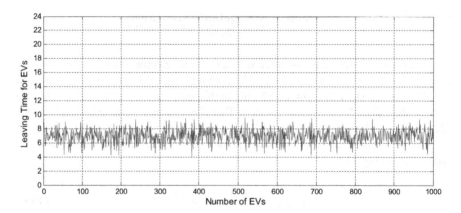

Fig. 15.7 Leaving time of different 1000 electric vehicles by normal distribution function with mean of 7 and standard deviation of 1

Where, $t_{leaving}$ determines the leaving time of electric vehicles. The mean and standard deviation of this distribution function are 7 and 1 respectively [46]. According to the number of electrical vehicles in the studied system (1000 cars), the normal distribution is applied to determine the time to reach the home, which is a number from 0 to 24. On the other hand, time greater than 10 or less than 4 are not acceptable due to the fact that they are not commonly used to leave the house and, if they exist, will be replaced at a later date. The leaving time of different 1000 electric vehicles is shown in Fig. 15.7.

Electric Vehicle Charging Model

The electric vehicle demand in charging time is defined as follows. It is related to charging time, battery charging point and initial soc0.

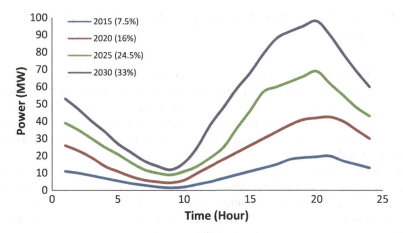

Fig. 15.8 Daily residential load growth by increasing the EV penetration

$$P(t, x, soc_0) = f(x - t)x_0 \leq x \leq 300 \tag{15.8}$$

Where, P shows the electric vehicle demand, x is the battery charging point. t is charging time and soc0 specifies initial state of charge. The charging time per day is in the range of 0–1440 minutes, 24 h a day. Therefore, the function f (x − t) is the same as shown in Fig. 15.1, as it corresponds to the time transmitted daily. This means that at any time of the day, if the car is connected to the charger, the above diagram, regardless of the connection time, requires 5 h of time for the full charge according to the curve displayed. The battery charge status will fit the curve during this time. x0, assuming soc0, is calculated from Eq. (15.3), where soc0 is obtained from Eq. (15.4).

15.3.3 Effect of Electric Vehicles on Smart Homes' Demand

The development of EVs has significant effect on demand in different places including workplace, shopping mall, home area and charging stations, which has been investigated in [49]. The EVs can be developed with high speed, which has an inevitable impact on increasing the demand. In this study, the role of EVs growth in the residential demand is analysed. It is an important issue to consider this effect comprehensively, and determine the amount of its impact by different penetration of EVs. Figure 15.8 shows the amount of increase in residential demand by increasing the EVs in the smart grids. As it can be seen in this figure, it will have an increasing trend by passage of time, which will be intensified in the future. As a result, the need to the accurate demand response program and energy management system can be confirmed through this rising trend.

15.3.4 Vehicle to Grid (V2G) and Grid to Vehicle (G2V) Effects

EVs can be applied in two different views base on their characteristics, so they can have two direction relation with network. They can be applied in two different modes vehicle to grid (V2G) and grid to vehicle (G2V). In G2V mode, the EVs are applied as a normal load which are charged through connection to the network. However, in the V2G mode, EVs can be used as an energy storage system and can be discharged to the network.

Air pollution in large urban areas, reliance on fossil fuels, climate change, and rising energy costs are all challenging in the current world. These significant issues have been raised by the transportation system and electricity generation sectors as major consumers of fossil fuels. In particular, in order to minimize reliance on conventional energy sources, many of the research activities have been done.

V2G technology is an emerging solution to these challenges. EVs act as an alternative to internal combustion engines, thus representing an attractive economic approach based on the mainstream of transport and manufacturing sectors. Recent studies have shown that EVs have a certain advantage over other conventional energy-saving technologies. They are easy to implement and easy to maintain, and they ensure environmental compatibility. As a result, with their higher efficiency, these EVs are likely to significantly increase market acceptance, especially in urban areas.

The V2G subject can improve the application of network in the efficiency, reliability and load flow. EVs can be applied as a load and energy storage system by the decision of energy management system. It can be applied especially in peak hours in order to inject electricity to the network. Network reliability, supply and demand balance, power transfer from source to buyer, all this can be maintained by bi-directional services. When the system provides better voltage regulation and frequency control, the grid can maintain peak power, manage load and do efficient rotation of reserves.

Challenges to the V2G system include battery degradation, total infrastructure changes, additional communication between electric vehicles and the supply network, the impact on distribution systems and its parameters, energy losses, and other technical barriers. The high number of overcharge cycles can reduce battery life and storage capacity. These barriers can be solved by using an economical and more efficient battery structure with an acceptable standard for operators and makers. Battery degradation depends on the amount of power drawn up, as well as the depth of discharge and the number of cycles it charges. Estimating the cost of battery degradation is difficult due to the fact that technology is constantly evolving. The only parameter that predicts battery life is its equivalent series resistance. Smart control minimizes the power of battery degradation to optimize time and demand.

In this study, the main aim is determination of demand of EVs on residential demand. It is important to assess the direct effect of EVs on residential demand in the present network. Considering all mentioned barriers, V2G structure needs comprehensive changes in the present infrastructure, so the G2V mode is considered in this study as the worst condition for EVs development in the residential areas.

15.4 Demand in Smart Homes

In order to demonstrate the efficiency of controlled charging on the residential demand and the benefits of intelligent energy management in the home for the development of electric vehicle, the results are examined under two scenarios. The profile of residential demand is shown in the following figure [23].

Demand in smart homes can be considered for two different seasons including hot and cold seasons, which have different peaks. The electricity demand increases in the hot seasons as it is demonstrated in the Fig. 15.9. In this study, the residential demand in hot season is considered.

In both scenarios, 1000 homes with the same load profile are considered, with the difference that in the first scenario the typical homes and in the second scenario are smart homes equipped with smart energy management system. The basis for this energy management system, as described in the second scenario, is based on minimizing the cost of domestic electricity.

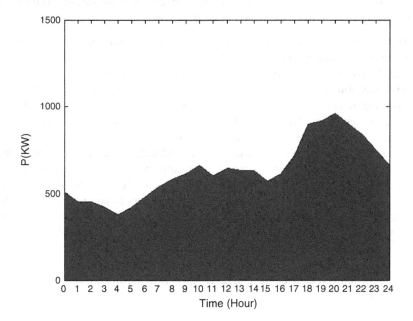

Fig. 15.9 Demand in smart home

In order to study the effect of the development of electric vehicles in these two scenarios, electric vehicles with the penetration level of 10%, 20% and 30% are added to the domestic system by 100, 200 and 300 cars.

The comparison index in this study is Peak demand and the standard deviation of the demand values in 24 h. Peak demand is the maximum value of demand in 24 h and demand's standard deviation, which determines the standard deviation of the data from the demand average in 24 h.

15.5 Charging Management

For hybrid vehicles, nightly battery charging is important at home, while electric vehicles should be able to be charged throughout the day and the ability to charge away from home is vital. The effect of a large number of electric vehicles on the total energy demand and demand in specific hours as well as the production and capacity of the lines should be well assessed. In this context, the ratio of charge per night to the day has an important role, where electricity price policy can also be effective in this regard.

The smart grid can play a key role in this area. All new charging methods (such as electric vehicle to grid, daily and nightly charging rates, charging in peak hours, etc.) are only possible with smart grids.

In this section, two different scenarios for charging an electric vehicle are considered, which will be explained in detail below:

1. Instant charging: When the infrastructure required to run the demand response issue is not provided, the consumer is reluctant to change the charging time of his car. As a result, the consumer charges his car as he arrives home from work. In the first scenario, electric vehicles are charged when they arrive at home, and charging time is not chosen based on energy management algorithm. In this case, the charging start time is based only on when electric vehicles received homes, so there is no optimal planning on it.
2. Charging with cost management: A cheap and simple way to postpone battery charging is to use a timer to connect your car to network while the consumer is charging, but starting a charging process with a time delay set by the timer. As a result, the charging process changes from peak hours to non-peak hours. In the second scenario, the time of charging an electric vehicle is selected based on the electricity price, so that household electricity costs are reduced to the minimum. In this case, it may not be necessary to start charging the car immediately after reaching the house.

The electric vehicle demand of all added cars to the smart grid is calculated as follows.

15 Optimal Charge Scheduling of Electric Vehicles in Smart Homes 375

$$P_{total}(t, x, soc_0) = \sum_{i=1}^{k} P_i(t_i, x_i, soc_{0i}) \tag{15.9}$$

Where, k is the number of electric vehicles, and i shows the ith electric vehicle. P_{total} specifies the total demand of electric vehicles.

Scenario 1 – Uncontrolled charging: In this case, there is no incentive to determine the time of charging an electric car; therefore, electric vehicles will be charged when they arrive at home. Arrival time of electric vehicles is determined by Eq. 15.6. In an uncontrolled vehicle charging scenario, once the vehicle is taken to the house, the vehicle's charging process will begin immediately according to the remaining battery charge, as shown in the Nissan Altra battery charge characteristics (Lithium-Ion). The results of this process are discussed in the next section.

Scenario 2 – Controlled charging: In this scenario, unlike the previous scenario, which do not have a method for determining the time of vehicle charging when it came to home, the main criterion for determining the charge time of an electric vehicle is the time of use (TOU). This criterion, based on off-peak time, mid-peak and on-peak, determines different prices for the consumer. Accordingly, the price for these three periods in Iran is 150, 300 and 600 Rials per kilowatt-hour (this price is based on the first stage of daily consumption, i.e., between 0 and 100 kWh) [22]. In the first half of the year, short, middle and peak hours are defined as 23–7, 7–19, and 19–23, respectively; Similarly, in the second half of the year, the off-peak hours are in the range of 21 up to 5, mid-hours are in the range of 5–17, and peak hours are in the range of 17–21 [22].

In the smart grid and in a smart home, the energy management program is set to minimize the cost of electricity consumed by the home, and this is important in electric vehicles by determining the time of electric vehicle charging. Therefore, in Eq. (, time should be chosen in order to minimize the cost.

$$\min \left(\sum_{t_0}^{t_1} P(t)M(t) \right) t_0 \le t \le t_1 \tag{15.10}$$

Where, P(t) is the electricity cost, P(t) shows the electric vehicle demand. t_0 and t_1 are initial charging time and charging timeout respectively.

15.6 Comparison Between Controlled and Uncontrolled Charging Effect on Demand Curve and Consumption Cost

In order to test the efficiency of proposed method, it is tested on the explained household demand. The case study includes different smart homes which have the various penetration level of electric vehicles. In order to study the effect of the development of electric vehicles in these two scenarios, electric vehicles with a value

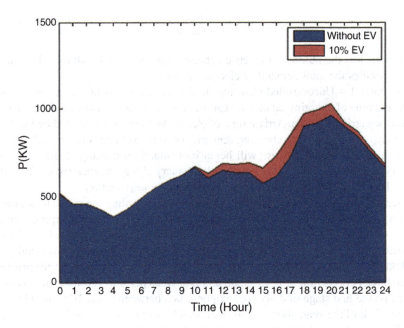

Fig. 15.10 Comparison of the effect of increasing the electric vehicle's penetration on household demand in an uncontrolled charging – penetration level = 10%

of 10%, 20% and 30% are added to the smart homes by 100, 200 and 300 cars. As it is explained in the previous section, the chosen comparison indices are peak demand and standard deviation of demand.

Scenario 1 – Uncontrolled charging

In the development of electric vehicles in the first scenario, the results of the development of 10%, 20% and 30% of the electric vehicles are shown in the following figures. The peak demand for this scenario are 1027.8, 1120.1 and 1217.8 kW, respectively, and the standard deviation of data are 190.14, 220.28 and 252.52 kW, respectively. As it is clear from the following figures, with the increase in electric car development, peak values and standard deviation increase, and this has a significant negative effect on the development of an electric vehicle in the absence of an electric energy management system.

As it is shown in Figs. 15.10, 15.11, and 15.12, the most significant effect on residential demand in the case of electric vehicles development is between 17 to 20 h. As these are typical time for reaching home, the charging process will be started in these hours. Although the peak time of each penetration level of electric vehicles is different, the rising trend can be seen by increasing the penetration level.

It is additionally can be seen in Figs. 15.10, 15.11, and 15.12 that the dispersion of data increases by raising the penetration level of electric vehicles. It is one of the system objective to decrease the standard deviation between demand values and

15 Optimal Charge Scheduling of Electric Vehicles in Smart Homes 377

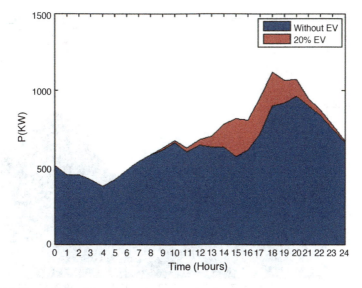

Fig. 15.11 Comparison of the effect of increasing the electric vehicle's penetration on household demand in an uncontrolled charging – penetration level = 20%

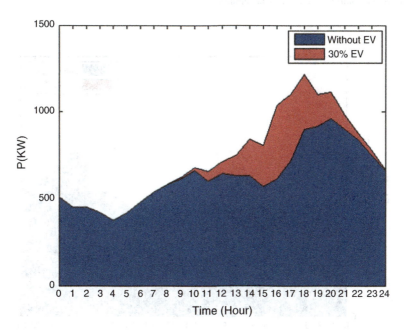

Fig. 15.12 Comparison of the effect of increasing the electric vehicle's penetration on household demand in an uncontrolled charging – penetration level = 30%

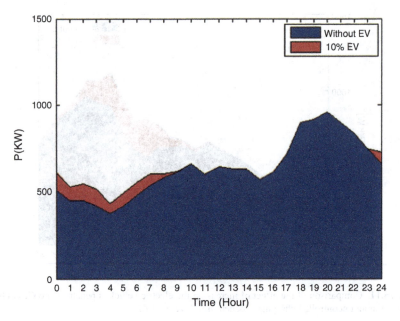

Fig. 15.13 Comparison of the effect of increasing the electric vehicle's penetration on household demand in a controlled charging – penetration level = 10%

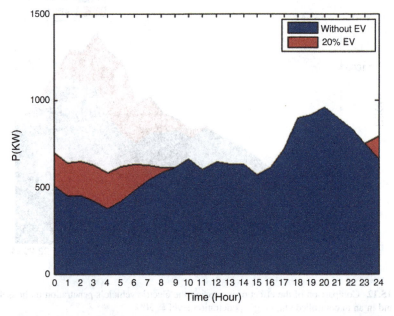

Fig. 15.14 Comparison of the effect of increasing the electric vehicle's penetration on household demand in a controlled charging – penetration level = 20%

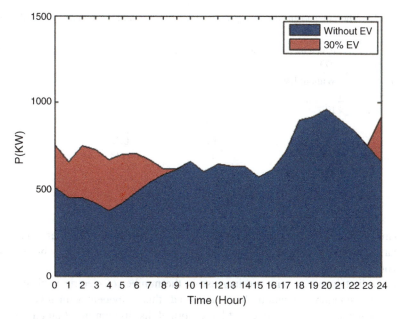

Fig. 15.15 Comparison of the effect of increasing the electric vehicle's penetration on household demand in a controlled charging – penetration level = 30%

flatten the demand pattern, which is not addressed by uncontrolled development of electric vehicles.

Scenario 2 – Controlled charging
The demand of smart houses is shown in the following figures (Figs. 15.13, 15.14, and 15.15) after the development of an electric vehicle with penetration level of 10%, 20% and 30%. Demand peak values for the three development factors are 960 kW; the standard deviation in these three modes is 139.42, 125.74, and 122.76 kW. As the values show, the peak value of these three scenarios is constant for the three cases, and the standard deviation values decrease with increasing development coefficient. The results indicate the superiority of the smart home and the existence of an electrical energy management system.

15.7 Conclusion

Development of electric vehicles in the smart grid needs different trend in the management of smart homes. The penetration level of electric vehicles increases considering both controlled and uncontrolled charging view in this study. The peak values and the standard deviation obtained from the two scenarios are summarized in the following Table 15.1. The values of the peak and the standard deviation of the

Table 15.1 The comparison between peak demand and standard deviation of demand in two scenarios

Scenario	Penetration level of EV (%)	Peak demand (MW)	Standard deviation of demand (MW)
Without scenario	Without EV	0.96	0.17
Scenario 1	10	1.027	0.19
	20	1.12	0.22
	30	1.218	0.25
Scenario 2	10	0.96	0.14
	20	0.96	0.13
	30	0.96	0.12

demand. In this study, the comparison criterion is chosen for the operation of the scenarios, which means that the lower of these values means more stability of the network.

In the first scenario, with the increase in the number of electric vehicles, peak values and standard deviations have increased, this is evident with regard to the vehicle's arrival time to a home that has a normal distribution of 17 and a standard deviation of 2.8. Therefore, increasing the number of electric vehicles without an electrical energy management system increases the possibility of system instability.

In the second scenario, with the increase in the number of electric vehicles, peak demand has been fixed and, on the other hand, the standard deviation is decreasing. The reason for this is the choice of TOU as the basis for determining the charging time and aiming to reduce household electricity consumption; given that electricity prices are minimum at off-peak hours, consumers are encouraged to charge their cars in these hours, and this reduces the standard deviation and makes the system more stable.

Acknowledgments Authors would like to thank the research council of Islamic Azad University, Damavand, Iran for financial support of this research project.

References

1. F. Rassaei, W.-S. Soh, K.-C. Chua, Distributed scalable autonomous market-based demand response via residential plug-in electric vehicles in smart grids. IEEE Trans. Smart Grid **9**(4), 3281–3290 (2018)
2. Q. Chen, F. Wang, B.-M. Hodge, J. Zhang, Z. Li, M. Shafie-Khah, J.P.S. Cataloe, Dynamic price vector formation model-based automatic demand response strategy for PV-assisted EV charging stations. IEEE Trans. Smart Grid **8**(6), 2903–2915 (2017)
3. L. Yao, W.H. Lim, T.S. Tsai, A real-time charging scheme for demand response in electric vehicle parking station. IEEE Trans. Smart Grid **8**(1), 52–62 (2017)
4. S. Pal, R. Kumar, Electric vehicle scheduling strategy in residential demand response programs with neighbor connection. IEEE Trans. Industr. Inform. **14**(3), 980–988 (2018)

5. O. Hafez, K. Bhattacharya, Integrating EV charging stations as smart loads for demand response provisions in distribution systems. IEEE Trans. Smart Grid **9**(2), 1096–1106 (2018)
6. J. Earl, M.J. Fell, Electric vehicle manufacturers' perceptions of the market potential for demand-side flexibility using electric vehicles in the United Kingdom. Energy Policy **129**, 646–652 (2019)
7. H. Zhao, X. Yan, H. Ren, Quantifying flexibility of residential electric vehicle charging loads using non-intrusive load extracting algorithm in demand response. Sustain. Cities Soc. **50**, 101664 (2019)
8. W. Yang, Y. Xiang, J. Liu, C. Gu, Agent-based modeling for scale evolution of plug-in electric vehicles and charging demand. IEEE Trans. Power Syst. **33**(2), 1915–1925 (2018)
9. K. chaudhari, N.k. kandasmy, A. krishnan, A. ukil, H.b. gooi, Agent-based aggregated behavior modeling for electric vehicle charging load. IEEE Trans. Industr. Inform. **15**(2), 856–868 (2019)
10. H. Lin, K. Fu, Y. Liu, Q. Sun, R. Wennersten, Modeling charging demand of electric vehicles in multi-locations using agent-based method. Energy Procedia **152**, 599–605 (2018)
11. J. Su, T. Lie, R. Zamora, Modelling of large-scale electric vehicles charging demand: A New Zealand case study. Electr. Power Syst. Res. **167**, 171–182 (2019)
12. Y. Lee, J. Hur, A simultaneous approach implementing wind-powered electric vehicle charging stations for charging demand dispersion. Renew. Energy **144**, 172–179 (2019)
13. S. zhang, m. chen, w. zhang, A novel location-routing problem in electric vehicle transportation with stochastic demands. J. Clean. Prod. **221**, 567–581 (2019)
14. M.B. Arias, S. Bae, Electric vehicle charging demand forecasting model based on big data technologies. Appl. Energy **183**, 327–339 (2016)
15. G. Berkelmans, W. Berkelmans, N. Piersma, R. van der Mei, E. Dugundji, Predicting electric vehicle charging demand using mixed generalized extreme value models with panel effects. Procedia Comput. Sci. **130**, 549–556 (2018)
16. H. moon, s.Y. park, c. jeong, j. lee, Forecasting electricity demand of electric vehicles by analyzing consumers' charging patterns. Transport. Res. D Transp. Environ. **62**, 64–79 (2018)
17. R. Iacobucci, B. Mclellan, T. Tezuka, Modeling shared autonomous electric vehicles: Potential for transport and power grid integration. Energy **158**, 148–163 (2018)
18. H. Jahangir, H. Tayarani, A. ahmadian, M.A. Golkar, J. Miret, M. Tayarani, H.O. Gao, Charging demand of plug-in electric vehicles: Forecasting travel behavior based on a novel rough artificial neural network approach. J. Clean. Prod. **229**, 1029–1044 (2019)
19. D. Hu, J. Zhang, Q. Zhang, Optimization design of electric vehicle charging stations based on the forecasting data with service balance consideration. Appl. Soft Comput. **75**, 215–226 (2019)
20. M. Pertl, F. Carducci, M. Tabone, M. Marinelli, S. Kiliccote, An equivalent time-variant storage model to harness EV flexibility: Forecast and aggregation. IEEE Trans. Industr. Inform. **15**(4), 1899–1910 (2019)
21. M. Shariful, N. Mithulananthan, D.Q. Hung, A day-ahead forecasting model for probabilistic EV charging loads at business premises. IEEE Trans. Sustain. Energy **9**(2), 741–753 (2018)
22. A. Chakrabarti, R. Proeglhoef, G.B. Turu, R. Lambert, A. Mariaud, S. Acha, C.N. Markides, N. Shah, Optimisation and analysis of system integration between electric vehicles and UK decentralised energy schemes. Energy **176**, 805–815 (2019)
23. D.V.D. meer, G.R.C. mouli, g.M.-E. mouli, Energy management system with PV power forecast to optimally charge EVs at the workplace. IEEE Trans. Industr. Inform. **14**(1), 311–320 (2018)
24. M.C. Kiasacikoglu, F. Erden, N. Erdogan, Distributed control of PEV charging based on energy demand forecast. IEEE Trans. Industr. Inform. **14**(1), 332–341 (2018)
25. H. Kikusato, K. Mori, S. Yoshizawa, Y. Fujimoto, H.A.Y. Ha, Electric vehicle charge–discharge management for utilization of photovoltaic by coordination between home and grid energy management systems. IEEE Trans. Smart Grid **10**(3), 3186–3197 (2019)
26. H.K. Nunna, S. Battula, S. Doolla, D. Srinivasan, Energy management in smart distribution systems with vehicle-to-grid integrated microgrids. IEEE Trans. Smart Grid **9**(5), 4004–4016 (2018)

27. F. Wu, R. Siochansi, A stochastic operational model for controlling electric vehicle charging to provide frequency regulation. Transp. Res. D Transp. Environ. **67**, 475–490 (2019)
28. K. Seddig, P. Jochem, W. Fichtner, Two-stage stochastic optimization for cost-minimal charging of electric vehicles at public charging stations with photovoltaics. Appl. Energy **242**, 769–781 (2019)
29. R. Lacobucci, B. Mclellan, T. Tezuka, Optimization of shared autonomous electric vehicles operations with charge scheduling and vehicle-to-grid. Transp. Res. C Emerg. Technol. **100**, 34–52 (2019)
30. S. Xu, Z. Yan, D. Feng, Z. Zhao, Decentralized charging control strategy of the electric vehicle aggregator based on augmented Lagrangian method. Int. J. Electr. Power Energy Syst. **104**, 673–679 (2019)
31. L. Wang, B. Chen, Distributed control for large-scale plug-in electric vehicle charging with a consensus algorithm. Int. J. Electr. Power Energy Syst. **109**, 369–383 (2019)
32. E. Pournaras, S. Jung, S. Yadhunathan, H. Zhang, Z. Fang, Socio-technical smart grid optimization via decentralized charge control of electric vehicles. Appl. Soft Comput. **82**, 105573 (2019)
33. F. E. R. C. Staff, Assessment of demand response and advanced metering, in *Federal Energy Regulatory Commission*, Aug 2006 to 2009
34. U. S. D. O. Energy, Energy Policy Act of 2005, U. S. Department of Energy, USA (Feb 2006)
35. Z. Wang, R. Paranjape, An evaluation of electric vehicle penetration under demand response in a multi-agent based simulation, in *2014 IEEE Electrical Power and Energy Conference*, (2014), pp. 220–225
36. K. Qian, C. Zhou, M. Allan, Y. Yuan, Modeling of load demand due to EV battery charging TOU price and SOC curve. IEEE Trans. Smart Grid **26**(2), 802–810 (2014)
37. A. Ovalle, A. Hably, S. Bacha, Optimal management and integration of electric vehicles to the grid: Dynamic programming and game theory approach, in *2015 IEEE International Conference on Industrial Technology (ICIT)*, (2015), pp. 2673–2679
38. R.P.Z. Wang, An evaluation of electric vehicle penetration under demand response in a multi-agent based simulation, in *2014 IEEE Electrical Power and Energy Conference*, (2014), pp. 220–225
39. M.A. López, S. de la Torre, S. Martin, J.A. Aguado, Demand-side management in smart grid operation considering electric vehicles load shifting and vehicle-to-grid support. Int. J. Electr. Power Energy Syst. **64**, 689–698 (2015)
40. K. Clement-Nyns, E. Haesen, J. Driesen, The impact of charging plug-in hybrid electric vehicles on a residential distribution grid. IEEE Trans. Power Syst. **25**(1), 371–380 (2016)
41. O. Sundstorm, C. Binding, Flexible charging optimization for electric vehicles considering distribution grid constraints. IEEE Trans. Smart Grid **3**(1), 26–37 (2012)
42. S. Shao, M. Pipattanasomporn, S. Rahman, Challenges of PHEV penetration to the residential distribution network, in *2009 IEEE PES Power & Energy Society General Meeting*, (2009), pp. 1–8
43. S. Shao, M. Pipattanasomporn, S. Rahman, Demand response as a load shaping tool in an intelligent grid with electric vehicles. IEEE Trans. Smart Grid **2**(4), 624–631 (2015)
44. Y. Cao, S. Tang, C. Li, P. Zhang, Y. Tan, Z. Zhang, J. Li, An optimized EV charging model considering TOU price and SOC curve. IEEE Trans. Smart Grid **3**(1), 388–393 (2014)
45. C. Madrid, J. Argueta, J. Smith, Performance characterization 1999 Nissan Altra-EV with lithium-ion battery, *Southern California EDISON*, 1999
46. Z. Wang, R. Paranjape, An evaluation of electric vehicle penetration under demand response in a multi-agent based simulation, in *Electrical Power and Energy Conference (EPEC)*, (2014), pp. 220–225
47. J. Argueta, A. Mendoza, Performance characterization-GM EV1 Panasonic lead acid battery, *Southern California EDISON*, 2000

48. S. Khemakhem, M. Rekik, L. Krichen, A flexible control strategy of plug-in electric vehicles operating in seven modes for smoothing load power curves in smart grid. Energy **118**, 197–208 (2017)
49. T.-H. Chen, R.-N. Liao, Analysis of charging demand of electric vehicles in residential area, in *International Conference on Remote Sensing, Environment and Transportation Engineering (RSETE 2013)*, (2013)

Index

A

Aggregator, 194, 236
  cost, 238
  distribution system, 236
  EVs parking, 234, 235
  operation cost, 244, 245
  V2G price, 236
Air pollution, 1
  all-electric vehicles, 6
  greenhouse gas emissions, 2
All-electric vehicles, 6
Artificial intelligence-based approaches, 22
  ANNs (*see* Artificial neural networks
    (ANNs))
  deep learning, 22
  Markov chain theory, 23
  MCS, 22
  Monte Carlo simulation, 22
  queuing theory, 23
  trip chain and origin-destination, 24
Artificial neural networks (ANNs)
  communication service, 26
  conventional ANN
    error back propagation learning
      method, 27, 28, 30
    LM method, 30, 31
  conventional forms, 25
  error criteria, 36
  EVs travel behavior, 24, 25
  forecasting results, 37, 39
  GPS, 24
  in handling large dimension tasks, 25
  overall configuration, 26
  with rough neurons
    and back propagation learning
      approach, 32, 33
    description, 31
    Levenberg–Marquardt training
      procedure, 34
  training procedure, 25

B

Batteries, 133
Battery Electric Vehicles (BEVs), 189
Battery swapping station (BSS), 249, 264
  advantages, 250
  bi-level model, 250
  dynamic operation strategy, 250
  microgrid, 250
  proposed method, 251
  PSI criteria, 251
  scenario-based approach, 252
BigM method, 278
Bi-level model, 250, 252, 257, 266
  with charging/discharging schedule, 81,
    82, 84
  with controlled charging, 77, 79, 80
  problem solving method, 86
  SDNO and PL owners, 78
  to single-level model, 117–121, 123
  structure, 77
Bi-level programming approach
  aggregator
    operation cost, 244, 245
    vs. EV parking, 235
    vs. parking, 243
  distribution network, 240–242
  EV parking, 235, 236, 239, 241

© Springer Nature Switzerland AG 2020
A. Ahmadian et al. (eds.), *Electric Vehicles in Energy Systems*,
https://doi.org/10.1007/978-3-030-34448-1

386   Index

Bi-level programming approach (*cont.*)
  load profile, 243
  mathematical model
    cost, aggregator, 238
    EVs parking profit, 236, 237
    single-level equivalent problem, 239
  parking-bidding price, 243
  parking profit, 245
  scenarios, 240, 241
  SOC, 243, 244
  Stackelberg leadership model, 236
  wholesale market prices, 241, 242
BSS to grid (B2G) mode, 256

C
Car efficiency, 2
Centralized control methods, 184
Charge bidding strategy, 341
Chargers, 191
Charging infrastructure, 4, 5
Charging management system (CMS), 352, 354
Charging management, EVs
  charging costs, 199
  cooperative control, 195
  emergency charging, 199
  EVSEs, 195
  fair charging/discharging rate, 196
  frequency stability, 197
  minimum SoC, 198
  power flow fluctuations, PCC, 197
  prevention, overcurrents, 198
  renewable-based DGs, 184
  SMG dependency, main grid, 197
  system ride-through-ability, 199
  voltage stability, 198
Charging power method, 353
Charging rates, EVs
  level 1 (about 2kW), 7, 8
  level 2 (about 7 kW), 7, 8
  level 3 (about 50 kW), 7, 8
Charging, EVs
  advantages and disadvantages, 10
  implementation, smart charging, 9
  purposes, optimal charging, 9
Combined heat and power (CHP), 269, 290
Conditional value at risk (CVaR), 251
Control architectures, 185, 187
Control frameworks, 184
Control requirements, 200
Controlled charging, 360, 366, 373, 375, 379
Conventional generation (CG), 150, 151, 153,
    156, 157, 159, 161

Conventional power grid, 130
Conventional vehicle, 233
Cooperative control, 189, 195–197, 200,
    201, 206
Cost function linearization, 167
Critical peak pricing (CPP), 331

D
Decentralized control methods, 185, 186
Decision-making model, bi-level programming,
    see Bi-level programming approach
Deep learning, 22, 37
Demand response (DR)
  application timescale, 363
  benefits
    additional, 363
    customer, 361
    network, 363
  categories, 362
  consumption change, 281, 282
  definition, 361
  DISCO, 270
  expected demand level, 282, 283
  network, 363–364
  probabilistic distribution network, 272
  renewable energy expansion, 272
  residential demand, 362
  risk strategy and level, 281
Demand response aggregator (DRA), 271
Demand response programs (DRPs), 213–216,
    222, 224, 291
  advantages, 327, 330
  categories, 329
  customer response, 331–332
  dynamic prices/incentive payments, 330
  electrical power systems, 330
  environmental concerns, 353
  EVs
    charging parking lots, 334
    charging residential area, 335
    discharging period, 333
    HAN strategy, 336
    implementation, 333
    NAN strategy, 335, 336
    smart charging, 332
    usage, 332
  implementation, 329
  incentive-based, 331, 342
  modeling, 337–338
  price-based
    CPP, 340
    RTP, 341–342

TOU, 338–340
type, 331
smart charging, 328, 354
V2G and G2V, 328
Demand side management (DSM), 333
Department of Energy's (DOE's), 192
Diesel generators (DGs), 215–217, 220–222, 224, 226, 228
Direct charging, 9, 11
Discharging management, 185, 186, 199, 200, 203, 204
Discrete temporal method, 297
Distributed controls, 185, 186, 189, 195, 199
Distributed energy resources (DER), 289
Distribution network (DN), 183
  characteristics, 214
  charge/discharge, EVs, 226
  demand, 222
  DGs data, 221
  DR program, energy management, 228
  DRAs, 224
  DRPs, 215
  energy exchange, UG, 226, 227
  energy management model, 215
  EVs aggregator, 222, 224
  EVs battery, 214
  formulation
    DGs, 216, 217
    DR programs, 219, 220
    EVs, 218, 219
    EVs aggregator, 215, 216
    power balance, 220
    PV, 216
    spinning reserve, 220
    UG, 217
    WTs, 216
  fuzzy method, 214
  LILs, 224
  multi-objective approach, 214
  objective function, 225
  operation costs, 228
  parameters, EVs, 222, 223
  PHEVs aggregator, 214
  power output, DGs, 226, 227
  PV output power, 224, 225
  radiation, solar, 223
  stochastic planning model, 214
  UG price, 222
  wind speed, 223
  WTs and PV data, 221
  WTs output power, 224, 225
DR programs, 219, 220
Dynamic operation strategy, 250

E
Economic approach, 372
Electric cars, 189
Electric engine, 134
Electric generators, 134
Electric mobility, 47
Electric motors, 189
Electric transportation fleet, 1, 2, 4, 5, 12, 13
Electric vehicles (EVs)
  advantages, 135
  aggregator, 21
  all-electric vehicles, 6
  batteries, 75
  battery capacity, 137, 138
  bi-level programming approach (*see* Bi-level programming approach)
  BSS (*see* Battery swapping station (BSS))
  charging and discharge limits, 171
  charging/discharging efficiency, 94
  control methods, 183, 184
  coordinated charge and discharge, 166
  cost, 4, 5
  as distributed ESSs, 195
  DN (*see* Distribution network (DN))
  DRP (*see* Demand response programs (DRPs))
  energy consumption, 2
  facilities, 4
  HEVs, 6, 135
  incentives, 4
  modeling, 75, 171
  penetration level, 166
  PHEVs, 7, 136
  and power system data, 36
  required time to fully charge, 140
  RERs, 73, 74
  revenue/cost, 171
  simulation results, 170, 172
  specifications, 7
  stochastic generation, WTs on EVs, 173, 175–178
  system without EVs and solar system, 94, 97
  technical specifications, battery, 140
  travel behavior, 21
  uncontrolled charging, 73
  unmanaged connection, 183
  vehicle fuel consumption, 2
  with/without the solar system
    charging/discharging, 104–111
    controlled charging, 98–105
Electric vehicles battery replacement stations (EVBRSs)

Electric vehicles battery replacement stations (EVBRSs) (*cont.*)
constraints, 318
consumer behavior, uncertainty, 314
demand, 321, 322
demand response and energy storage, 314
demand variation range, 322
driver behavior and load effects, 314
exchanged power, 321
generation of PV, 321
market price, 320, 322
objective function, 318
PV system, 316
renewable energy, 314
station performance, 315
station structure, 315
transportation industry, 313
uncertainties, 316
Electrical power systems, 330
Electricity consumers, 213
Energy consumption, 2
Energy dependence, 190
Energy dispatch strategy, 290
Energy efficiency, 189
Energy hubs
CHP and ESS technologies, 291
CHP implementation, 305
demand response programs, 291
energy consumption behavior, 304
energy dispatch strategy, 290
energy storage technologies, 297
GHG emissions, 306–307
inputs, 298
model, 295–296
Monte Carlo simulation, 298
multi-objective approach, 294
operating and optimization strategies, 290
operating costs analysis, 305–306
optimization, 289, 300, 301
plug-in electric vehicles, 292–294
residential, 301–302
simulation scenarios, 301
transmission and distribution infrastructure, 308
Energy management method, 360
Energy management system (EMS), 75
Energy storage system (ESS), 47, 194, 195, 250, 251, 290
dedicated simulation models, 59
EV-BCS, 56
off-board EV-BCS, 61
power converter to interface, 59
and RES, 57

Environmental pollution, 213
Environmentally friendly, 190
EV batteries
chargers, 191
charging and auxiliary applications, 190
technical features, 190
EV battery charging systems (EV BCS)
and advanced operation modes, 49
H2V mode, 51
off-board, 48
on-board, 48
power limits, 48
EV charging, 193
EV-charging management, 189
EV parking, 234, 235
EV Supply Equipment (EVSE), 191, 195–201, 203

F
Fossil fuels, 233
consumption, 2
passenger transportation, 3
transportation fleet, 2
Fuel cells (FCs), 269
Fuel consumption, 2

G
Gasoline engine, 134
Global warming, 233
Greenhouse gas (GHG), 289, 292, 300, 302
Grid security, 130
Grid to BSS (G2B) mode, 256
Grid to vehicle (G2V), 269, 372

H
Home area network (HAN), 335, 354
Home energy management system, 143
Home general controller (HGC), 146–148
Hybrid electric vehicles (HEVs), 6, 135, 365
Hybrid vehicle, 133
batteries, 134
electric engine, 134
electric generators, 134
gasoline engine, 134
performance, 135
vs. pure-electric vehicles, 134

I
IEEE 33-bus distribution system, 93–95, 114

Index
389

Incentives for EVs, 1, 4, 16
Indirect charging, 11
Information-gap decision theory (IGDT), 314, 316, 317, 323
Integrated dynamic method, 360
Intelligent grid, 142
Intelligent Residential Energy Management System (IREMS), 143
Internal combustion, 133
International energy agency (IEA), 327
Interruptible Loads (IL), 329
Iterative approach, 290

K
Karush–Kuhn–Tucker (KKT), 239, 270

L
Lagrangian method, 276, 361
Latin hypercube sampling (LHS), 253
Levenberg-Marquardt (LM) method, 30, 31
Linear equivalent, 167
Linear power flow, 115, 116

M
Markov chain theory, 23
Matrix transposition, 202
Microgrid, 15
    bi-level model, 266
    BSS operation, 256–257
    characteristic of MTs, 262
    characteristics of wind turbine, 262
    EV model, 255
    formulation
        lower-level, 259–260
        real-time pricing mechanism, 260, 261
        upper-level, 257–259
    load demand, 255
    power generation, 263
    and profit, 265
    PV power, 254–255
    small/medium-scale distribution network, 250
    and upstream network, 264
    wind power, 252–254
Minimum up and down time constraints, 169
Mixed integer linear programming (MILP) problem, 93, 290, 295
Mixed integer non-linear programming (MINLP) model, 290
Mixed integer quadratic programming, 294

Monte Carlo method, 298, 314
Monte Carlo simulation (MCS), 22–25, 37–39, 130, 142, 152, 162, 253, 345
Multilayer perceptron (MLP) network, 27
Multi-objective approach, 294

N
Neighbourhood area network (NAN), 335, 354
Non-linear bi-level model, 86
Nonlinear expression, 167
Non-polluting vehicles, 6
Non-prediction-based control methods, 183, 184
Normal charge, 143
Novel approach, 272

O
Objective function, 225
Off board EV BCS
    and advanced operation modes, 49
    dc-dc converter, 56
    G2V/V2G/H2V/V4G modes, 55
    power quality functionalities, 56
    smart home, 55, 56
    smart homes and grids, 59
    three-phase, 53
On board EV BCS
    analysis, 52
    benefits, 53
    G2V/V2G/H2V modes, 52
    and off-board, 55
    possibility, 53
    smart home, 48, 49, 51, 54
    V4G mode, 53
Operating and optimization strategies, 290
Operational scheduling, SDN
    bi-level model
        with charging/discharging schedule, 81, 82, 84
        with controlled charging, 77, 79, 80
        problem solving method, 86
    risk management
        bi-level model, 91
        single-level model, 92
        strategies, 90
    single-level model
        with charging/discharging schedule, 88–90
        with controlled charging, 87
Optimal charging
    approaches, 22

Optimal charging (*cont.*)
communication services, 26
equations, 34
grey-Markov chain theory, 23
Optimal charging equations, 34
Optimal operation, 322
Optimization
DR programs, 343
parking lots, 350–353
SDSs
constraints, 347–348
DGs uncertainties, 344
EVs uncertainty, 345–346
objective function, 347
problem solving, 348
renewable energy sources, 344

P
Parking-bidding price, 243
Parking lot, 234
electricity market uncertainty, 350
EV uncertainty, 350
objective function, 351
optimal charging scheduling, 351
Particle swarm optimization, 162
Peak load shifting strategy, 335
Performance benefits, 189
PHEVs aggregator, 214
Photovoltaic cells (PVs), 269
Photovoltaic system (PV), 214, 216, 221,
222, 224, 229
Piecewise method, 275
Plug-in electric vehicles (PEVs), 6, 213, 214
battery, 274
bigM method, 278
charge station, 269, 270, 272, 273,
275, 279, 284
CVaR, 279
distribution network planning, 270, 272
DR (*see* Demand response (DR))
Lagrangian method, 276
piecewise method, 275
proposed method, 278
risk assessment methods, 277
risk strategy, 279
two-step screening method, 271
upstream network, 283
usage, 269
virtual electricity-trading process, 271
Plug-in hybrid electric vehicle (PHEV), 365
availability, 142
batteries, 7, 130

battery capacity, 137
battery charging, 139
battery charging rate, 141
charge specification, 144, 145
charging management, 130
charging program, 152
consumption and battery capacity, 137
coordinated V2G mode, 149
cost, PHEV batteries, 136
description, 7, 136
energy demand, 140
energy resources to drive, 132
feature, 7
as flexible loads, 152
vs. HEVs, 136
managed charge, 146
performance, on power grid, 136
and photovoltaic, 143
random charge, 145, 146
renewable resources, 156
specification, 144
stored energy, batteries, 138
Power balance constraint, 169
Power line carriers (PLCs), 183
Power quality, 51–57, 65
Prediction-based methods, 183, 184
Probability distribution functions (PDF), 345
Probability of successful islanding (PSI), 251
Problem constraints
minimum up and down time, 169
power balance, 169
ramp-up and ramp-down, 168
start-up and shut-down cost, 169
Proposed method, 251, 278, 366, 375

Q
Queuing theory, 23, 24

R
Ramp rate constraints, 168
Real-time pricing (RTP), 271, 331
Real-time pricing mechanism, 260, 261
Renewable energy resources (RERs), 251, 252,
290, 347
Renewable energy sources (RES)
dc-dc converter, 56
electrical appliances, 61, 62
EV optimization, 50
future off-board EV-BCS, 60
G2V/V2G flexibility, 50
Renewable resources

Index

output power, solar panels, 132
PHEVs charging, 150
residential smart grid, 131
solar energy, 131
Rough neuron, 31–34

S

Scenario-based approach, 252, 345, 350
Single-level model
  bi-level model, convertion
    charging/discharging schedule, 120,
     121, 123
    controlled charging, 117–119
  charging power, EVs, 99
  maximum profit of SDNO, 98, 99
  power purchasing, 101
  risk-based, 92
  with charging/discharging schedule, 88–90
  with controlled charging, 87
Smart charge, 144, 328, 331, 332, 349, 354
  centralized charging control method, 9, 11
  centralized *vs.* decentralized
    structures, 12, 13
  decentralized charging control method, 12
  direct charging, 9
Smart distribution network (SDN)
  modeling of operational scheduling
    (*see* Operational scheduling, SDN)
  optimal operation, 74
  risk level, strategies, 90
Smart distribution systems (SDSs), 344
Smart grids
  and smart homes, 49, 56–58
  G2V/V2G modes, 48
  G2V/V2G operation, 52
  off-board EV-BCS, 66
  power quality functionalities, 56
  purposes, 130
  V2G/V2V mode, 50
Smart home, 130, 131, 143, 152
  battery charging modelling, 368–369
  behaviour, 369–370
  charging management, 374–375
  controlled charging, 379
  demand, 373–374
  demand response method, 359
  electric vehicle charging model, 370
  electric vehicle power curve, 367
  energy management system, 359
  EVs, 364, 371
  HEV, 365
  integrated dynamic method, 360

Lagrangian method, 361
  peak and standard deviation, 379
  PHEV, 365
  power system, 366
  proposed method, 375
  uncontrolled charging, 376–379
  V2G and G2V effects, 372–373
Smart Microgrids (SMGs), 334, 354
  cooperative control-based EV-charging
    management, 196
  distributed controllers, 186
  ESS, 199
  frequency, 197
  ride-through-ability, 189
  scheduling EV charging, 186
  setups, 206
  stand-alone micro-grids, 194
  VREs, 195
Socio-Techno-Economic-Political (STEP)
  factors, 189
Solar based-electric vehicle parking lots
  (PLs), 74
Solar panels, 131, 132, 135
Solar system, modeling, 76
Stackelberg game theory, 271, 272
Stackelberg leadership model, 236
Start-up and shut-down cost, 169
State of charge (SOC), 274
Stochastic methods, 293
Stochastic programming, 167
Swarm optimization approach, 293

T

Technical aspects, 190
10-bus MG test system, 252, 261, 263, 266
Time of use (TOU), 270, 331, 338, 375
Traffic control policies, 3
Transformer general controller (TGC), 146, 148
Transportation, 233
Transportation systems, 22
Trip chain modeling approach, 24
Two-step screening method, 271

U

Uncertainty, 314, 316, 317, 320, 323
Upstream grid (UG), 226, 227

V

Variable renewable energies (VREs), 195

Vehicle charging management, 147, 148, 156, 161
Vehicle electrification, 1, 47
Vehicle to building (V2B), 335
Vehicle to grid (V2G), 234, 269, 293, 332, 372
  for ancillary services, distribution network, 15, 16
  charging without vehicle management, 153
  coordinated V2G mode, 149
  and discharge mode, 150
  G2V/V2G power interaction, 48
  managed charge, 159
  management system, 50
  microgrids, implementation, 15
  mode as virtual power plants, 14
  power system security and resiliency, 14
  smart home management system, 52
  V2V, 50
  vehicle managed charging without V2G capacity, 156, 157

Vickrey-Clarke-Groves mechanism, 271
Virtual power plants, 14

W
Weibull distribution function, 171, 172, 178
Wind
  generated power, 178
  integration, EVs and wind units, 166
  turbines, 172, 173, 176, 177
  units, 165
  Weibull probability distribution function, 171
Wind turbines (WTs), 214–216, 220–222, 224, 226, 229, 269
Wind unit, 166, 169, 172, 178

Z
Zero pollution, 6